Paleogeography and Gas Exploration of Marine Multiple Formations in Central Sichuan Basin

Qiufen Li · Tongshan Wang

Paleogeography and Gas Exploration of Marine Multiple Formations in Central Sichuan Basin

With Contributions by Congsheng Bian, Tong Lin, Jie Zhao, Jinghai Dong, Shiwei Huang, Xiaobin Tian, Li Jun, Baomin Zhang, Weihua Lu, Qisen Sun, Xiaorong Zhang, Shipeng Huang, Cong Tan, Xuan Chen, Yanhu Chen, Xiaodong Lan, Hengfei Wei, Zhonggui Hu, Shengfei Qin, Anna Xu, Xiufen Zhai, Zhaohui Xu

石油工业出版社
PETROLEUM INDUSTRY PRESS

Springer

Qiufen Li
Research Institute of Petroleum Exploration and Development
Beijing, China

Tongshan Wang
Research Institute of Petroleum Exploration and Development
Beijing, China

ISBN 978-981-99-2527-8 ISBN 978-981-99-2528-5 (eBook)
https://doi.org/10.1007/978-981-99-2528-5

Jointly published with Petroleum Industry Press
The print edition is not for sale in China (Mainland). Customers from China (Mainland) please order the print book from: Petroleum Industry Press.
ISBN of the Co-Publisher's edition: 978-7-5183-5715-4

This Springer imprint is published by the registered company Springer Nature Singapore Pte Ltd.
The registered company address is: 152 Beach Road, #21-01/04 Gateway East, Singapore 189721, Singapore

Preface

Sichuan Basin is a large superimposed petroliferous basin developed above the Yantze craton with two evolution stages: craton basin and foreland basin. Sinian-Triassic marine carbonate rocks are widely developed in the basin, with a thickness of more than 6000 m. The Central Sichuan area is located in the central Sichuan Basin, with stable structures and well-developed marine strata and favorable petroleum geological conditions and exploration potential. At present, strata of Sinian, Cambrian, Permian, and Triassic in the Central Sichuan area have been explored, which indicates that the multiple marine strata in Central Sichuan area have good exploration prospects.

The study of lithofacies paleogeography is the base of oil and gas exploration, and it is of great significance for oil and gas exploration and development. The predecessors have done a lot of effective researches relating to marine strata in Sichuan Basin, such as the series of lithofacies paleogeography in southern China (Xia Wenjie, et al., 1994), lithofacies paleogeography map of Sinian-Triassic in Yangtze Region (Ma Yongsheng, et al., 2007), lithofacies paleogeography map of Cambrian-Ordovician in Sichuan Basin (Yang Wei, et al., 2012), lithofacies paleogeography map of Permian-Triassic complied by sequence stratigraphic units (Zhao Zongju, et al., 2012), lithofacies paleogeography map of Sinian-Cambrian in Sichuan Basin (Du Jinhu, et al., 2016). These studies play an important role in oil and gas exploration by clarifying the Sinian-Triassic lithofacies paleogeographic patterns of Sichuan Basin. With the development of exploration and the deepening of understanding, there are many problems restricting exploration and research, such as how to depict lithofacies paleogeographic pattern of different periods more precisely by combining seismic data with drilling and outcrop data, and how to make detailed reservoir research under the constraints of lithofacies paleogeography.

In 2018, PetroChina Exploration and Development Research Institute began to carry out fine lithofacies paleogeography research and natural gas exploration evaluation for multiple marine strata in the Central Sichuan area. After three years of research, the project team built a more detailed paleostructure restoration and evolution of multiple marine strata in Central Sichuan area with comprehensive application of outcrop data and 2D, 3D seismic data finely calibrated by drilling and well logging data, clarified the paleogeographic pattern of each sedimentary period in the basin, reconstructed the fine Sinian-Triassic lithofacies paleogeographic pattern, revealed the genesis and distribution characteristics of reservoirs under the control of the paleogeographic pattern, and evaluated the key exploration areas and favorable zones, which provided detailed basic geological evidence for oil and gas exploration.

The book is divided into six chapters. The preface was written by Wang Tongshan and Li Qiufen, Chap. 1 Outline of "Regional Geological Setting" was written by Bian Congsheng, Li Qiufen, Lin Tong, Zhao Jie, and Dong Jinghai, Chap. 2 Structural Characteristics and Evolution was written by Wang Tongshan, Li Qiufen, Lan Xiaodong, Lu Weihua and Tan Cong, Chap. 3 Lithofacies Paleogeography and Sedimentary Evolution was written by Li Qiufen, Wang Tongshan, Chen Xuan, Hu Zhonggui, Zhang Baomin, Zhai Xiufen, Huang Shiwei and Zhang Xiaorong, Chap. 4 Reservoir Characteristics and Distribution Prediction was written by Li Qiufen, Wang Tongshan, Zhang Baomin, Chen Yanhu, Wei Hengfei, Tian Xiaobin and Xu Anna, Chap. 5 Evaluation of Natural Gas Accumulation Conditions was written by Bian Congsheng, Li Qiufen, Wang Tongshan, Huang Shipeng, Qin Shengfei and Xu Zhaohui, Chap. 6 Recent Key Exploration Fields and Zones was written by Wang Tongshan, Li Qiufen, Bian Congsheng, Li Jun, Lintong and Sun Qisen. The whole book was edited by Li Qiufen and Wang Tongshan.

The book was strongly supported by PetroChina Exploration and Development Research Institute, PetroChina Daqing Oilfield Company, PCL Southwest Oil and Gas Branch, Changjiang University, China University of Geosciences, Beijing, Research Institute of Zhongheng Lihua Petroleum Technology, Beijing. At the same time, it received guidance and help from Prof. Hu Suyun, Prof. Li Jianzhong, Prof. Luoping, and Prof. Bi Jianjun. We would like to express our heartfelt thanks.

This book contains the latest achievements in the petroleum geology research of multiple marine strata in the central area of Sichuan Basin in recent years. It has both scientific and practical significance for exploration guidance that may help advance basic geological research. This book is a reference book integrating teaching, production, and scientific research. It is liable to mistakes or omissions, so we welcome suggestions and criticisms.

Beijing, China

Qiufen Li
Tongshan Wang

References

Xia W J, Du S G, Xu X H, et al. 1994. Sinian lithofacies, paleogeography and mineralization in South China [M]. Beijing: Geological publishing house.

Ma Y S, et al. 2007. Marine oil and gas exploration in China [M]. Beijing: Geological publishing house.

Yang W, Xie W R, Wei G Q, et al. 2012. Sequence lithofacies paleogeography, favorable reservoir distribution and exploration zones of the Cambrian and Ordovician in Sichuan Basin, China [J]. Acta Petrolei Sinica, 33(s2):21–34.

Zhao Z J, Zhou H, Chen X, et al. 2012. Sequence lithofacies paleogeography and favorable exploration zones of the Permian in Sichuan Basin and adjacent areas, China [J]. Acta Petrolei Sinica, 33(s2): 35–51.

Du J H, Wang Z C, Zou C N, et al. 2016. Discovery of intra-cratonic rift in the Upper Yangtze and its coutrol effect on the formation of Anyue giant gas field [J]. Acta Petrolei Sinica, 37(1): 1–16.

Contents

Chapter 1
Outline of Regional Geological Setting

Sichuan Basin is located in the northwest margin of the Yangtze plate, bordered by Longmenshan Mountains in the west, Qiyaoshan Mountains in the east, Micangshan-Dabashan Mountains in the north, and Daliangshan-Daloushan Mountains in the south. The basin area is about 18×10^4 km^2. Sichuan Basin is developed with strong tectonic activities as it has long been in the transition-transform zone between Gondwana and Laurasia. Multiple regional unconformities are developed in the craton of Sichuan Basin, and the craton margin was strongly transformed due to orogenic deformation in the later stage.

The Leshan-Longnvsi paleo-uplift is located in the Central and Western Sichuan Basin, and the plunging zone is located in the gentle structural belt in Central Sichuan. It is the most prominent structural unit in Sichuan Basin, with an irregular shape. The core of the paleo-uplift is formed by the absense of Silurian, covering an area of 6.25×10^4 km^2. The outcrops in Central Sichuan are mainly composed of Middle and Upper Jurassic, in Weiyuan and Laolongba anticline they are mainly composed of the Triassic, which are relatively old, and in Western Sichuan are mainly composed of Cretaceous and Quaternary. This study focuses on the Central Sichuan at the eastern end of the Leshan-Longnusi paleo-uplift, located on the west side of Huayingshan tectonic belt.

1 Regional Geotectonics

Geographically, the research area is located within the inner Upper Yangtze craton. The development and evolution of marine sedimentary sequences in the study area must be consistent with the whole Upper Yangtze craton. According to the theory of "basin controlled by tectonics", the analysis of characteristics and influence of

Q. Li and T. Wang, *Paleogeography and Gas Exploration of Marine Multiple Formations in Central Sichuan Basin*, https://doi.org/10.1007/978-981-99-2528-5_1

regional and areal tectonic movements is an important link in the process of sedimentary sequence filling. The following is a phased summary of the regional tectonic characteristics of the craton evolution stage in the Middle and Upper Yangtze Region.

The basement of Yangtze platform was formed by the Jinning movement in the Late Neoproterozoic, which made the Yangtze landmass enter the evolution stage of craton basin under plate movement mechanism since the Nanhua period. The evolution and development are closely related to the expansion and contraction evolution stages of the Tethys ocean during the reconstruction process of the mainland China and the interaction of the adjacent landmasses. In general, from the Early Paleozoic to Middle Triassic, the Yangtze landmass experienced the development and evolution history of differential action with adjacent landmarks (Cathaysian landmass, North China landmass, Indosinian landmass, etc.) in different periods of the process of South-to-North rotational drift in the proto-Tethys and paleo-Tethys ocean (Fig. 1). The Late Indosinian movement lasted to the Middle Triassic resulted in the closure of the paleo-Tethys ocean, the withdrawal of seawater, the tectonic inversion and the foreland orogeny, which ended the development and evolution stage of the Yangtze craton basin. As a result, the Yangtze landmass has entered a new stage of basin evolution, which includes intracontinental orogeny and foreland basin formation.

(1) Stage of Jinning-Tongwan movement. The accretive fold belts in the southeast margin of the Yangtze landmass and the trench-arc-basin system in the margin of the Cathaysia continent were formed by the subduction of South China Sea to the Yangtze landmass. About 850–880 Ma ago, between the Cathaysia and the Yangtze, the ancient South China Sea in the eastern part of the Yangtze landmass disappeared after the movement of the Jinning II episode, and the South China residual Sea basin in the western part remained to the Caledonian period. The basement of the Yangtze platform was formed after the Jinning movement. During the period of Chengjiang-Tongwan movement, the Rodinia supercontinent began to break up (Fig. 2), and the Yangtze block was under the influence of extensional tectonic action. The rift trough in the platform margin and the intercontinental rift developed together with the Sinian sedimentary caprock.

(2) The stage of the Caledonian movement. It is said to be the development stage of plate movement, including the extension-rifting stage and the convergence-matching stage, corresponding to the development of rifting-passive continental margin basin, the Formation of "great uplift to big depression" and foreland basin on the Yangtze craton.

The Caledonian period tectonic movements are mainly composed of the Tongwan movement, Xingkai movement, Yunan movement, and Duyun movement (Beiliu movement). Among them, ① Tongwan movement at the end of Sinian had a great impact on the Western Yangtze landmass, resulting in the uplift of the Western Sichuan Basin, which formed the prototype of paleo-uplifts in Leshan-Longnvsi, Longmenshan, Hannan-Dabashan, Xuefeng, and Central Guizhou. At that time, the Deyang-Anyue rift trough had begun to develop on the west margin of the Yangtze landmass. During the development period of

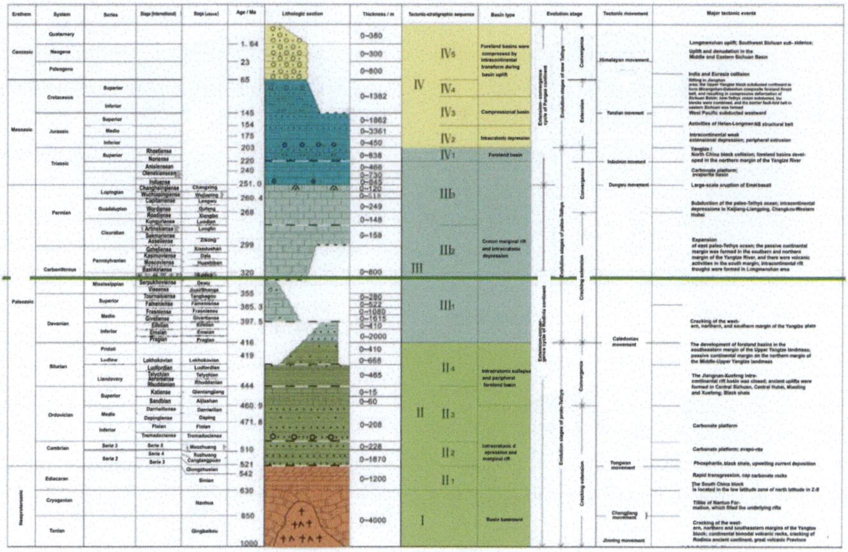

Fig. 1 Tectonic-stratigraphic sequences and evolution stages of Sichuan Basin (He et al., 2011)

the Dengying Formation, the mound-bank bodies in the platform margin were developed on both sides of the rift trough. ② The Xingkai movement at the end of the Early Cambrian had little effect on the Yangtze Region, characterized by uplift. ③ The Yunan movement at the end of Cambrian also had no significant influence on the Yangtze Region, and it was mainly developed as uplift, resulting in local unconformity. ④ The Duyun movement at the end of Ordovician was relatively strong, with intense activity in the west and weak in the east, strong in the margin and weak in the interior. Leshan-Longnvsi area uplifted early and intensely, Hannan and Dabashan uplifts developed inherently, and then Xuefeng and Central Guizhou uplifts rose sharply, which changed the tectonic pattern of the Yangtze Region from high in the west and low in the east to the uplift of the southeast margin. ⑤ Under the background of early tectonic pattern, at the end of Silurian, the Guangxi movements were dominated by uplifts, and caused the main uplift of the whole region to form a unified South China uplift.

Overall, the tectonic characteristics during the period of Caledonian movement were affected by the subduction of the paleo-Asian Ocean and the branches of proto-Tethys ocean, as well as the matching and collision of Chinese landmasses after the first expansion. Affected by the northward subduction of the proto-Tethys ocean, the northern margin of the Yangtze landmass formed a back-arc expansion zone and affected by the northward subduction of the ancient South China Ocean in the southern margin in the Yangtze landmass, the Hunan-Guangxi landmass overlapped on the Yangtze continental margin in the form of flexure basin.

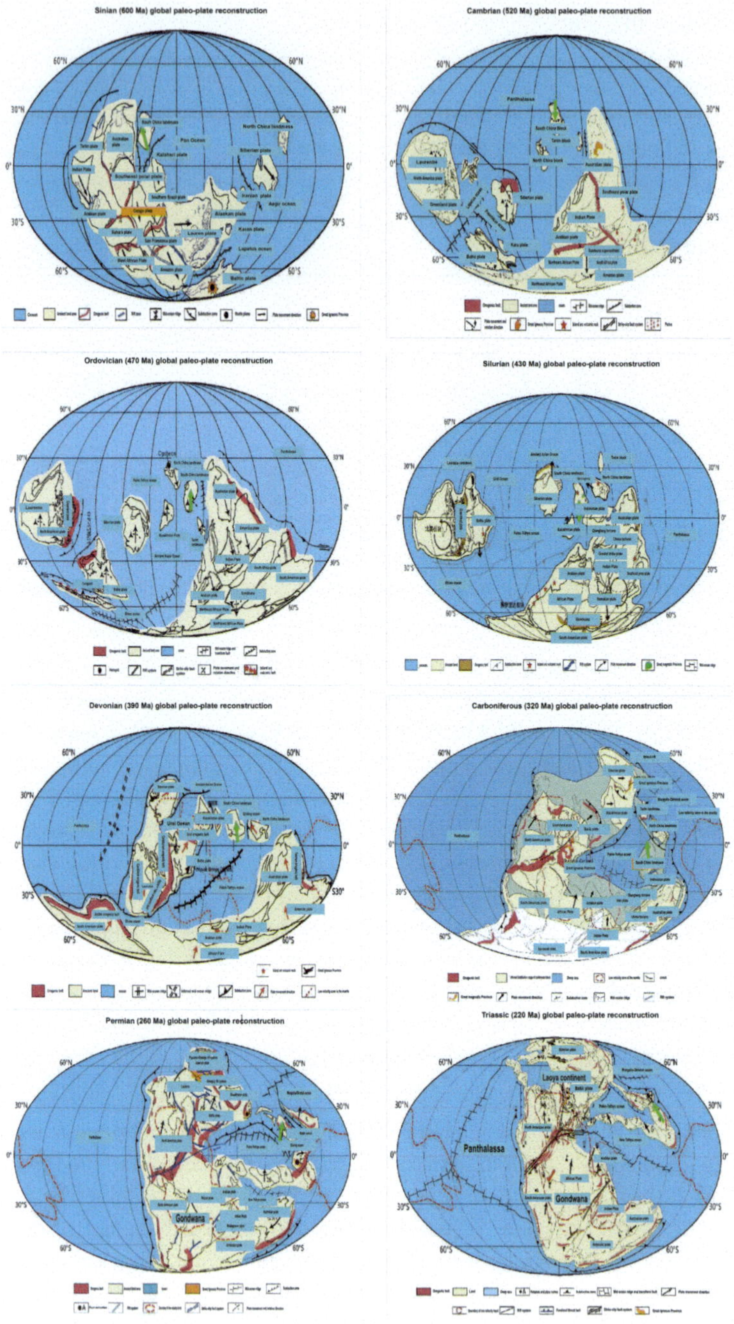

Fig. 2 Tectonic-stratigraphic sequences and evolution stages of Sichuan Basin (Li et al., 2014)

(3) Hercynian-Indosinian period (D-T_2). It is the development stage of intraplate activities, which includes the two geotectonic stages of convergence-extension and convergence orogeny-extension. In the stage of convergence-extension, the Nanpanjiang Basin, Hunan-Guangxi-Jiangxi rift basin and small Mianlue ocean basin were formed by the continental rifting. In the late stage, the Qinfang Trough closed. In the convergent orogeny -extension stage, a continental rift formed in the early period was further developed. In the late stage, the intracontinental convergence occurred, and the whole region changed from sea to land. In the Hercynian-Indosinian period, under the tectonic background of the Guangxi movement, many tectonic movements were further developed, including Ziyun movement, Yunnan movement (Sichuan-Hubei movement), Guizhou-Guangxi movement, Dungwu movement and the first episode of Indosinian movement. Among them, ① Guangxi movement (the end of Silurian): had a profound impact on the development and evolution of the whole Middle and Upper Yangtze Region in the late stage. The tectonic pattern of relatively stable main uplift and relatively active marginal rifting was remained until to the late Early Permian. ② Early Hercynian (D-C): most of the Yangtze platform continued to uplift into the land, and the platform margin (the front of Longmenshan) was sharply rifted and received stable-transitional sediments with huge thickness. The Ziyun movement (at the end of Devonian) occurred in this period, which was active in the Sichuan-Hubei area, resulting into unconformities and overlapping between Carboniferous and Devonian. In the late Early Carboniferous, the unconformity between the Late and the Early Carboniferous was formed by the Yunnan movement (Sichuan-Hubei movement), and the Wudang uplift was formed. At the end of the Carboniferous, the Guizhou-Guangxi movement was dominated by the vertical movement, resulting in interval unconformities. ③ Dungwu movement (the end of P_2): the tectonic environment was relatively stable in the Middle and Early Permian. A large-scale gentle carbonate slope was developed and intensely stretched during the late stage. Under the influence of "Emei Taphrogenesis" at the end of the Middle Permian, the paleo-Tethys was opened, and the Emei hot mantle plume uplifted. In the early Late Permian, large-scale basalt magmatic effusion occurred in the Wolong-Panxi rift, forming an erosion gap between the Late and the Middle Permian, and the South China region had been raised and turned into a denudation zone. ④ Early Indosinian (T_1–T_2): Songpan-Ganzi trough was developed with back-arc extension, subsidence and uncompensated active sediments. Beach islands were arranged around the platform margin, and the stable marine evaporation deposition of the Upper Yangtze was received in the platform.

In short, in the Hercynian-Indosinian period, the Yangtze landmass's tectonic characteristics were affected by the expansion and contraction of the paleo-Tethys ocean. The effects are shown as: ① two EW sub-branch oceans, Mianlue ocean and Qinfang trough, were developed by the expansion of northern and southern margin of Yangtze platform in Carboniferous; ② in Late Permian, the Mianlue ocean in northern margin subducted to North China, the Yuehai Ocean in the southern part subducted from east to west and south, and the paleo-Tethys ocean subducted from west and south to north. ③ In the Middle Triassic, by inheriting the previous tectonic background, the Yangtze landmass was shown as a subducting collision in whole. Foreland basins and foredeep basins were formed around South China landmass, and Xuefengshan, Dabashan, Jiangnan orogenic belt and corresponding foreland uplift areas developed in the landmass. That is, the northern margin of the landmass was affected by the northward collision of the Mianlue ocean to the North China landmass, the southern part was affected by the southward subduction of Yuehai ocean, and the western margin was affected by the west to east subduction of small Ganzi-Litang ocean basin.

(4) Late Indosinian period (T_3): it belongs to the Indosinian-Yanshan-Himalayan cycle. Paleo-Tethys ocean closure: in the northern part, the South China and North China landmass collided and converged into one piece; in the southern part, South China landmass and Sanjiang region were unified as shallow sea area; in the western part, Songpan-Ganzi trough and Longmenshan platform margin depression returned, tectonic inversion was developed, orogeny basins were formed. The foreland basin in T_3x of Western Sichuan was formed and evolved and entered a new intracontinental orogeny and foreland basin development stage.

The above analysis shows that from Sinian to Middle Triassic, the Middle and Upper Yangtze Region was developed with multi-stage structural evolutions, the transformation of multi-period tectonic movement, multi-type basin superposition, multi-cycle sedimentary filling, and multi-stage hydrocarbon accumulation.

2 Stratigraphic Development Characteristics

The drilling wells reveal that the sedimentary caprocks in Central Sichuan are Sinian, Cambrian, Ordovician, Silurian, Carboniferous, Permian, Triassic and Jurassic from old to new, without Devonian. The basement stratum is the pre-Sinian crystalline rock (Fig. 3). The Sinian unconformity overlays on the granite basement of pre-Sinian.

(1) Pre-Sinian. The pre-Sinian strata in Central Sichuan are mainly developed metamorphic rocks and magmatic rocks. According to drilling data of Well Nvji, the pre-Sinian volcanic rocks developed in the basement of Longnvsi anticline, with the lithology of dacite. Well Wei 15 in the Weiyuan area reveals that the basement was developed with biotite quartz diorite, while Well Wei 28 and Wei 117 reveal that the basement was composed of granite mixed with thin diabase.

(2) Sinian. It is the oldest sedimentary overlying layer in Sichuan Basin, which is unconformable overlaid on the pre-Sinian. Doushantuo and Dengying Formation developed in Sinian, and Doushantuo Formation mainly developed in the Lower Sinian. The sedimentary sequence of Sinian can be divided into four

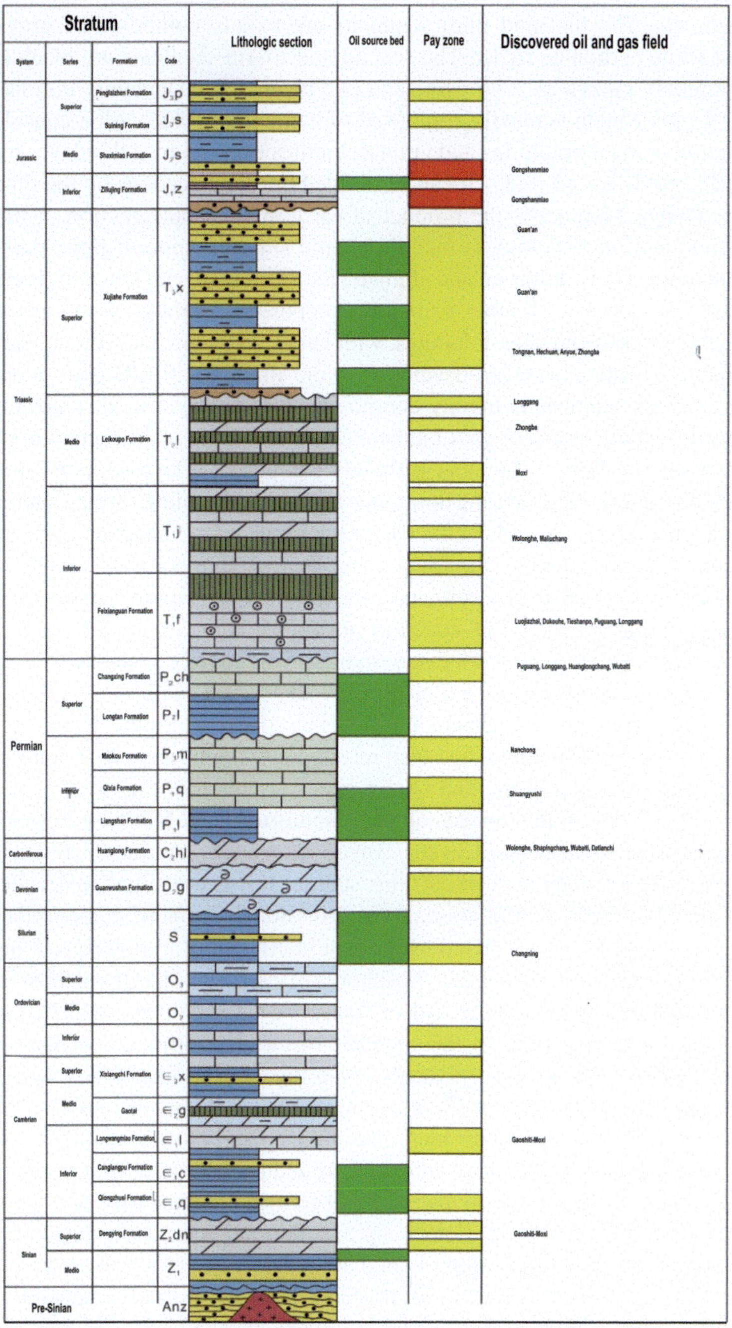

Fig. 3 Comprehensive stratigraphic column of the study area

segments. The first and third segments are mainly composed of gray-white and white carbonate rocks. The second and fourth segments are black shales, commonly known as "two white and two black". The Dengying Formation of the Upper Sinian is mostly composed of carbonate rocks, with a considerable amount of algal dolomites and grain dolomites overlain by a thin layer of sand, shale, and siliceous rocks. It can be divided into four sections by the lithology. The Deng1 Member is the product of early marine transgression of the Late Sinian, and the lithology is mainly layered silty dolomite of light–dark gray, containing sandy dolomite and alginate, locally mixed with siliceous bands and chert. The Deng 2 Member is mainly composed of light-gray and gray-white algal dolomites and silty dolomites, with laminated, snowflake, block and grape structures, paleoalgaes are developed in the stratum with siliceous dolomites. The Deng3 Member is mainly composed of dark-gray and gray layered fine silty dolomites, and dark-gray mudstones are developed in the bottom of Central Sichuan. The Deng 4 Member is mainly composed of layered, pore-dissolved silty dolomites, sand-bearing dolomites and algal dolomites, locally intercalated with siliceous bands and cherts. Under the influence of Tongwan movement in Sinian Dengying period, the Deng 2 and Deng 4 Member were weathered and denuded to form regional pore and cave layers, which are the main production layers of platform marginal zone in Gaoshiti-Moxitai.

(3) Cambrian. It is a set of relatively stable layers mainly composed of marine carbonate rocks intercalated with clastic rocks, and it can be compared regionally. The Tongwan movement uplifted the structures in the Upper Yangtze Region, strata denudation, and then resulted into the formation of non-uniform landform. The Maidiping and Qiongzhusi Formation of the Lower Cambrian were deposited on this uneven ancient landform. The Maidiping Formation is mainly distributed in the Deyang-Anyue rift trough, and the Well Gaoshi 17 in the western study area reveals this statum. It was developed under the sedimentary background of large-scale transgression. It is mainly composed of dark fine-grained sediments, including laminar interbedding argillaceous dolomites, silicophosphate dolomites and phosphorus black shale. The Qiongzhusi Formation, which is mostly composed of transgression sediments with fine grains, is a major source rock in the Sichuan Basin. The lithology of Qiongzhusi includes fine sandstones and sandy shale mixed with fine sandstones. In the lower part of the study area, Canglangpu Formation is mainly composed of fine-coarse-grained quartz sandstones, and in the upper part, it is mainly composed of gravel quartz sandstones with high roundness. The stratum is developed with relatively high mineral maturity and composition maturity, and it deposits gravel sandy tidal-flat. The Longwangmiao Formation is mainly developed with oolitic limestones, lime dolomites and dolomites, most of which are striped and mottled, and mixed with wavy argillaceous bands. Fossils are relatively rare in Longwangmiao Formation. The stratum deposits shallow water gentle slope. The reservoirs in the gas field of the Longwangmiao Formation in the Moxi-Gaoshiti area are mainly composed of a set of porous dolomites in beach facies,

and the reservoir spaces are mainly intergranular solution pores and intergranular solution pores. The reservoirs are porous reservoirs with good reservoir performance. The lithology of the Gaotai Formation in the Middle Cambrian is divided into upper and lower sections: the lower part is mainly composed of a set of argillaceous dolomites with aubergine mudstones, and the upper part is a set of aubergine argillaceous siltstones and dolomitic siltstones mixed with aubergine mudstones. The lithology of Xixiangchi Formation of Middle and Upper Cambrian is relatively single, which is a large set of thick massive fine-grained dolomites and silty dolomites, bioclastic dolomites, and dissolved dolomites can be seen in some layers with conodont fossils.

(4) Ordovician. It is mainly composed of terrigenous clastic rocks and carbonate rocks. The Ordovician on the uplift was denudated in varying degrees by the Caledonian movement. In the high part of the uplift, the Ordovician was denudated and missing, and the residual thickness in the two wings of the paleo-uplift increased. The middle and lower part of the Lower Ordovician Tongzi Formation is argillaceous shale with bioclastic limestones of thin layer, and the upper part is shale. Honghuayuan Formation is mainly composed of thick massive coarse-grained bioclastic limestones. Meitan Formation is mainly composed of a set of shale and siltstones, occasionally intercalated with limestones. The lower part of the Middle Ordovician Shizipu Formation is mainly composed of limestones, where as the upper part is dominated by argillaceous shale mixed with argillaceous limestones, containing conodonts, chitins and other fossils. Baota Formation and Linxiang Formation of Upper Ordovician are mainly composed of limestones with more mud. The former is characterized by unique "turtle crack" and the latter is characterized by "pimple limestone". The Wufeng Formation is mainly developed with a set of black carbonaceous shale, and the upper part is siliceous shale.

(5) Silurian. It comprises deposited black and grayish-green shale, intercalated with thin layers of siltstones, biological limestones and calcareous shale. The Silurian was denuded entirely in the high portion of the Leshan-Longnvsi uplift, and the residual thickness progressively thickened to both sides and the lower section of the uplift. The Silurian consists of Longmaxi Formation, Xiaoheba Formation, Hanjiadian Formation, jintaiguan Formation and Chejiaba Formation from bottom to top. The lithological characteristics are as follows: the lower part of Longmaxi Formation of Lower Silurian is black graptolite shale, which contains a large amount of graptolites with varying thickness, and the upper part is mainly composed of gray, grayish-yellow, or greenish-yellow shale mixed with siltstones or lime tumor cores, containing a small amount of graptolites and it is a continuous deposit with the overlaying Wufeng Formation. The lower part of the Xiaoheba Formation is mainly composed of medium and thick green-gray siltstones, locally intercalated with thin layers of bioclastic limestone, and the upper part is yellow-green, gray-green shale and silty shale, intercalated with thin layers of biological limestone or limestone lens. The lower part of Shiniulan Formation (i.e. Qiaogou section) is interbedded with black lime shale, thin layers of nodular limestone and limestone, and the upper part (i.e. Shiniulan

section) is mainly composed of gray nodular marlites and argillaceous limestones, occasionally intercalated with yellow-green mudstones. The Hanjiadian Formation in Middle Silurian is mainly developed with a set of gray-green and blue-gray mudstone and silty mudstone, and the lower part is mainly developed with yellow-green and blue-gray mudstones.

(6) Carboniferous. The study area's Carboniferous layers are concentrated mostly in the eastern part. They are mainly developed with Huanglong Formation composed of thick dolomites, superimposed on the Silurian and pseudo-integrals under the Permian Liangshan Formation. They contain fusulinid belts, which are generally tens of meters thick. The Carboniferous is an important natural gas reservoir in the study area.

(7) Permian. The Permian strata are draped on Leshan-Longnvsi paleo-uplift. In Central Sichuan, the lower part of Liangshan Formation in Lower Permian is mainly composed of light gray and purple clay shale and iron-bearing chlorite clay rocks. It is locally enriched with oolitic and pisolitic hematites. The middle part is mainly composed of gray and gray-white ferrolite mudstones, dense and bean-shaped bauxites. The upper part is mainly composed of grayish-brown and black carbonaceous shale, which is mixed with coal lines or thin anthracite coal seams. Qixia Formation of Middle Permian is mainly deposited by carbonate platform comprising a large set of dark-gray bioclastic micrite limestones intercalated with argillaceous rocks and bioclastic limestones. Maokou Formation is mostly composed of dark-gray, gray and gray-white thick to massive micritic limestones, micritic bioclastic limestones and algal limestones, with a small amount of dolomites containing more flint limps and strips. Longtan Formation in Upper Permian is mainly developed with a set of black-gray shale and brown-gray limestones interbedded by unequal thickness. The lower part of Changxing Formation is mainly developed with gray thick-layer micritic limestones, skeletal fragment limestones and algal limestones, and the middle and upper parts are mainly developed with gray-white medium-thick layer strip limestones containing flints, dolomitic limestones and biological limestones. The top is mainly developed with a thin layer of micritic limestones, dolomitic limestones and clay rocks interbedded with unequal thickness, intercalated with siliceous layers and tunnel flints.

(8) Triassic system. It is parallel integrated above the Permian. The Lower Triassic includes Feixianguan Formation and Jialingjiang Formation, the Middle Triassic is mainly developed with Leikoupo Formation and the Upper Triassic is mainly developed with Xujiahe formation. The Feixianguan Formation of the Lower Triassic is a set of alternating deposition of open sea platform facies limestones and intertidal-supratidal red clastic rocks. Carbonate rocks and evaporite rocks are deposited in the closed bay of Jialingjiang and Leikoupo Formation. The Leikoupo Formation was denudated by the Indosinian movement that occurred at the end of the Middle Triassic, and the denudation degree gradually intensified from northwest to southeast. Xujiahe Formation in Late Triassic is a set of sand and mudstone deposits of marine-continental alternative deposition.

3 Tectonic Evolution

In general, a set of extremely thickened sedimentary caprock was developed above the Presinian basement in Sichuan Basin, which further divided into two sets of marine and continental strata, with a total thickness of 6000–12,000 m. The Middle Triassic Xujiahe Formation and above are mainly developed as clastic rock strata with a thickness of 2000–5000 m. The strata of the Sinian-Middle Triassic are dominated by marine carbonate rocks, with a thickness of 4000–7000 m, among which, the study area is located above the Central Sichuan paleo-uplift, and the east transition area is the Luzhou paleo-uplift (Fig. 4). The sedimentation and structures in the study area are controlled by the evolution of Central Sichuan paleo-uplift and Luzhou paleo-uplift. According to the development characteristics of strata above the paleo-uplift in Central Sichuan, 7 regional unconformities are mainly developed in the marine strata. (1) Sinian/pre-Sinian: Is the product of Jinning movement, and the Sinian sedimentary rocks are unconformable on the pre-Sinian metamorphic rocks or intrusions. (2) Cambrian/Sinian: It represents the Tongwan movement. The Cambrian and Sinian are in parallel unconformity contact. From the East and west sides, the strata of Cambrian are thinned towards the direction of Leshan-Longnvsi paleo-uplift in Central Sichuan. The bottom boundary of Lower Cambrian is onlapping, and the top boundary of Cambrian is eroded. It shows that Leshan-Longnuvsi paleo-uplift is developed with both synsedimentary and denudation characteristics. The Leshan-Longnvsi paleo-uplift is surrounded by a depression area, and the altitude difference between uplift and depression is about 3000 m, indicating that Sichuan Basin has presented the structural pattern of great uplift and depression in Cambrian. (3) Ordovician/Cambrian: It is the product of the middle Caledonian movement, which is mainly characterized by the onlaping deposition on strata of Ordovician in the slope. The top of Ordovician was truncated from east to west. (4) Silurian/Ordovician: The onlap of Silurian strata can be seen on both sides of Leshan-Longnvsi uplift. (5) Permian/pre-Permian: there is a regional angle unconformity interface between Permian and pre-Permian, which resulted due to late Hercynian movement. The lower Permian is missing in a large range of the basin, and the Middle Permian mainly overlaps above the Carboniferous, Silurian or Ordovician. (6) Upper Permian/Middle Permian: The Middle and Upper Permian is also a parallel unconformity contact, which is called the Dungwu movement in the Upper Yangtze Region. In the Late Permian, affected by the activity of "Emei mantle plume", Emeis basalt erupted on a large scale and normal faults in the basin were frequently developed. (7) Upper Triassic/Middle Triassic. It is the angular unconformity interface between the Middle and Upper Triassic and is the product of early Indosinian movement, indicating the end of marine sedimentation in the basin. The top Leikoupo Formation of the Middle Triassic was truncated, and the thickness of Leikoupo Formation in the whole study area was increased from southeast to northwest.

3.1 Tectonic Evolution of Central Sichuan Paleo-Uplift

(1) Original stage. The basement of Sichuan Basin was formed after the Jinning movement and before the Sinian sedimentation, where as for tillites were deposited in Nantuo period of pre-Sinian, which was got distributed in Northeast Sichuan, Longmenshan, Dabashan and Xichang-Ganluo areas. This set of tillite sedimentation is missing in Sichuan Basin. After the glacial period, the deposition of black shale, siltstone and thin-layer carbonate rock of Doushantuo Formation was received by Sichuan Basin, with the thickness of 0–250 m. However, the stratum thickness in Weiyuan-Longnvsi and other Central Sichuan areas is only 9–20 m, which is the thinnest developed area in the basin, indicating that the Central Sichuan area may have been the uplift at that time, and the uplift is developed with low tectonic amplitude, which is mainly related to the basement activity.

Dengying Formation was deposited after Doushantuo Formation, and at the initial stage of deposition, transgression occurred in Sichuan Basin. Dengying Formation is mainly deposited with a large set of algal dolomite, a series of onlap pinchout points were developed in the southeast wing of the paleo-uplift in the Deng1 Member. Combined with the comprehensive well data and seismic data, during the deposition period of Dengying Formation, the southeast wing of Leshan-Longnvsi paleo-uplift was relatively low, and the Central Sichuan may be the uplift. Well Nvji and Well Gaoshi 1 are located on the slope of the uplift. At the end of the deposition of the Deng1 Member, the sea flooded the study area and further extended to the Longmenshan area. In this period, the Sichuan Basin received the deposition of the Deng2 Member. The Deng 2 Member was relatively stable with large deposition thickness. At the end of the stratigraphic deposition, the first episode of Tongwan tectonic movement began to move, the paleo-uplift was differentially uplifted, and the denudation thickness gradually increased from east to west. After the tectonic movement, the paleo-uplift subsided again, and the Deng 3 Member was deposited. The gentle slope was developed inside the basin. There were only two onlap pinchout points in the Deng 3 Member on the southeast wing. After the deposition of the Deng 4 Member, the overall uplift could be lifted again and denudated by the

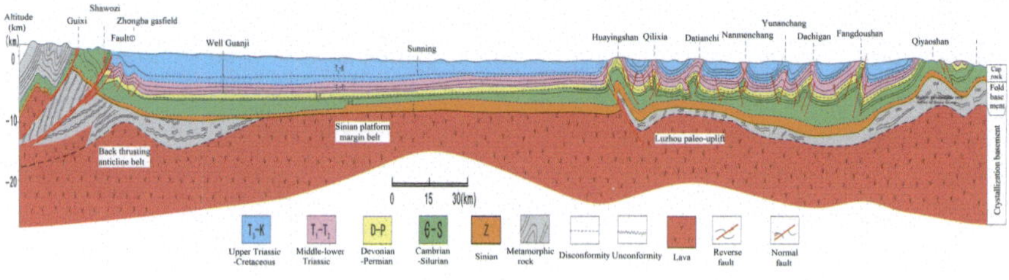

Fig. 4 Structural cross section of Sichuan Basin

second episode of Tongwan movement. Well Gaoshi 17 may be located at the high part of the paleo-uplift, the third and the Deng4 Member were completely denuded.

Based on the above analysis, the paleo-uplift was lifted on large-scale during the Tongwan movement, and the stratum was strongly denuded. The residual thickness of Dengying Formation in the Weiyuan-Quxian area was 500–700 m, and it could be 800–1200 m to both sides, indicating that the paleo-uplift was characterized by both sedimentation and uplift denudation during the deposition of Sinian Dengying Formation, which was the original development period of the paleo-uplift.

(2) Development stage. After the Tongwan movement, a nearly NS rift trough was developed in Central Sichuan, resulting in the uneven shape of the paleogeomorphology of the paleo-uplift. A series of onlap pinchout points were identified on the southeast wing of the paleo-uplift, as well as inside the rift trough. The Central Sichuan was mainly developed as filling-leveling up sedimentation in the NE-SW section. A rift trough developed in the Weiyuan-Gaoshiti area with near NS direction. Maidiping and Jianzhusi Formation of Lower Cambrian with large thickness deposited in the trough. At the end of the Early Cambrian, the paleo-uplift of Leshan-Longnvsi was uplifted again, and denudation pinchout points were developed in the northwest wing, while the denudation in Central Sichuan was affected by the late Caledonian movement. After the tectonic movement of this period, transgression occurred. The onlap pinch points can be identified in the Middle and Upper Cambrian at the margin of the paleo-uplift. Under the influence of the late Caledonian movement, in most parts of the core in the paleo-uplift, the Middle and Upper Cambrian were denuded. The tectonic activities in the Late Cambrian may have occurred only in Longmenshan area and Western Sichuan. These areas were uplifted, as a result, the Cambrian was seriously denuded, while the southeast wing of the paleo-uplift was still underwater, so that the Ordovician continued to deposit. From the well correlation section, the thickness of the Middle and Lower Cambrian was significantly decreased towards the Central Sichuan, indicating the existence of transgression. The structural form of the paleo-uplift at the end of the Cambrian was inherited by Ordovician, which later became an uplift in the northwest part, highpoint in the central part, and depression in the southeast part. At the same time, the sediments were coarse in the northwest and fine in the southeast. The Ordovician in the core of Leshan-Longnvsi paleo-uplift had been denuded, which is only distributed at the core margin. The southeast wing of the paleo-uplift is relatively intact, and multiple onlap pinchout points can be identified on the section. From the well data, only the Lower Ordovician remains at the well location on the paleo-uplift. However, the Ordovician is relatively intact at the wing of the paleo-uplift, especially the southeast wing. It can be seen that the thickness of the stratum gradually decreases from the southeast wing to Central Sichuan. It is judged that the paleo-uplift of Leshan-Longnvsi area may be a synsedimentary uplift during the deposition of the Ordovician. In

the Late Ordovician, the study area was uplifted again, and the amplitude in the southeast was the smallest, only the Upper Ordovician was denuded. From the seismic section, it can be seen that, in most areas, the Lower Ordovician was basically in direct contact with the Silurian, indicating that the Middle and Upper Ordovician may be denuded in this period.

After the tectonic action of the Late Ordovician, the paleo-uplift subsided again and received sedimentation. In the seismic section, there is an overlapping phenomenon in the Silurian, indicating that the strata of the study area have entered the stage of slow transgression again.

(3) Stage of strong uplift and denudation. At the end of Silurian, large-scale uplift and strong denudation occurred in the Upper Yangtze Region, and the core of Leshan-Longnvsi uplift was seriously denuded. The denudation continued until to the Sinian Dengying Formation in Southwest Sichuan, and was relatively weak in the wings of the paleo-uplift. The denudation period may last until before the deposition of Permian.

In summary, the Leshan-Longnvsi paleo-uplift developed on the pre-Sinian basement, and the original form was developed at the end of Sinian through the Tongwan movement. Then the multicycle tectonic uplift and subside was developed in the paleo-uplift. Finally, the paleo-uplift was flattened after the strong uplift and denudation of the Late Caledonian movement and was finalized as an inherited paleo-uplift of the Caledonian period before the deposition of Permian.

(4) Stable burial stage. After the deposition of Permian, Leshan-Longnvsi paleo-uplift entered the tectonic stage of later adjustment and transformation. In Permian, Sichuan Basin entered the stable craton basin development stage, and the paleo-uplift received stable deposition. At the end of the Early Permian, the paleo-uplift was lifted and short-term denudated again by the Soochow movement. Subsequently, the thickness of the strata in the paleo-uplift was stable by the Middle and Lower Triassic deposition. By the end of the Middle Triassic, tectonic inversion occurred in Sichuan Basin, Luzhou and Kaijiang paleo-uplift developed, under the effect, the southeast Leshan-Longnvsi paleo-uplift was uplifted, and the northwestern part was subsided. The structural axes of the Paleozoic strata in the paleo-uplift migrated to the southeast, and the Ziyang paleo-uplift was formed at this time. The Middle Triassic of Permian in the paleo-uplift was developed as monoclinic, and the Leikoupo Formation was strongly denuded in the southeast of the study area.

(5) Development stage of West Sichuan Foreland Basin. Until the Late Triassic, the tectonic pattern in Sichuan Basin is still developed as depression in West Sichuan, relative uplift in Southeast Sichuan, and foreland basin in West Sichuan. The pre-Sinian detachment layer of the paleo-uplift was developed, the structure axes of the Paleozoic strata continued to migrate to the southeast, and the monoclinic structure of Permian-Middle Triassic was more prominent. The detachment layer of the Jialingjiang formation was only developed in the

front of Longmenshan area, and the tectonic stress was not transmitted to the Middle Sichuan paleo-uplift on the west side of Huayingshan.

(6) Adjustment and finalization stage. At the end of the Jurassic, the structural layer under the detachment layer of the Jialingjiang Formation inherited the characteristics of previous structural deformation, and the Western Sichuan area was still developed as a depression basin. In the Himalayan period, the tectonic stress continued to transfer to Central Sichuan through the gypsum detachment layer of the Jialingjiang Formation, and Longquanshan continued to uplift. A series of thrust faults and recoil faults were also developed on the gypsum detachment layer in Central Sichuan. During the period of 8–13 Ma, affected by the expansion and uplift of the eastern margin of the Qinghai Tibet Plateau, a large-scale structural wedge was developed in the deep layers of the study area, resulting in the rapid uplift of Longmenshan. The Southwest Sichuan was compressed and deformed. The paleo-uplift axes continued to migrate to the southeast, the Weiyuan anticline uplifted rapidly, and the paleo-uplift of Leshan-Longnvsi evolved into the current structural form.

3.2 *Tectonic Evolution of Luzhou Paleo-Uplift*

The development period of Luzhou paleo-uplift crosses almost the whole Triassic period, with sustainable development and change. As we all know, the Middle-Late Triassic is the period for the transformation of the Sichuan Basin from the stage of craton basin to the stage of compressional or foreland basin, especially the sedimentary period of Leikoupo Formation in the Middle Triassic, which is the last period of marine sedimentation in Sichuan Basin and the prelude to the formation of foreland basin. The Sichuan Basin in this period was dominated by marine carbonate deposits. Influenced by the paleogeographic pattern of the basin by the NE-SW trending of Luzhou Kaijiang paleo-uplift belt, the sedimentary filling of the basin was developed with prominent differentiation characteristics. In the uplift area, barrier island sedimentation was the main sedimentary environment, evaporation platform and calcareous lagoon were the main sedimentary environment on the northwest side of the paleo-uplift, and carbonate rocks were widely developed While in the southeast side of the paleo-uplift, tidal basin was the main sedimentary environment. From the Southeast to Sangzhi area on the northwest side of Jiangnan Xuefeng paleo-uplift, the strata were mainly gray, aubergine mudstones, silty mudstones and cinereous silty-fine sandstones. It shows that the content of clastic particles gradually increases from northwest to southeast.

From the perspective of geotectonic background, the southeast margin of Sichuan Basin is the Jiangnan-Xuefeng orogenic belt, which was comprehensively affected by plate matching, intracontinental deformation and deep power in the Middle Triassic, and continued to advance to the northwest, forming a series of thrust nappes. The Jiangnan-Xuefeng orogenic belt gradually expanded and uplifted in the Middle Triassic and became an important provenance area in the Sichuan Basin, which led to the formation of a piedmont depression on the southeast side of Luzhou paleo-uplift, with an overall terrain of high in the east and low in the west, and the sedimentary center gradually migrated to the west. Longmenshan area in the west of Sichuan Basin was still dominated by tensile environment in the Middle Triassic. It was a low and gentle underwater uplift area controlled by normal faults, mainly developed in the sedimentary environment of the salinized shallow sea and rich in algae. The west of Longmenshan was developed as the Songpan-Ganzi passive continental margin basin, which was still dominated by the tensile environment. A series of alternating rift zones and tilted uplift fault blocks were developed. Generally, it was developed with the geomorphic characteristics of low in the west and high in the east, and was a set of marine carbonate formation. The Songpan-Ganzi area was developed with strong tectonic activities in the early-middle period of the Late Triassic, and deposited a large thickness of flysch formation, mainly in the passive continental margin deep-sea slope and shallow sea shelf environment. Until the late period of Late Triassic, the strong compression of the rift zone was developed in this area by the closure of the paleo-Tethys ocean. Under the regional dynamic metamorphism, the fold deformation was developed in the early strata. As a result, the Western Sichuan foreland basin was formed by compression and matching between the rift zone and Eastern Sichuan basin in Longmenshan area.

On the NW–SE seismic section of the core area of Luzhou paleo-uplift, the Middle Triassic was generally characterized by high in the middle part and low on both sides. It is close to the margin of the basin. Above the pre-existing Paleozoic strata with weak deformation, Early and Middle Triassic strata were deposited, especially Leikoupo Formation, which remained on both sides of the paleo-uplift with thin thickness. By the Late Triassic, compressive tectonic activities were developed in a large scale of the Upper Yangtze Region, especially the compressive stress from the Jiangnan-Xuefeng area in the southeast, which gradually lifted the southeast region to form a terrain low in the west and high in the east. In the Late Triassic, the thickness of the Xujiahe Formation gradually decreased from west to east, and the strata in Lower Cambrian composed of mudstone, shale and gypsum salt were folded and deformed under the influence of tectonic activities. Thrust faults occurred along the Cambrian detachment layers in some areas, cutting through the upper Cambrian strata and spreading upward. It can be seen that the dynamic origin of Luzhou paleo-uplift mainly comes from the Jiangnan-Xuefeng belt on the southeast margin. The front uplift belt on the west side of its piedmont depression was developed by the long-distance stress transmission effect resulted due to the Early Indosinian orogeny. At that time, the Western Sichuan was not developed as a foreland basin, while it was characterized by a depression basin.

During the Jurassic-Early Cretaceous sedimentary period, the basin was developed with weakened stress in the eastern part, entered the evolution stage of the inner craton depression basin, and deposited thicker Cenozoic strata. The eastward propagating faults along the Cambrian detachment layer were developed by the tectonic compression transmitted eastward from the western margin of the basin. In the Late Cretaceous, the tectonic compression activities were strengthened again, and the Jurassic and Late Cretaceous deposited in the early stage were uplifted and denuded in large scale. The fault activities developed along the Cambrian detachment layer were strengthened. The faults cut through the Silurian mud and shale layers continued to spread upward, as a result, the fold shape in the early period was formed. After the Late Cretaceous, affected by the continuous action of tectonic compressive stress from the EW direction, the faults spread upward, and the folds continued to uplift and expand. This resulted, in the current tectonic pattern to be gradually formed under the long-term weathering and denudation.

4 Regional Paleogeographic Pattern

4.1 Sinian Lithofacies Paleogeography in Sichuan Basin and Around Areas

A large-scale transgression occurred in the deposition period of the Deng1 Member, and an epicontinental sea deposit was formed. In this period, with the gentle slope of the basin, there was little difference in the depth of the sedimentary water body. A set of argillaceous dolomite was deposited, and a small amount of algal sheet dolomites can be seen locally. Leshan, Mianzhu, Guangyuan and other areas in the west of the basin near the land margin were mainly deposited with mud-dolostone flat deposits, while the central and eastern parts of the basin were deposited with algal-dolostone flat and grain beach. The distribution range of grain beach was relatively limited. Gypsum salt flat was deposited in Changning area of the southern basin, and a large set of gypsum rocks were deposited in Well Ning 1, reflecting the relatively restricted environment of the Deng1 Member. The basin is close to Xiaoji of Yuqing, Guizhou to Nangao of Danzhai, and was deposited by muddy lime flat. It was mainly deposited by micrite limestones, dolomitic limestones and mudstones with obviously decreased thickness. During the deposition period of the Deng 2 Member, large-scale regression occurred, the water depth was shallow, and algal dolomites were developed in a large scale. Tidal flat sedimentary structures such as bird eye and pane can be seen on the field outcrops. The sediments were mainly composed of grape algal dolomites, laminated dolomites, granular dolomites and muddy to micritic dolomites. The basin was characterized by the large-scale distribution of algal mounds and grain beaches, among which the grain beach subfacies were mostly developed in the Weiyuan and Gaoshiti-Moxi areas. The Well Dingshan 1 and Well Lin 1 in the southeast of the basin revealed that the thickness of Dengying Formation was large, and the algal laminar dolomites and algal arenaceous dolomites were developed, and the strata were developed with grain beach facies, especially on the sections of Qingzhen, Jinsha Yankong and Songlin Zunyi. Gaomo area in Central Sichuan was mainly deposited as algal mound facies in platform marginal, and eastern Gaomo area was deposited by algal mound inner the platform.

During the deposition of the Deng 4 Member, the water depth was relatively shallow as a whole. Due to the influence of the Tongwan movement, the basin uplifted in the west and subsided in the east, showing a regressive process. Mud-dolostone flat facies were deposited in the west of the basin, algal mound and grain beach facies were deposited in the middle part of the basin, and slope-basin facies were deposited in the east of the eastern margin of the basin. Gaomo area in Central Sichuan were mainly developed with algal mound facies of platform marginal, and eastern Gaomo area was developed with algal mound deposition of inner platform.

4.2 Lithofacies Paleogeography of the Cambrian Longwangmiao Formation in Sichuan Basin and Around Areas

Based on lithofacies paleogeographic pattern study, the grain beach produced in the Weiyuan-Gaoshiti-Moxi-Yilong area has a direct connection with the development of gypsum lagoon. The grain beach was located between the mixed tidal flat and gypsum lagoon, that is, the development area of grain beach was far away from the interference of terrigenous materials and lagoon. In addition, the sedimentary environment of grain beach was formed as a result of strong hydrodynamic force, it can be observed in the lithofacies paleogeographic map, in development location of grain beach between the mixed tidal flat and the gypsum Lake (Bao et al., 1998; Chen & Li 1994). In Gaomo area, the grain beach was distributed from NE to SW. Large-scale grain beaches were distributed in the Moxi area, and there were back beach lagoons between some grain beaches. The back beach lagoons were developed from reservoirs with relatively poor scale. For example, Well Moxi 205, Well Moxi 18 and Well Moxi 203 are located in the development area of the back beach lagoon. The development scale of grain beach in Gaoshiti area is slightly worse than that in the Moxi area, however, grain beach in the direction of Well Gaoshi 6-Well Gaoshi 10 is still relatively developed. Between the structures of Moxi and Gaoshiti, there is an intertidal sea developed with the direction of NE-SW, which is mainly developed with muddy argillaceous dolomite and poor reservoirs. For example, Well Moxi 21, Well Gaoshi 3 and Well Gaoshi 2 are located in intertidal subfacies, and reveal relatively poor reservoirs. Well Anping 1 and Well Moxi 20 are located at the intertidal sea margin and reveal the slightly poor reservoirs (Dai et al., 1999; Li ei al., 2012; Zhang et al., 2010). In general, the Longwangmiao Formation of Gaomo gas field in Central Sichuan is a large-scale continuous platform grain beach facies, with small distribution and thickness of grain beach in the east.

4.3 *Lithofacies Paleogeography of the Cambrian Xixiangchi Formation in Sichuan Basin and Around Areas*

Xixiangchi Formation was denuded in the west of the Sichuan Basin. From the perspective of lithofacies paleogeographic pattern, the open-semi limited platforms were developed in Central Sichuan and Northeast Sichuan, the limited evaporation platforms were developed in Southeast Sichuan, and the platform margin basin facies were developed in the Dabashan in the northern part and eastern Western-Hubei area. Sporadic intra-platform grain beach deposited in the Gaomo area of Central Sichuan, but the thickness and migration law of intra-platform beach is still unclear.

4.4 *Lithofacies Paleogeography of the Permian Qixia Formation in Sichuan Basin and Around Areas*

On the coastal marsh facies of Liangshan Formation, a large-scale transgression occurred, forming the clear water carbonate platform deposition of Qixia Formation, most of which were dark-gray argillaceous limestones micritic limestones, micritic bioclastic limestones and bioclastic limestones, and dolomitic bioclastic limestones in some parts. The light-gray massive crystal arenaceous limestones (local dolomitization) were developed in Hongkou, Shangsi Changjiang Valley and Beichuan Tongkou outcrops in Dujiangyan, Western Sichuan, representing medium-gentle slope facies deposition (platform margin). In the Suining-Ziyang-Ya'an area of Central Sichuan, three platform beaches with large area were developed.

4.5 *Lithofacies Paleogeography of the Permian Maokou Formation in Sichuan Basin and Around Areas*

The sedimentary period of the Maokou Formation inherited the sedimentary pattern of Qixia Formation. The overall sedimentary water body in the sedimentary period of Maokou Formation was deeper than that of the Qixia period, and the lithology was mainly composed of eyeball limestone and dark-gray micritic bioclastic limestone. In the southwest of the basin, breccia limestone of mixed accumulation can be seen at Qiaozhitoudao bridge in Baoxing County, which is the sediment of the upper slope (the upper part of the outer gentle slope). It can be inferred that there is a shallow gentle slope platform margin in the eastern part. Platform beach was also developed in the Suining-Ziyang-Ya'an area, showing strip distribution near EW, mainly distributed in Kaijiang paleo-uplift and Luzhou paleo-uplift.

4.6 Lithofacies Paleogeography of the Permian Changxing Formation in Sichuan Basin and Around Areas

During the sedimentary period of Changxing Formation, the sedimentary appearance generally inherited the pattern of the early stage. However, with the continuation of tensional rift and differential lifting, the transgression range was larger, the distribution of basin facies in Kaijiang-Liangping was significantly increased, while the sedimentary area of terrigenous clastic rock in Southwest Sichuan was significantly reduced, a large and stably distributed intra-platform depression subfacies were developed in Central Sichuan, and a large number of biological reefs were developed on the platform margin and in the platform. In the southeast of the inner platform depression of the surrounding rocks in Gaomo area, there were differentiation of the sedimentary facies between the inner platform depression and the reef-bank at the margin of the platform in Hechuan-Tongnan area.

4.7 Lithofacies Paleogeography of the Triassic Leikoupo Formation

The basin's interior was dominated by limestones intercalated with dolomites in semi-restricted platform facies. Relatively depressed areas such as the Suining-Yingshan area in Central Sichuan evolved into the sedimentary environment of gypsum salt lagoon, and was mainly deposited by micrite limestones and gypsum salt rocks, which was the subtidal low-energy products with relatively deep water and low energy. In Wanxian-Yunyang-Fengjie areas and other Eastern Sichuan areas, terrigenous sand and argillaceous matters increased, and clastic tidal flat deposited. One particular concern is that four sets of inner-platform porous shoal grain dolomite were developed in the Lei 3 Member of Moxi-Hechuan-Guang'an area in Central Sichuan. The strata were mainly composed of sandy and bioclastic dolomites and were locally developed with oblique beddings. With the apparent cross-well correlation, the response characteristics of logging curves are reflected as low resistance, low natural gamma and relatively high acoustic time difference.

References

Bao Q, Chen H D, Xiao M, et al. 1998. Evaluation and prediction of Sinian-Cambrian reservoirs in Caledonian paleouplift, Sichuan Basin [R]. Geological Exploration and Development Research Institute of Sichuan Petroleum Administration.

Chen X B, Li E. 1994. Research report on sedimentary facies and reservoirs of Sinian, Cambrian, and Ordovician in Leshan-Longnvsi ancient uplift in Sichuan Basin [R]. Geological Exploration and Development Research Institute of Sichuan Petroleum Administration.

Dai H M, Wang S Y, Wang H Q, et al. 1999. Formation characteristics of natural gas reservoirs and favorable exploration areas in Sinian Cambrian petroleum system of Sichuan basin [J]. Petroleum Exploration and Development, 05: 16–20.

He D F, Li D S, Zhang G W, et al. 2011. Formation and evolution of multi-cycle superposed Sichuan Basin, China [J]. Chinese Journal of Geology, 46(3): 589–606.

Li W, Yu H Q, Deng H B. 2012. Stratigraphic division and correlation and sedimentary characteristics of the Cambrian in central-southern Sichuan Basin [J]. Petroleum Exploration and Development, 39(6): 681–690.

Li J H, Wang H H, Li W B, et al. 2014. Discussion on global tectonics evolution from plate reconstruction in Phaneroroic [J]. Acta Petrolei Sinica, 35(2): 207–218.

Zhang M L, Xie Z Y, Li X Z, et al. 2010. Characteristics of lithofacies paleogeography of Cambrian in Sichuan Basin [J]. Acta Sedimentologica Sinica, 28(1): 128–139.

Chapter 2
Structural Characteristics and Evolution

1 Structural Interpretation and Structural Characteristics

1.1 Regional Stratigraphic Correlation

The accurate regional stratigraphic framework is of great significance in clarifying the distribution and spatial configuration of various sedimentary bodies in the stratigraphic framework. The rules are: stratigraphic division from large to small, divide first and then compare. The marine stratigraphic sequence framework is established following such ideas and steps: firstly, based on the previous stratigraphic division and correlation results, the drilling wells sensitive to the changes of marine sedimentary water bodies in the study area are selected as standard wells, in order to identify marker layers and vertical sedimentary cycles; secondly, based on the selected marker layers for standard wells, drilling of the adjacent wells are extended, and the strata are divided by extension in turn. Finally, based on the stratigraphic division of standard wells, the sections of stratigraphic division and correlation for well tie are established based on the typical sections selected.

Based on the stratigraphic division and correlation of single well and well ties, through the accurate well-seismic horizon calibration, seismic reflections corresponding to the marker layers are found in the seismic data volume, and reflectors are tracked and closed. Finally, a spatial stratigraphic sequence framework is established based on the seismic data volume providing stratigraphic framework constraints for the detailed description of sedimentary bodies and reservoirs.

(1) **Marker layer selection and stratigraphic division**

Eleven marker layers were identified from the marine sedimentary strata of Central Sichuan through the analysis of drilling electrical properties and lithologic characteristics, with the mutation of lithology and lithofacies on the top interfaces (Fig. 1).

Q. Li and T. Wang, *Paleogeography and Gas Exploration of Marine Multiple Formations in Central Sichuan Basin*, https://doi.org/10.1007/978-981-99-2528-5_2

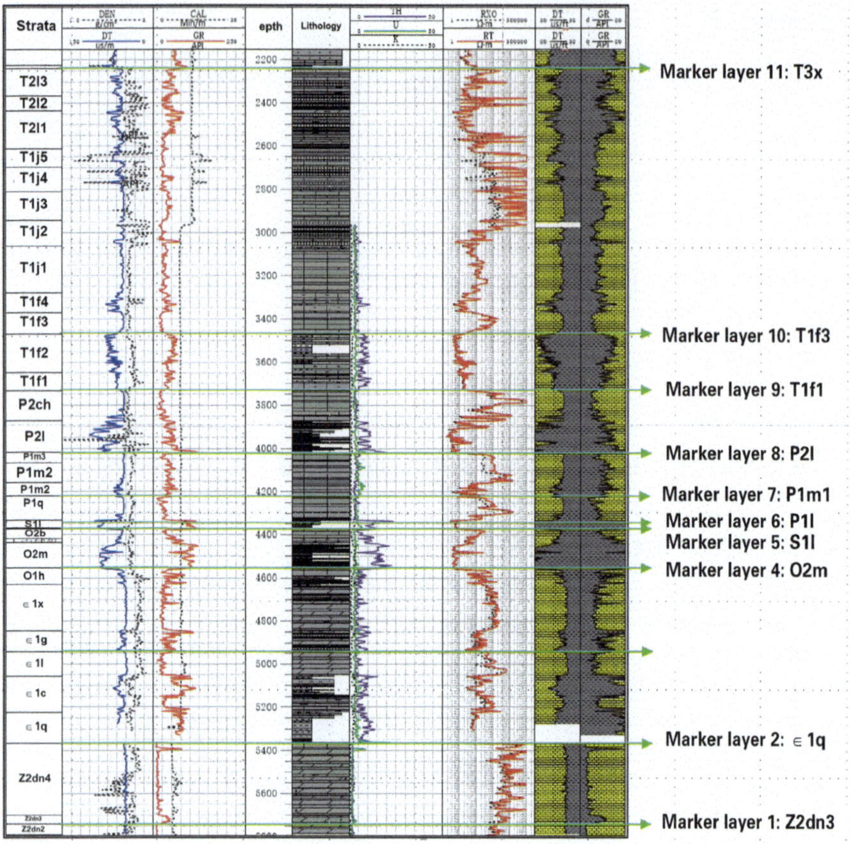

Fig. 1 Marker layers and stratsigraphic division of Well Hetian 1

Marker layer 1—Deng 3 Member. It is a set of marine flooding sedimentary layer as a whole, which has good contrast on the drilled wells in the whole area. Mudstones are mainly deposited in the Gaoshiti-Moxi platform margin belt on the western study area, and the content of dolomite increases with the change of sedimentary geomorphology to the east. Generally speaking, Deng 3 Member is mainly composed of a set of black shale intercalated with thin layers of siliceous rock and siltstone, with a thickness of 30–80 m. The Moxi area develops with thin-layered shales and is mainly composed of interbedded dark-gray siltstones and gray dolomitic siltstones. In terms of electrical characteristics, the value of the GR logging curve drastically increased between Deng 3 and Deng 2 Members, however it progressively changes between the top Deng 3 Member and Deng 4 Member.

Marker layer 2—Qiongzhusi Formatio: The Qiongzhusi Formation was formed by a large-scale transgression after the exposure and erosion following up to the deposition of the Dengying Formation,, with good correlation in the whole region. The Qiongzhusi Formation is mainly composed of black shales, carbonaceous shales and siltstones, with phosphorus at the bottom. Also with regard to electrical characteristics, the GR, AC and RT logging curves indicate that the Qiongzhusi Formation is in connected to the underlying Dengying Formation, in which the values of GR and AC logging curves increase suddenly, and the value of RT logging curve decrease suddenly. However, the logging curves between Qiongzhusi and overlying Canglangpu Formation change gradually.

Marker layer 3—Gaotai Formation: The Gaotai Formation is mainly composed of a set of dolomites and sandy dolomites, intercalated with brownish-red mudstones and shales. In terms of electrical properties, there are also obvious differences compared with the logging curves of the dolomites in underlying Longwangmiao Formation and overlying Xixiangchi Formation, in which the values of GR and AC logging curve become larger with bigger tooth amplitude, and the value of RT curve decreases.

Marker layer 4—Meitan Formation: It is a set of regional marine flooding layer as a whole, with abrupt contact at the bottom. It develops with good correlation in the whole region, and is denuded locally on the Leshan-Longnvsi uplift. Meitan Formation is composed of a set of dark-gray and black shales mixed with biologenic limestones, and more limestones in the upper part. Regarding electrical characteristics, the GR and AC logging curve increases suddenly at the bottom, and the RT logging curve decreases suddenly, showing an obvious stepped shape. And the GR, AC, and RT logging curves in the top part gradually change.

Marker layer 5—Longmaxi Formation: It is a set of regional marine flooding layer as a whole, with abrupt contact at the bottom. It develops with good correlation in the whole region. It developed as an onlap on Leshan-Longnvsi uplift, and was partially denuded on Leshan-Longnvsi uplift. It is mainly composed of a set of black shales, rich in graptolites, and the upper part is dark-gray to grayish-green shales and silty shales. In terms of electrical characteristics, the GR and AC logging curves at the bottom increase suddenly, and the RT logging curve decreases suddenly, showing an obvious stepped shape.

Marker layer 6—Liangshan Formation: The bottom of Liangshan Formation forms as a regional angular unconformity, which is marine flooding sediments of Permian with good regional correlation. The Liangshan Formation mainly comprises gray and gray-black shales, aluminous mudstones mixed with thin marls and thin coal seams.

Marker layer 7—Mao1 Member: It is mainly composed of a set of dark-gray argillaceous micrite limestones and micrite limestones with eye spherical structures. In terms of electrical properties, there are also clear differences between the limestones of the underlying Qixia Formation and the overlying Mao 2 Member. The value of GR and AC curves becomes larger and the toothing amplitude, and the value of RT curve becomes smaller.

Marker layer 8—Longtan Formation: It is a regional marine-flooding surface as a whole with good regional correlation, and the bottom is in disconformity contact with Maokou Formation. Longtan Formation in the study area is mainly composed of dark-gray and gray shales, sandstones with coal seams, and the content of marl layer increases upward. Compared with Maokou Formation, the GR, AC and RT in Longtan Formation show an abrupt increase.

Marker layer 9—Fei1 Member: It develops with good contrast in the study area and is mainly composed of grayish-brown and gray marls mixed with purple mudstones. Compared with Changxing Formation, the GR, AC and RT curves in Fei1 Member show a sudden change.

Marker layer 10—Fei 3 Member: It develops with good contrast in the study area, and is mainly composed of grayish-brown limestones. Compared with overlying Fei 2 Member, the curves of GR and AC show a rapid change.

Marker layer 11-Xujiahe Formation: It develops with regional correlation interface. Terrestrial clastic rocks deposited above the interface, and marine carbonate rocks were deposited under the interface. The lithologic interface is developed with high correlation.

Based on the development characteristics of marine marker layers, the strata in the drilling wells are finely divided to lay a foundation for stratigraphic correlation.

(2) **Stratigraphic correlation of well tie**

Based on the division of marine sedimentary strata of typical wells in Central Sichuan, the well connections are selected for stratigraphic correlation.

The strata are completely developed in the study area, from Sinian to Triassic. Two layers are missing, which are Ordovician–Silurian in west of Central Sichuan, and top Leikoupo Formation in south of Central Sichuan. For example, the section of Well Gaoshi 1—Guangtan 2 is located in the south wing of Leshan-Longnvsi uplift, parallel to the hinge of Leshan-Longnvsi uplift as a whole. From Fig. 2, it can be seen that there are many missing strata in Ordovician–Silurian. Near Well Guangtan 2, the residual strata of Ordovician–Silurian are developed with gradually increasing thickness. On this section, Permian strata are well developed with good correlation. The top Leikoupo Formation of the Triassic is a truncated unconformity. The Lei 4 Member in the line of Well Moxi 18—Guangtan 2 is little remaining, but it is missing on the line of Well Gaoshi 1—Moxi 8.

The profile of Well Gaoshi 17—Lai 1 is located on the south wing of the Leshan-Longnvsi uplift, parallel to the uplift's hinge. According to Fig. 3, some strata of Ordovician Silurian are missing. On this section, Permian strata are well developed with good correlation. The top Leikoupo Formation of Triassic is a truncated unconformity, and the Lei4 Member is missing as a whole in this section.

1.2 Well-to-Seismic Calibration

Based on the horizon calibration by regional VSP logging, the geological horizons of each r reflection event are calibrated by comparing the acoustic synthetic seismogram with the cross-well seismic section. It can be seen from the synthetic seismogram map (Fig. 4) that the waveform characteristics, wave group relationship and time

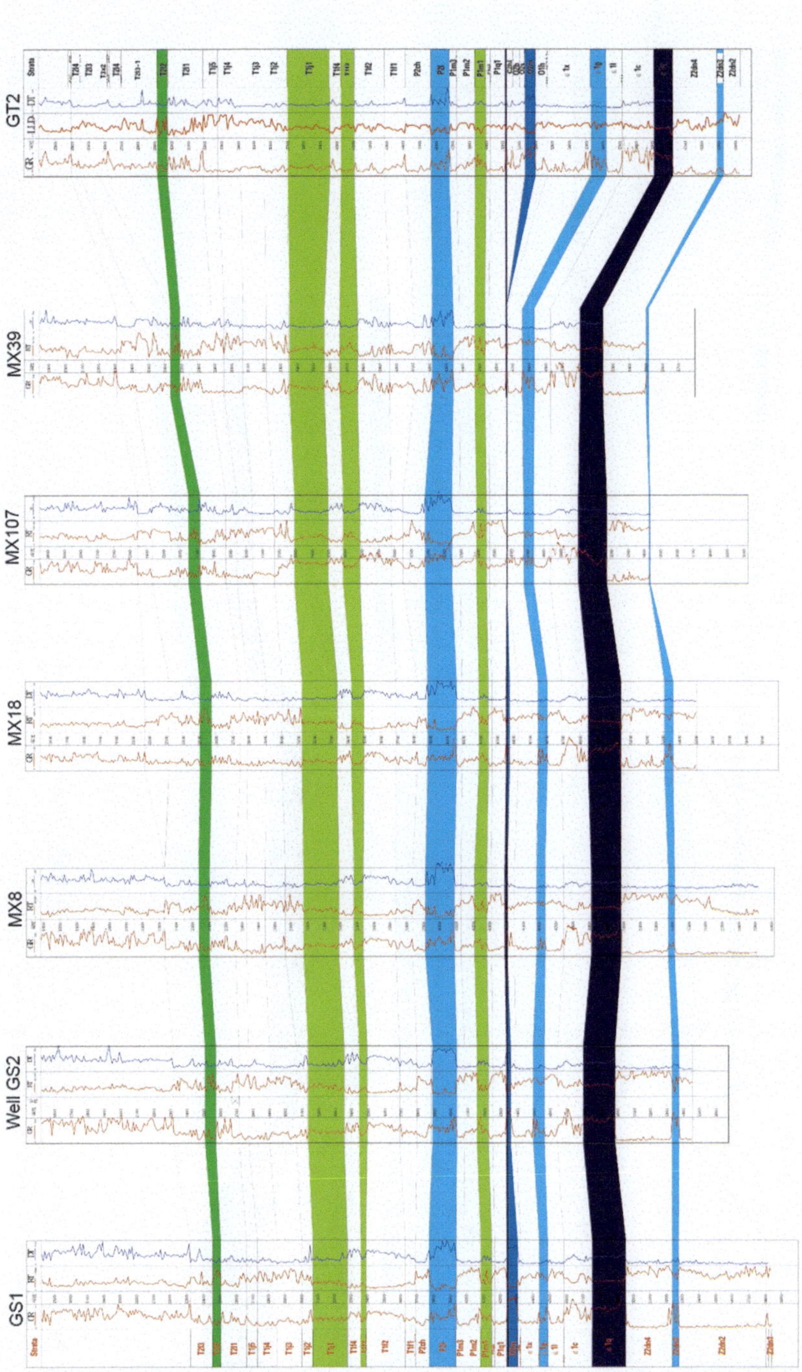

Fig. 2 Stratigraphic correlation of Well Gaoshi 1—Gaoshi 2—Moxi 8—Moxi 18—Moxi 107—Moxi 39—Guangtan 2

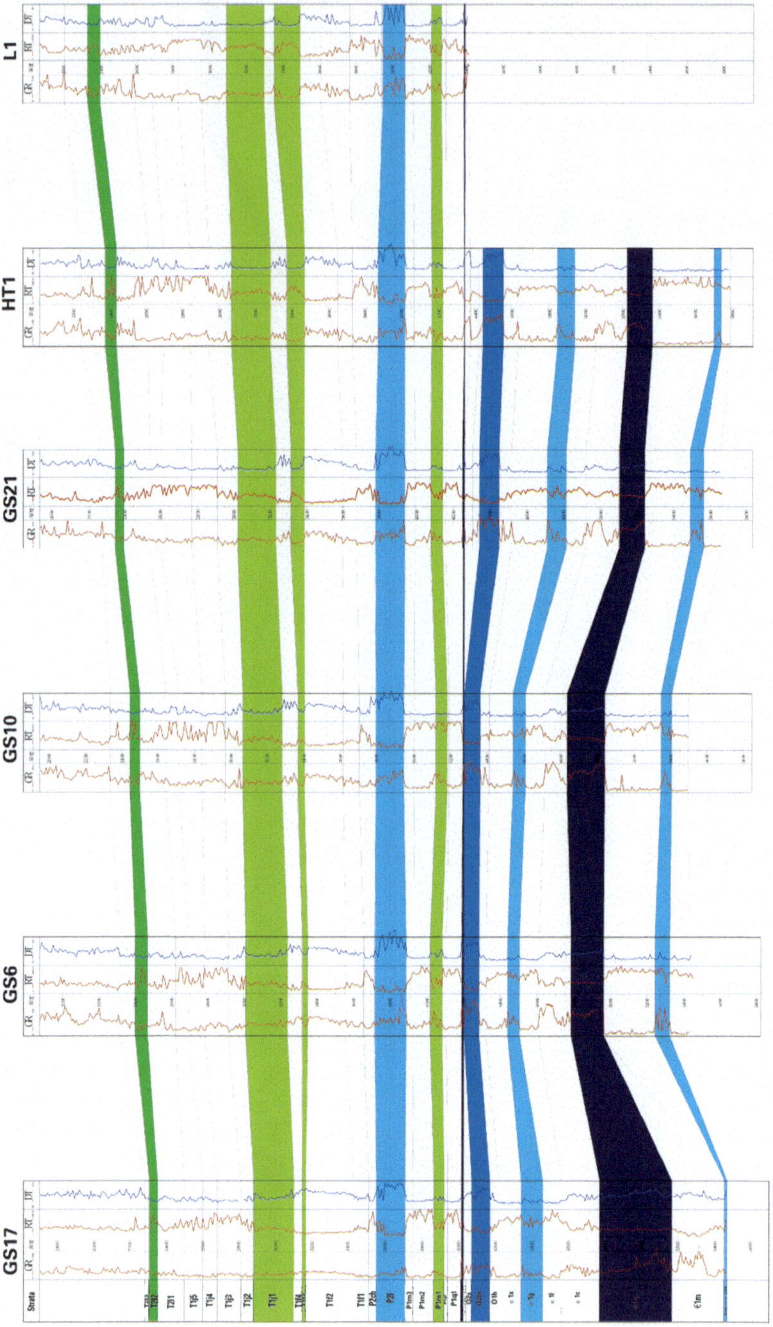

Fig. 3 Stratigraphic correlation of Well Gaoshi 17-Gaoshi 6-Gaoshi 10-Gaoshi 21-Hetian 1-Lai 1

difference of each reflection event in the actual seismic traces are relatively consistent with the synthetic seismogram, indicating a high matching degree, with a correlation coefficient more than 0.7. As a result, the seismic horizons can be calibrated for geologic horizon by this time-depth relationship.

The calibration results of seismic-geological horizon show that the top Sinian is the contact surface between the mudstone at the bottom of Cambrian Qiongzhusi Formation and the dolomite in Deng 4 Member, and the bottom of Deng 3 Member is the contact surface between the mudstone, argillaceous siltstone of Deng 3 Member and the dolomite in the Deng 2 Member. The wave impedance difference of the lithology across the interface is large, which indicates an interface with positive strong reflection coefficient developed by the contrast of low and high wave impedance. Seismic reflection events are characterized by strong and stable wave crests and

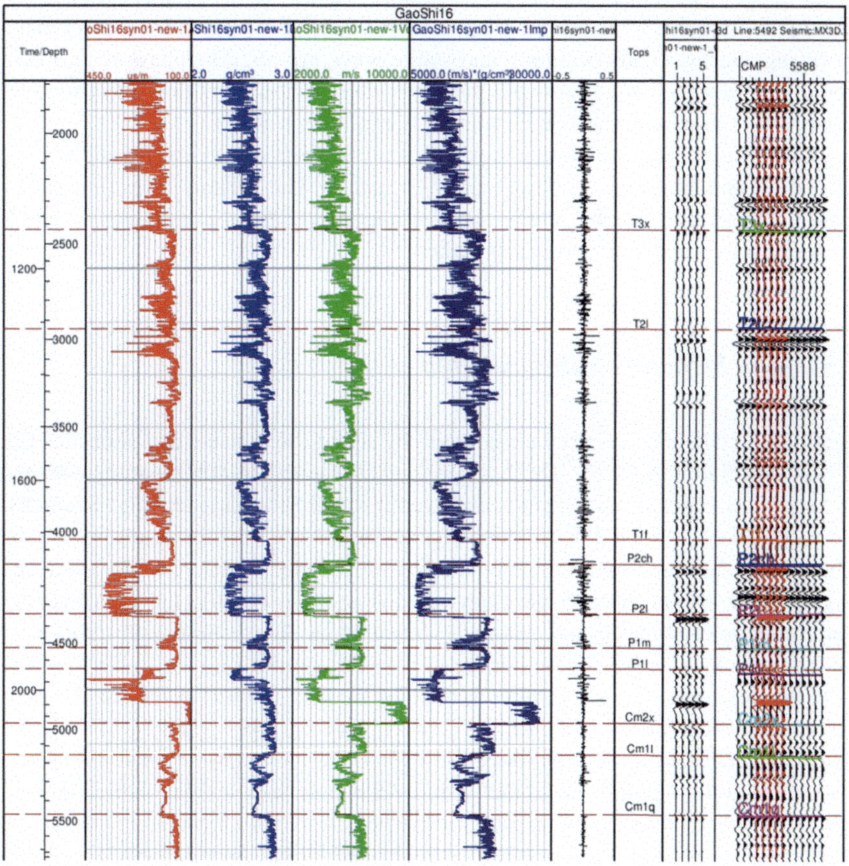

Fig. 4 Synthetic seismogram of Well Gaoshi 16

good lateral continuity, which is favorable to comparison and interpretation. The bottom Dengying Formation is characterized by stable reflectors with continuous and strong wave troughs, as well as good lateral continuity. It is also easy to conduct comparison and interpretation.

The reflect events of the bottom in the Upper Permian is characterized by single strong phases (wave crest), which is the regional marker reflection event, with strong energy, good continuity and stable wave groups. The marker lays are characterized by strong phase. The reflect events of the bottom in Maokou Formation is generally characterized by medium-strong phases (wave crest), varying energy in lateral direction, good continuity and stable wave groups. The marker lays are characterized by strong phase. The reflect events of the bottom in Qixia Formation is generally the zero point from the wave crest to the wave trough about 35–40 ms down from the bottom of Maokou Formation, and the wave trough with continuity characterizes the local part. The reflect events of the bottom in the Lower Permian consists of two phases, the former is weak and the latter is strong. The reflect events are characterized by strong energy, good continuity and stable wave groups. The phase is the wave trough between the two phases, and the time difference between the reflective layer of Upper Permian and Maokou Formation is relatively stable.

The top Leikoupo Formation is the contact surface between mudstones of Xujiahe Formation and limestones of Leikoupo Formation. The seismic reflection shows a strong wave crest, stable characteristics and good lateral continuity, which is favorable for comparison and interpretation.

1.3 Fine Structure Interpretation

Firstly, the synthetic seismogram is made by using the density and acoustic data of wells in the study area. The horizons are calibrated in combination with the previous horizon calibration results and overall waveform characteristics. Secondly, the vertical and horizonal distribution characteristics of structures and faults in the whole region are preliminarily understood by browsing the seismic exploration data of the entire region through the interpretation system. Finally, the interpretation framework is established and gradually densified by the sections with prominent wave group characteristics based on the cross well sections.

In this research, the faults are interpreted by making full use of seismic data and extracting the seismic volumes such as coherent, ant and phase. Faults with different levels are identified in multiple directions. First, the clear development model and reasonable combination style of faults are established based on the main strike and occurrence characteristics by investigating structural evolution characteristics in the whole region. Second, the faults are adjusted and closed from large to small and from primary to secondary. Third, the individual dissenting faults can be identified by analyzing the formation mechanism combined with coherent volume and 3D visualization technology. As a result, the reliability of fault interpretation can be ensured.

After the seismic interpretation, according to the fault characteristics in the section, the fault polygons can be drawn in layers by the breakpoint combination with reference to the coherent volume data, ant body data and edge detection data. The drawn fault polygons of each layer are stacked together for inspection, and the same fault cannot intersect.

In the process of interpretation, the formation occurrence and its changes should be carefully considered, and the details of small faults should be noticed. As a result, geophysical exploration data can be closely combined and penetrated with geology and drilling data, and the geological understanding can be increasingly deepened.

Finally, the interpretation of main reflective layers and faults in Central Sichuan is completed.

1.4 Structural Mapping

Under the influence of the surface in the study area, it is difficult to obtain the velocity and implement the structural high points. As a result, the method of variable speed mapping is adopted. In the process of mapping, pay attention to the following points: first, the rationality of isolines of the two walls of the fault; second, the rationality of isoline correlation in the region with structural morphological changes; third, the rationality of isolines near the mapping boundary. For the horizons of the 2D study area, with the help of 3D method, the "pseudo" 3D survey network can be established for the variable speed mapping. The horizons of the 2D study area can be imported with the coordinates, then meshed, and then the wells in the 2D study area can be calibrated by seismic data, the velocity field can be established, and the time-depth conversion can be carried out, so as to achieve the purpose of variable speed mapping.

Structural Characteristics of the Deng 2 Member

The reliability of structural results can be verified by analysing structural map accuracy with the comparison between actual drilling data and the structural results interpreted by seismic data. The altitude of Deng 2 Member interpreted by seismic data is in good agreement with the actual drilling altitude, with an absolute error range of 0–7.31 m and a relative error range of 0–0.14%. For example, the altitudes of Well Gaoshi 1, Gaoshi 10, Gaoshi 16, Gaoshi 18, Guangtan 2, Hetan 1, Moxi 10, Moxi 107, Moxi 11 and Moxi 41 are in good agreement with the values on the structural map. The overall structural form of the Deng2 Member is high in the west and low in the east (Fig. 5), with the structural altitude of − 5200 to − 5950 m and the main structure altitude on the inner uplift of above − 5600 m. Eight structural traps are found in Deng2 Member, with a cumulative area of 265.69 km^2.

Structural Characteristics of the Top Dengying Formation

The altitude interpreted by seismic data of Dengying Formation is in good agreement with the actual drilling altitude, with an absolute error range of 0–7.31 m and a

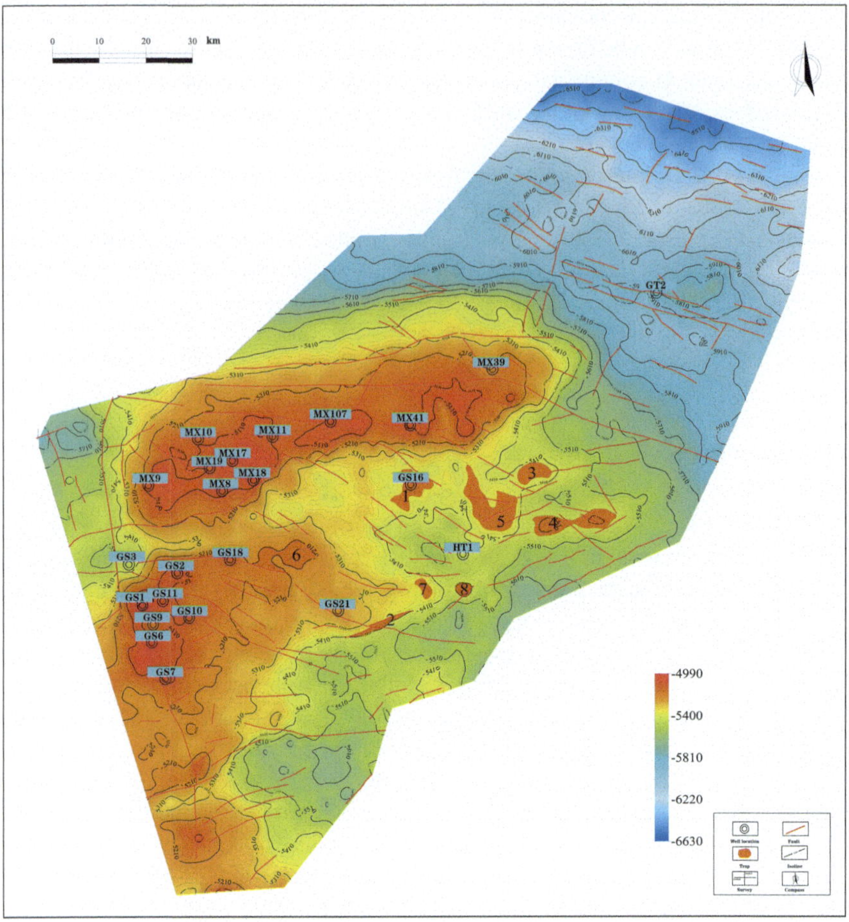

Fig. 5 Structure of top Deng2 Member in Central Sichuan

relative error range of 0–0.14%. For example, the altitudes of Well Gaoshi 1, Gaoshi 10, Gaoshi 105, Gaoshi 18, Gaoshi 19, Gaoshi 2, Gaoshi 21, Gaoshi 6, Gaoshi 7, Guangtan 2, Hetan 1, Moxi 117, Moxi 129H, Moxi 13 and Moxi 8 are in good agreement with the values on the structural map. The top Dengying Formation is generally high in the west and low in the east (Fig. 6), with the structural altitude of – 4900 to – 5800 m, and the main structure altitude in the palaeo-uplift of above – 5500 m. Eleven structural traps are found in Dengying Formation, with a cumulative area of 338.8 km^2.

Structural Characteristics of Longwangmiao Formation

The altitude interpreted by seismic data of Longwangmiao Formation is in good agreement with the actual drilling altitude. The absolute error range is 0.03–7.48 m

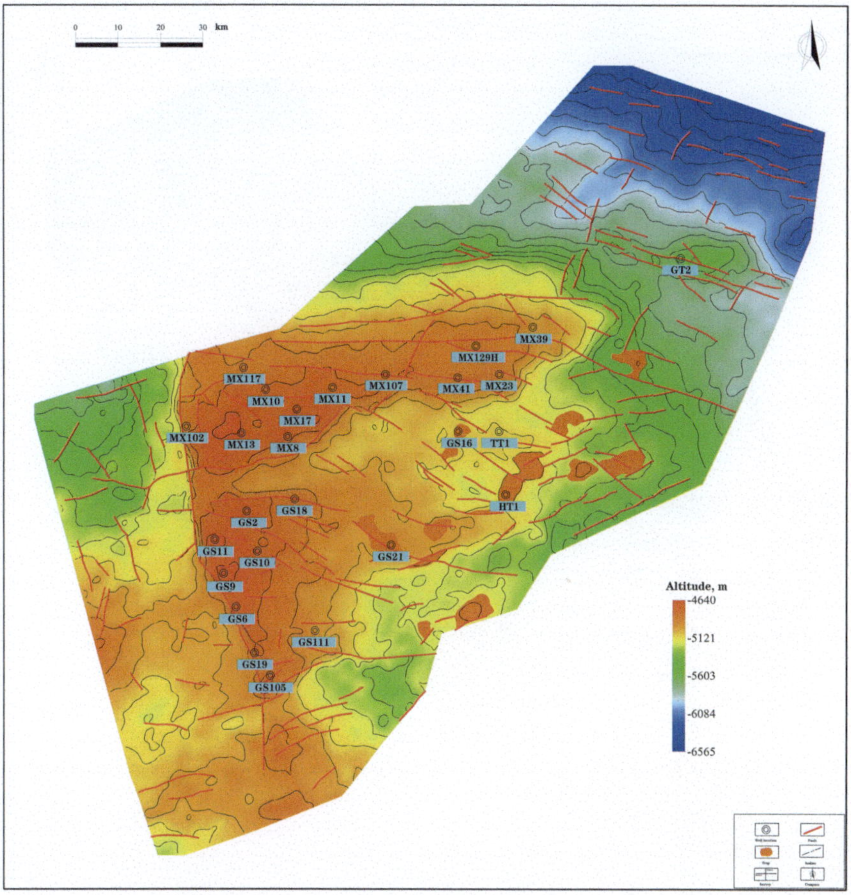

Fig. 6 Structure of top Dengying Formation in Central Sichuan

and the relative error range is 0–0.18%. For example, the altitudes of Well Gaoshi 2, Gaoshi 21, Moxi 11, Moxi 117 and Moxi 53 are in good agreement with the values on the structural map. The overall structural form of Longwangmiao Formation presents as high in the west and low in the east (Fig. 7), with the structural altitude of − 4500 to − 5500 m, the main structure altitude in the palaeo-uplift of above − 5200 m. Nine structural traps are found in Longwangmiao Formation, with a cumulative area of 255.5 km^2.

Structural Characteristics of Xixiangchi Formation

The altitude interpreted by seismic data of the Xixiangchi Formation is in good agreement with the actual drilling altitude, with an absolute error range of 0.04–33.13 m and a relative error range of 0–0.77%. For example, the altitudes of Well Gaoshi 111, Gaoshi 112, Gaoshi 113, Gaoshi 16, Gaoshi 18, Hechuan 12, Hetan

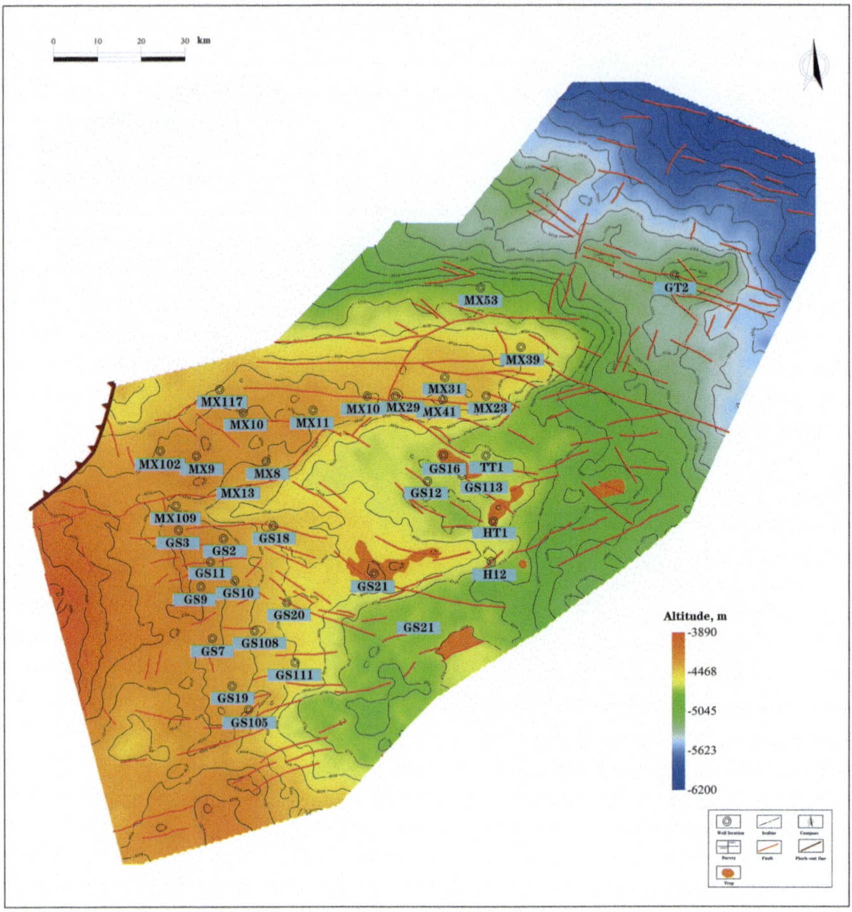

Fig. 7 Structure of top Longwangmiao Formation in Central Sichuan

1 and Moxi 107 are in good agreement with the values on the structural map. The overall structure of Xixiangchi Formation presents as high in the west and low in the east (Fig. 8), with the structural altitude of − 4200 to − 5100 m, and the main structure altitude in palaeo-uplift of above − 4600 m. Five structural traps are found in Xixiangchi Formation, with a cumulative area of 364 km^2.

Structural Characteristics of Qixia Formation

The altitude interpreted by seismic data of Qixia Formation is in good agreement with the actual drilling altitude, with an absolute error range of 0.03–9.46 m and a relative error range of 0–0.25%. For example, the altitudes of Well Gaoshi 1, Gaoshi 21, Guangtan 2, Hetian 1, Hualai 1, Tongtan 1 and Wangjia 1 are in good agreement with the values on the structural map. The overall structure of Qixia Formation is a monoclinal structure dipping from southwest to northeast (Fig. 9). The structural

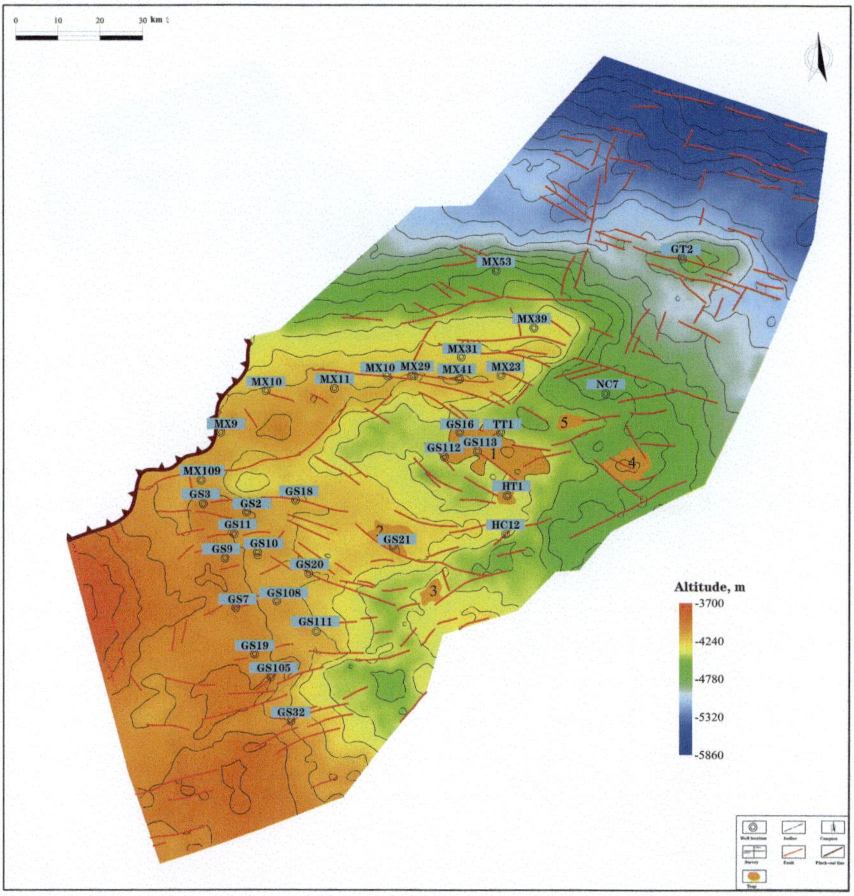

Fig. 8 The top view Structure of Xixiangchi Formation in Central Sichuan

altitude is – 3650 to – 4450 m, and the altitude of main structure in palaeo-uplift is above – 4300 m. Seven structural traps are found in Qixia Formation, with a cumulative area of 48.9 km^2.

Structural Characteristics of Maokou Formation

The altitude interpreted by seismic data of Maokou Formation is in good agreement with the actual drilling altitude, with an absolute error range of 0.03–7.65 m and a relative error range of 0–0.2%. For example, the altitudes of Well Gaoshi 1, Gaoshi 10, Guang3, Huaxi 2, Lai1, Moxi 10, Nanchong 7 and Tong 4 are in good agreement with the values on the structural map. The overall structure of Maokou Formation is a monoclinal structure dipping from southwest to northeast (Fig. 10). The structural altitude is – 2700 to – 5365 m and the altitude of main structure in palaeo-uplift is

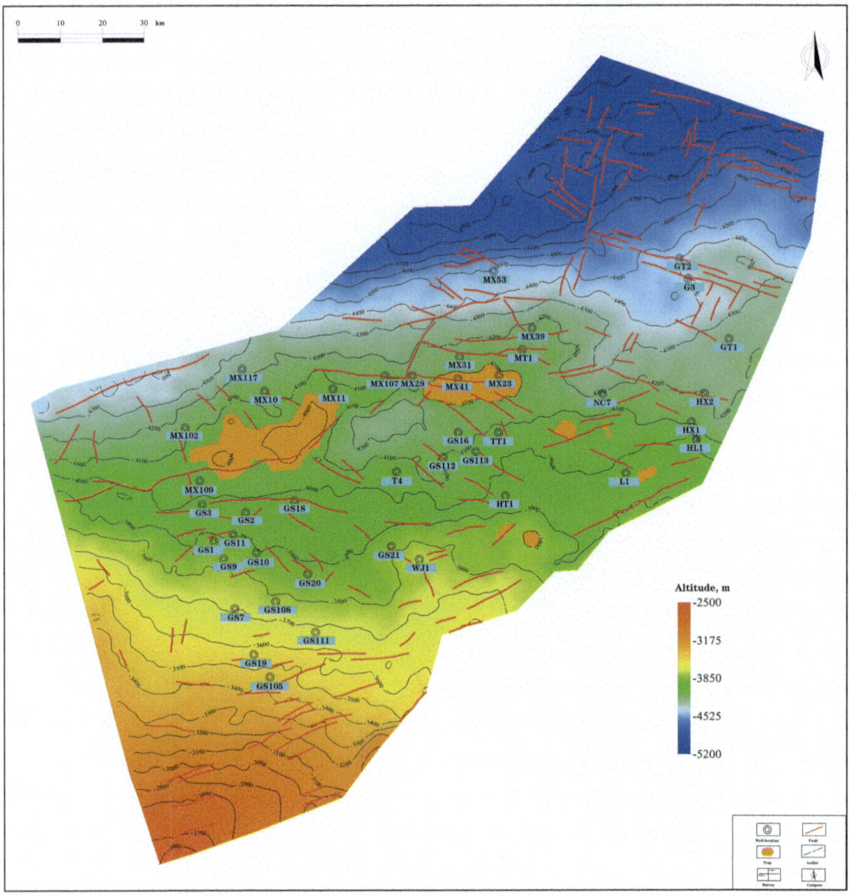

Fig. 9 Structure of top Qixia Formation in Central Sichuan

more than – 5000 to – 3800 m. Five traps are found in Maokou Formation, with a cumulative area of 69.6 km^2.

Structural Characteristics of Changxing Formation

The altitude interpreted by seismic data of Changxing Formation is in good agreement with the actual drilling altitude, with an absolute error range of 0.01–17.5 m and a relative error range of 0–0.46%. For example, the altitudes of Well Gaoshi 11, Guang3, Hetian 1, Lai 1, Moxi 29, Tongtan 1 and Wangjia 1 are in good agreement with the values on the structural map. The overall structure of Changxing Formation presents as a monoclinal structure dipping from southwest to northeast (Fig. 11). The structural altitude is – 3200 m to – 4000 m. The area of the structural traps in the Changxing Formation is 5.8 km^2.

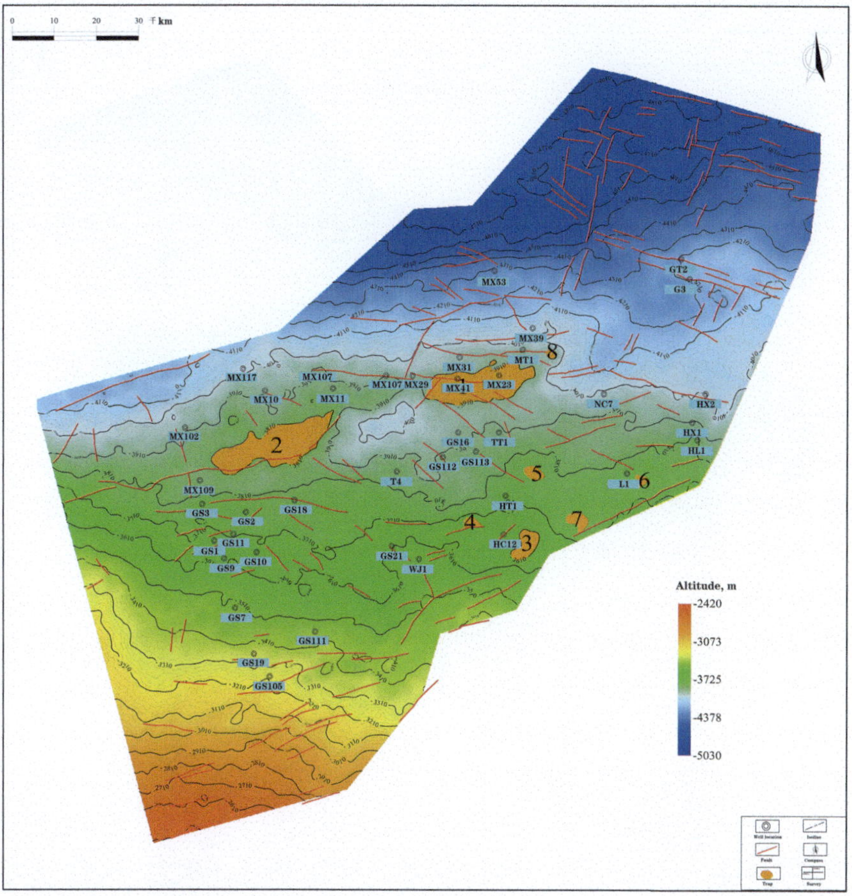

Fig. 10 Structure of top Maokou Formation in Central Sichuan

Structural Characteristics of Bottom Leikoupo Formation

The altitude interpreted by seismic data of Leikoupo Formation is in good agreement with the actual drilling altitude. The absolute error range is 0.02–8.69 m and the relative error range is 0–0.36%. For example, the altitudes of Well Gaoshi 111, Guangtan 2, Hetian 1, Moxi 41, Tong 4 and Tongtan 1 agree with the structural map values. The bottom Leikoupo Formation is characterized by a monocline dipping from southwest to northeast as a whole (Fig. 12). The structural altitude is − 1320 m to − 3782 m. Five structural traps are found in Leikoupo Formation, with a cumulative area of 88 km^2.

Structural Characteristics of Bottom Leikoupo Formation

The altitude interpreted by seismic data of top Leikoupo Formation is in good agreement with the actual drilling altitude, with an absolute error range of 0.02–8.43 m

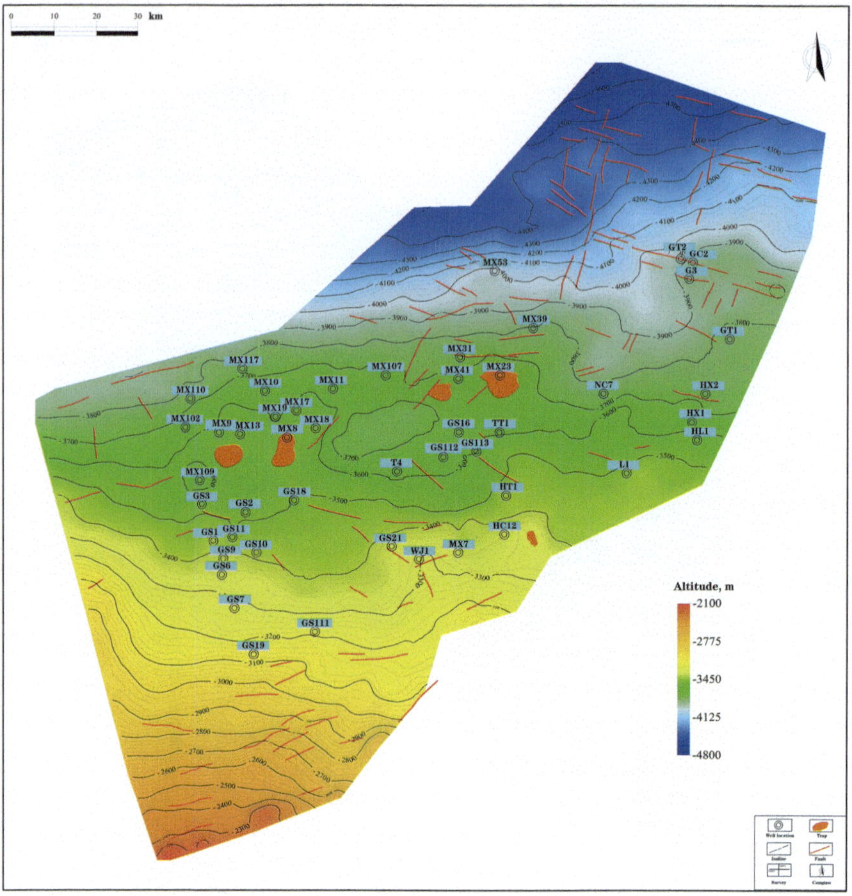

Fig. 11 Structure of top Changxing Formation in Central Sichuan

and a relative error range of 0–0.42%. For example, the altitudes of Well Gaoshi 113, Gaohetian 1, Lai 1, Moxi 29 and Wangjia 1 are in good agreement with the values on the structural map. The structure of top Leikoupo Formation is characterized by a monocline dipping from southwest to northeast as a whole (Fig. 13). The structural altitude is – 1210 to – 2810 m. Seven structural traps are found in top Leikoupo Formation, with a cumulative area of 1199.8 km^2.

1.5 Fault Development Stages

Tectonic movements with multiple stages and cycles are developed in Central Sichuan. Each tectonic cycle develops under both extension and compression, and

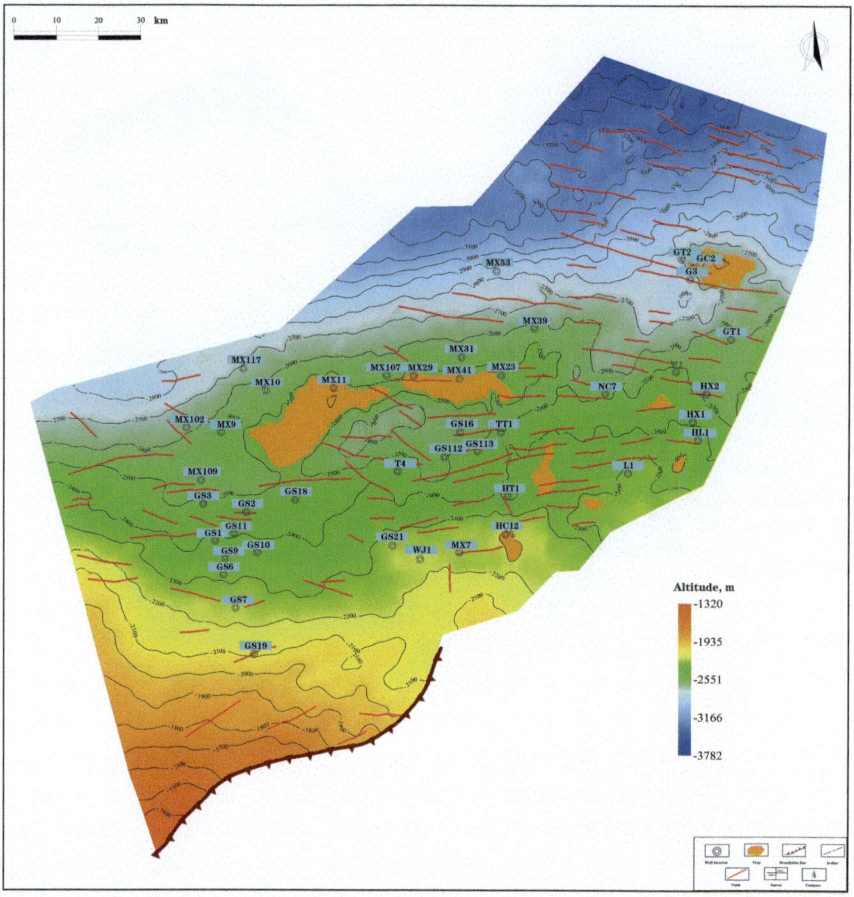

Fig. 12 Structure of botton Leikoupo Formation in Central Sichuan

different directions of tectonic stress. As a result, it is difficult to divide the fault stages. In addition, due to the multi-cycle tectonic evolution of Sichuan Basin, the faults formed in the early stage are often activated during the later tectonic movement, and characterized by the result of multi-stage activities. For such faults, it is difficult to determine the final activity stage. Therefore, it's difficult to determine the activity stages of large-scale faults that vertically break through multiple horizons. The study area is located in the south wing of Leshan-Longnvsi uplift, and the overall structure is relatively flat. By analysing horizon characteristics by comparative interpretation, faults with different scales are mainly developed in the study area, with the strikes close to NW, EW, NE and NS which are distributed in rows and belts. The following are the criteria used in this study to evaluate fault activity: First, the stages can be determined by the horizon cut by the fault in the seicmic section. The formation period of the latest horizon cut by the fault is the latest period of fault activity, that

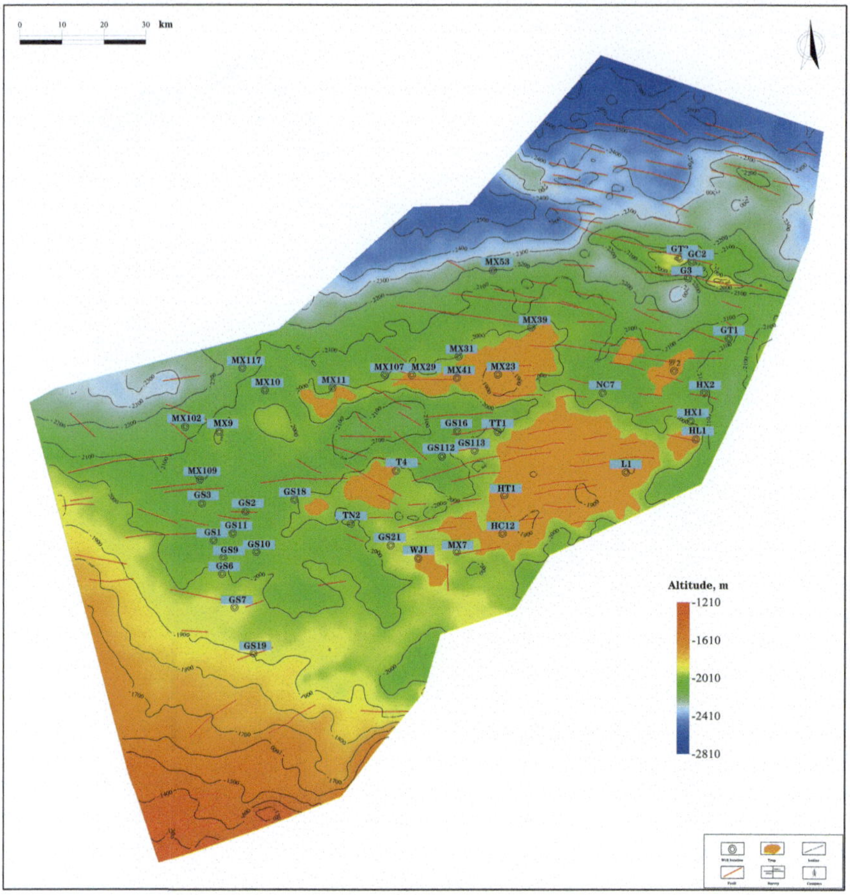

Fig. 13 Structure of top Leikoupo Formation in Central Sichuan

is, the fault forms before the formation period of the latest horizon to which it cut. Second, based on the study results of regional tectonic stress and tectonic evolution, properties of a fault in specific stage and itsextension direction in the plane can be determined.

According to the fault interpretation results of the study area, five stages of faults are developed in the study area (Fig. 14). The first stage is Tongwan period. The faults in this stage are mainly normal faults, mainly developed on the eastern Deyang-Anyue rift trough (Fig. 15) and mainly arrive at the bottom of Cambrian Qiongzhusi Formation. On the plane, the fault strikes are mostly near NS, and there are also normal faults developed with near EW direction locally. The development of faults in this stage is related to the development of Deyang-Anyue rift trough. The second stage is the Caledonian period. Most of the faults in this stage are normal faults with NW trending (Fig. 15), mostly reached the Lower Paleozoic. Some faults are activated in the Caledonian period and reach to the strata of Upper Paleozoic. The third stage is Dongwu period. The fault strikes in this stage are mostly EW in the southern study area and NW in the northern study area, which are parallel to the strikes of Pengxi-Wusheng trough and Kaijiang-Liangping trough formed in the sedimentary period of Changxing Formation. The fourth stage is Indosinian period. In the southern part of the study area, the plane fault strikes in this stage are mostly NE, while in the northern part, the fault strikes are mostly NNE. The formation of the faults in this stage is related to the formation of Luzhou palaeo-uplift formed in Indosinian period, which is related to the compressive stress with NW–SE direction. The fifth stage is Yanshanian period. The Faults of Yanshanian period are mainly developed in the Middle and Upper Triassic and Jurassic with the development of folds. Jurassic strata were formed by folds. So the faults are considered to be the product of tectonic movement after Jurassic (Fig. 16). The faults in this period are mainly developed with plane strikes of near EW and NW, and are mainly reverse faults. It is speculated that the formation of the faults is related to the rapid uplift and basin thrust of the Micangshan-Dabashan area in the northern part of the Late Jurassic. Under the action of SN compressive stress, a detachment layer is formed in the gypsum salt development zone from Leikoupo Formation to Jialingjiang Formation, at the same time, faults and folds are formed on the detachment layer.

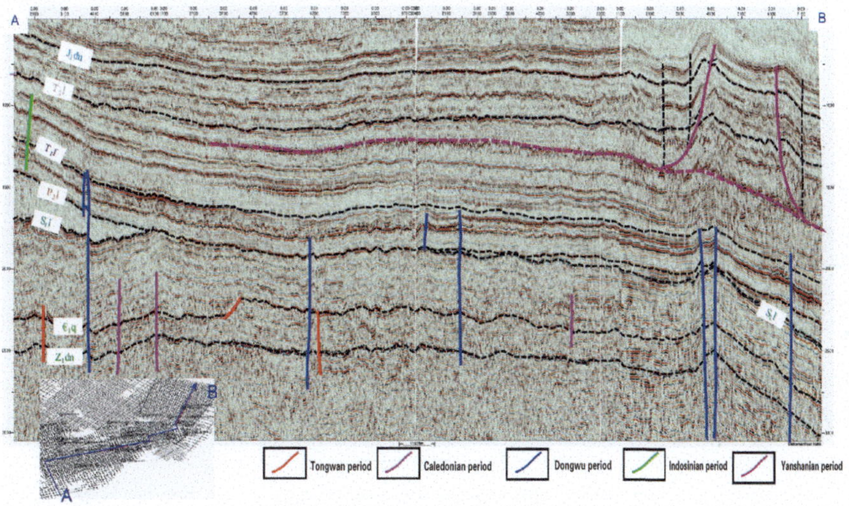

Fig. 14 Characteristic of faults in each stage of Central Sichuan

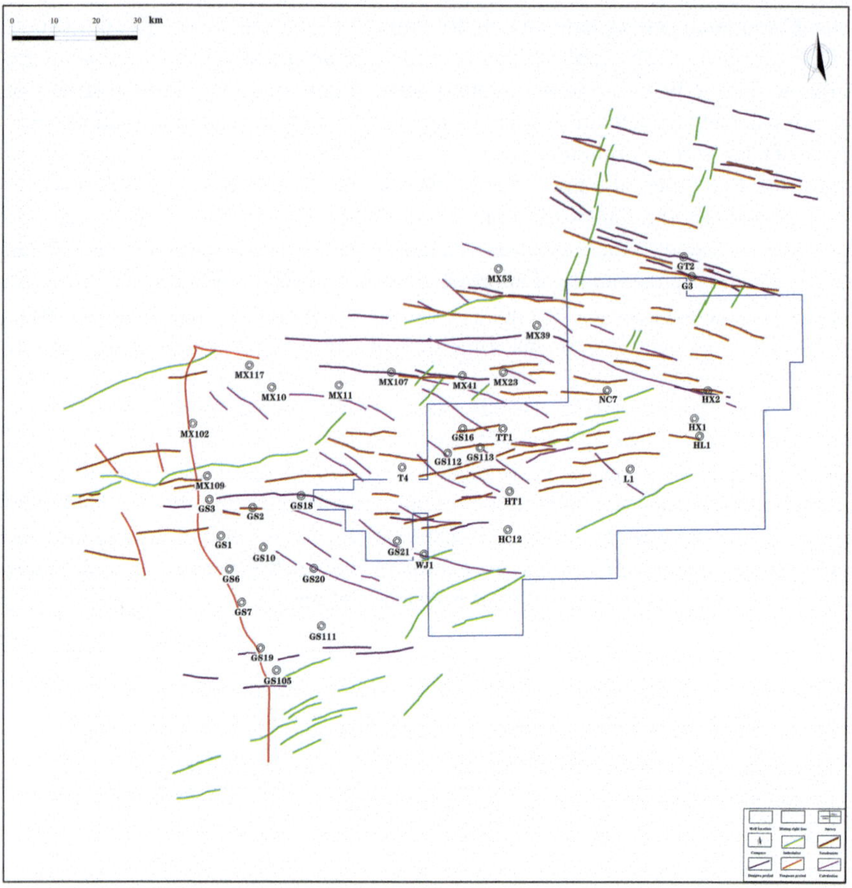

Fig. 15 Plan division of main faults in each stage of Central Sichuan

2 Erosion Recovery of Key Periods

2.1 Overview

The erosion recovery is very important, as the erosion parameters need to be provided in many aspects of oil and gas exploration research, such as stratigraphic burial history, basin heat flow history, hydrocarbon generation history of source rock, oil and gas filling history, resource calculation, dynamic evolution of caprock sealing and evaluation of oil and gas preservation conditions. There are many methods for erosion recovery, such as the method of vitrinite reflectance, apatite fission track, formation correlation (analogy), porosity, acoustic time difference method, deposition rate, deposition wave analysis, etc. For the erosion recovery of Sichuan Basin, predecessors (Zeng 1988; Lu et al., 2005, 2006; Yuan et al., 2008, 2013; Deng et al.,

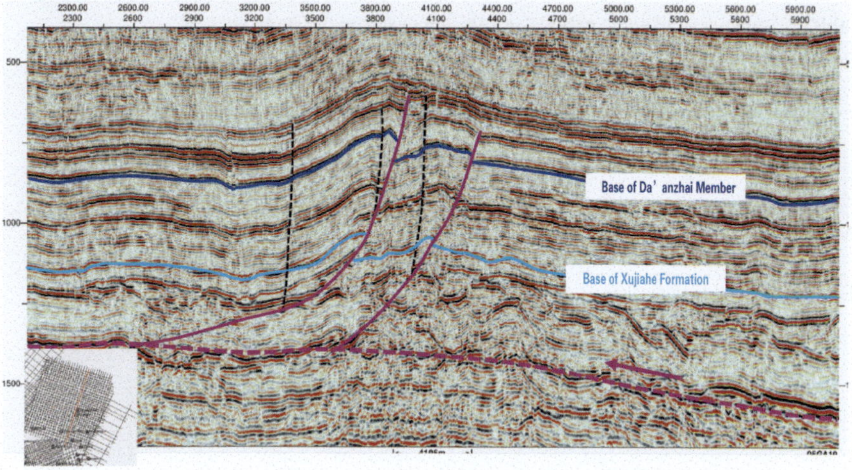

Fig. 16 Fault characteristics of Yanshanian period in Central Sichuan

2009; Jiang et al., 2012; Zhang et al., 2013) preliminarily restored the erosion of different strata in different periods by using sedimentation rate ratio, inversion and correction of vitrinite reflectance and inversion of paleotemperature scale. The above research results have quantitatively constrained the erosion of each stratum in Sichuan basin to a certain extent or within a certain period. However, all the existing erosion recovery methods have their own inherent limitations and applicable conditions, and the number of wells that can be used for the erosion recovery is always limited due to the limitation of data sources. One of the major issues is how to reasonably draw the plane map of erosion in the study area through the recovery results of erosion from some representative drilling wells or sections.

Firstly, based on the sedimentary strata and previous studies, the regional tectonic events and occurrence time are clarified, and the regional tectonic evolution process is established (Table 1). The credible recovery result of stratum erosion must be consistent with the regional geological background conditions and sedimentary structural evolution characteristics. Multicycle development and evolution process are formed in Sichuan Basin, including rifting (Sinian) → craton (Late Sinian-Silurian) → re-rifting (Devonian and Carboniferous) → craton (Permian-Middle Triassic) → foreland basin (Middle Triassic-Paleogene). Regional unconformities are developed between each cycle, and there is a certain degree of stratigraphic erosion on the unconformity surface. Generally speaking, after the deposition of Sinian Dengying Formation, Tongwan period (I, II and III), Yunnan-Guizhou period, Caledonian period, Hercynian period (Yunnan), Dongwu period (I and II), Early and Late Indosinian period, Yanshan period and Himalayan period have been developed in the study area.

Secondly, based on the characteristics of existing regional data, structural evolution and unconformity distribution, the method for erosion recovery is determined.

Table 1 Statistics of tectonic periods and erosion events in the study area

Strata		Beginning of deposition (Ma)	End of deposition (Ma)	Beginning of denuration (Ma)	End of denuration (Ma)	Periods of movement
Cretaceous		145	100	100	0	Yanshanian, Himalayan
Middle Jurassic		170	145			
Late Jurassic	Lianggaoshan Formation	174	170			
	Da'anzhai Formation	183	174			
	Maanshan Formation	191	183			
	Dongyuemiao Formation	199	191			
	Zhenzhuchong Formation	201	199			
Upper Triassic	Xujiahe Formation	237	208	208	201	Late Indosinian
Middle Triassic	Leikoupo Formation	247	241	241	237	Early Indosinian
Lower Triassic	Jialingjiang Formation	251.2	247			
	Feixianguan Formation	251.9	251.2			
Upper Permian	Changxing Formation	256.1	253.9	253.9	251.9	DOngwu II
	Longtan Formation	259.1	256.1			
Middle and Lower Permian	Maokou Formation	273	266.7	266.7	259.1	Dongwu I
	Qixia Formation	290	273			
	Liangshan Formation	299	290			
Upper Carboniferous	Huanglong Formation	323.2	307	307	299	Hercynian (Yunnan)
Lower Carboniferous	Hezhou Formation	358.9	323.2			
Upper Devonian				425.6	358.9	Caledonian
Middle Devonian	Hanjiadian Formation	433.4	425.6			

(continued)

Table 1 (continued)

Strata		Beginning of deposition (Ma)	End of deposition (Ma)	Beginning of denuration (Ma)	End of denuration (Ma)	Periods of movement
Lower Devonian	Longmaxi Formation	443	433.4			
Upper Ordovician	Wufeng Formation	458.4	443			
Middle Ordovician	Baota Formation	467.3	458.4			
	Shizipu Formation	470	467.3			
Lower Ordovician	Meitan Formation	477	470			
	Honghuayuan Formation	480	477			
	Tongzi Deng4 Member (Nanjinguan Formation)	485.4	480			
Middle and Upper Cambrian	Xixiangchi Formation	497	494	494	485.4	Yeli
Middle Cambrian	Gaotai Formation	500	497			
Lower Cambrian	Longwangmiao Formation	514	509	509	500	Yunnan-Guizhou
	Canglangpu Formation	521	514			
	Qiongzhusi Formation	526.5	521			
	Maidiping Formation	535.2	531.2	531.2	526.5	Tongwan III
Sinian	Deng4 Member	538.2	535.5	535.5	535.2	Tongwan II
	Deng3 Member	539.6	538.2			
	Deng1, 2 Member	551.1	541	541	539.6	Tongwan I

That is, the method of Milankovitch cycle for the carbonate rock erosion in single-stage tectonic movement, the method of multi-layer accumulation and seismic section extension for the denudation caused by multi-stage structure movements, with the combination of previous researches.

(1) Milankovitch cycle

The Milankovich cycle analysis can be used to effectively record the characteristics of the sedimentary cycles within the stratum. It can also restore the erosion of carbonate strata lost in the local region through mutual comparison and verification of the difference in stratum deposition rate and erosion thickness. According to Milankovitch's theory, the summer solar radiation in the middle and high latitudes of the southern and northern hemisphere is changed by the periodic changes of the three elements, including the earth's orbit eccentricity, inclination and precession. And the sea (Lake) level is regularly changed under the effect of the periodic fluctuations of the climate. As a result, the sedimentary sequences are developed regularly, and the lithology and lithofacies show rhythm and cyclicity (Table 2). If the Milankovitch cycle in the stratum is identified by spectrum analysis technology, the sedimentary time limit and the characteristics of sedimentary and erosion in the stratum can be calculated according to its inherent cycle. The change of relative sea level in carbonate sedimentary environment is generally analyzed with the Th/K curve by scholars at home and abroad. However, due to the high price of gamma ray spectral logging, the logging work on this kind of curve is not carried out in many wells. By comparing and analyzing the Milankovitch cycle characteristics of Th/K curve and gamma ray curve in the same well, the practical analysis shows that, in the absence of Th/K curve, the gamma ray curve can be used to replace the Milankovitch cycle, and the error is within an acceptable range. The main work of Milankovitch cycle analysis is to carry out spectrum analysis. The essence lies in that a logging curve is considered a comprehensive superimposed signal reflecting the regular stratum changes, which is formed under the comprehensive action of various geological elements on the time scale (or depth scale). The spectrum curve can be formed by the transformation of the superimposed signal (logging curve) from the depth scale (or time scale) to frequency scale by Fourier transformation.

The analysis process is as follows. ① Original curve analysis is used to remove abnormal values; ② Wavelet transform is used to decompose the gamma ray spectrum curve or gamma ray curve with fixed sampling interval, in order to remove the low-frequency background noise and high-frequency noise signal of the curve, and then reconstruct the logging data containing regular low-frequency components; ③ Fourier transform method is used to transform the reconstructed data from time domain to frequency domain, in order to form a 1D spectrum curve; ④ The cycle period parameters calculated by the spectrum curve are compared with the Milankovitch astronomical parameters; ⑤ The stratum erosion is figured out with the correlation parameters and deposition rate.

The specific methods are as follows. ① Data selection and processing. The logging data of Ln(Th/K) and GR from key wells are used for data resampling, with the original sampling interval of 0.125 m, in order to eliminate data with low-frequency, strong interference and high-frequency. ② Spectrum analysis. Redfit software is used for spectrum analysis and calculation, and the spectrum curve is recorded in the form of frequency. The high point frequency in the research's periodic frequency is selected within 95% confidence. The high point of periodic frequency represents

Table 2 Sequence level division and sequence stratigraphic terms

Level	Term	Formation time limit (Ma)	Formation mechanism	
First-order	Megasequence	200–400	Tectonic change of sea level caused by plate movement	Change of global sea level caused by the formation and disintegration of the pan continent
Second-order	Supersequence	10–40		Change of global sea level caused by the spread of mid-ocean ridges
Third-order	Sequence	1–10		Effect on the change of regional sea level caused by global sea level change and tectonic subsidence and uplift inner plate by mid-ocean ridge change and continental glacier growth and decline
Fourth-order	Subsequence set	0.4	Change of glacial sea level caused by Milankovitch cycle	Long eccentricity cycle
Fifth-order	Subsequence	0.1		Short eccentricity cycle
Sixth-order	Rhythmity/meter-scale cycle	0.02 or 0.04		Precession cycle or yellow red angle cycle
Seventh-order	Rhythm layer	0.002–0.005	Glacier melting and geoid change	

more frequent cycles of this frequency appearing in the stratum. ③ Comparation and analysis of orbital period and spectrum. The average cycle thickness and its ratio relationship corresponding to the frequency high point of each cycle in the spectrum curve is repeatedly verified and compared. The closer the ratio is to the natural ratio of the earth orbit cycle in the sedimentary period (Table 3), the more it can reflect that the stratigraphic cycle is affected by the Milankovitch cycle period. The average cycle thickness and deposition rate of the layers can be figured out by the value of the periods. ④ Wavelet analysis and verification. Morlet algorithm is used for wavelet transform to verify the vertical change of the period. ⑤ Erosion calculation. The total sedimentary thickness without erosion can be obtained by comparing the existing complete sedimentary strata (non-denuded) or sedimentary cycles, and by referring to the deposition rate and cycle thickness. And the estimated denuded thickness of the formation (member) can be figured out by deducting the residual stratum thickness (Fig. 17) from the total sedimentary thickness without erosion.

The functional relationship used are as follows: $\lambda = 1/f$, $Hc = \lambda L$, $Nc = Hr/Hc$, $Tr = NcT$, $Vd = Hc/Tf$ represents the frequency; λ represents the wavelength, that is, the number of cycles contained in the formation with the thickness of each sampling interval; L represents the sampling interval of logging curve; Hc represents the average cycle thickness of the formation, m; Nc represents the number of cycles

Table 3 Milankovitch cycle period

Geologic age	Age (Ma)	Precession cycle (ka)		Slope period (ka)		Eccentricity period (ka)
		S1	S2	X1	X2	E
Quaternary	0	19.00	23.00	41.00	54.00	E1 = 95 E2 = 125 E3 = 413
Paleogene	50	18.80	22.60	39.90	52.10	
End of Cretaceous	72	18.62	22.45	39.28	51.06	
Changxing Formation	254	18.08	21.67	37.03	47.38	
Maokou Formation	270	17.55	20.90	34.78	43.70	
Early Permian	298	17.39	20.68	34.16	42.74	
Late Devonian	380	16.92	20.02	32.39	40.00	
Early Silurian	440	16.57	19.53	31.13	38.10	
Late Ordovician	450	16.30	19.20	30.30	36.80	
Middle Ordovician	470	16.26	19.13	30.13	36.58	
Early Ordovician	480	16.24	19.10	30.05	36.47	
Xixiangchi Formation	497	16.22	19.06	29.97	36.35	
Late Cambrian	500	16.21	19.03	29.89	36.25	
Longwangmiao Formation	510	16.09	18.90	29.38	35.45	
Deng4 Member	537	16.03	18.82	29.20	35.19	

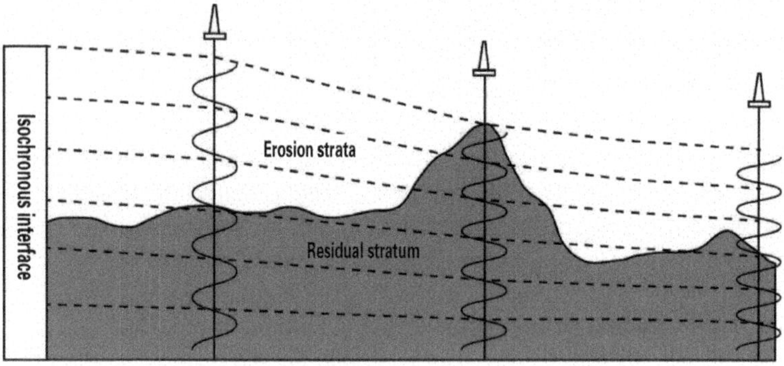

Fig. 17 Schematic of erosion recovery by spectrum analysis

contained in the residual formation; Hr represents the thickness of residual stratum, m; Tr represents the deposition time of residual strata, Ma; T represents cycle period, Ma; Vd represents the formation sedimentation rate, m/Ma.

(2) Stratigraphic extension method

The erosion unconformity trend and erosion thickness are restored by stratigraphic structure extension and related deformation analysis methods. The deformation axis and relevant deformation are inferred from the development of known strata and faults. Based on the data from the existing strata, the erosion strata can be interpreted according to the stratum development trend.

With the comprehensive analysis of well-seismic data in Central Sichuan, it can be seen that unconformities formed by three stages of large-scale erosion are mainly developed in the study area. In the process of erosion recovery, the erosion trend of unconformity is mainly restored by the methods of stratigraphic extension and related deformation analysis (Fig. 18), and then the erosion thickness is estimated. The basic principle is to identify the main unconformity based on fine seismic section interpretation and restore the pattern, development characteristics and distribution of the stratum before erosion, based on the stratum characteristics of non-denuded stratum close to the unconformity, such as development characteristics, extension trend, fault cutting relationship and fold deformation characteristics, as well as related deformation characteristics and extension trend. The distance between the restored stratum interface and the unconformity surface (Erosion surface) is the thickness of the denuded stratum (Erosion thickness). The specific implementation steps are as follows. Firstly, fine interpretation of unconformities, faults and folds on seismic sections. Secondly, the non-denuded stratum closest to the unconformity is selected, and according to the extension pattern and characteristics of fold and fault, the stratigraphic morphology of the unconformity before the erosion is restored and is paralled to the selected stratum. Thirdly, the thickness of the denuded stratum is estimated, and the denuded paleogeomorphic plan according to the erosion value is drawn.

Through a thorough analysis of drilling and seismic data in Central Sichuan, in combination with previous research, it is found that the studied region had experienced three distinct stages of erosional unconformity caused by fold uplift.. The unconformity formed in the first stage is the erosion unconformity on the palaeo-uplift of Leshan-Longnvsi. Before the deposition of Permian, the strata on the palaeo-uplift of Leshan-Longnvsi deposited before the Permian suffered serious erosion, forming the angular unconformities between the Permian and the underlying strata. According to the stratum extension trend in Fig. 19, the strata of Cambrian-Ordovician gradually narrows towards the direction of Leshan-Longnvsi uplift, and the erosion thickness gradually increases. In the northwest of the study area, the maximum erosion thickness before the deposition of Permian is 1300 m (Fig. 20), and the erosion thickness gradually decreases towards both the south wing and the dip direction of Leshan-Longnvsi palaeo-uplift.

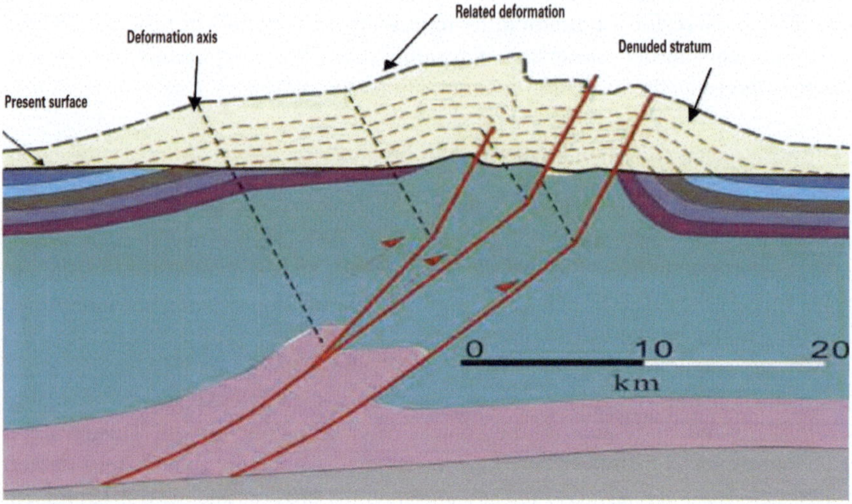

Fig. 18 Schematic of unconformity erosion trend restored by stratigraphic structure extending and related deformation analysis

The unconformity formed in the second stage is the erosional unconformity of the top Leikoupo Formation. The unconformity in this stage is formed by the formation of Luzhou palaeo-uplift in the eastern part. Due to the tectonic movement in this period, the deposition of marine carbonate rock in the study area is ended and transformed into the deposition of continental clastic rock. On the whole, the strata of Leikoupo Formation in the study area gradually thins to the southeast, which is angular unconformity with the overlying Xujiahe Formation, and the erosion thickness gradually increases to the southeast. In the south of the study area, the maximum erosion thickness of Leikoupo Formation reaches 240 m (Fig. 21). The erosion thickness gradually decreases northwardly towards Well Moxi 53—Moxi 39—Nanchong 7—Hualai 1, and gradually increases from Well Moxi 53—Moxi 39—Nanchong 7—Hualai 1 to the east of the study area.

The third stage is the differential erosion of Jurassic on the present surface. Due to the Yanshanian movement, some anticlines are developed in the Jurassic of the study area, as a result, the strata of Jurassic are uplifted and denuded (Fig. 22).

2.2 *Erosion Recovery of Tongwan Movement*

The Tongwan movement was first proposed by Liu Guochang, which refers to the unconformity formed between the Early Cambrian Wulipai Formation and Tillite of Nantuo Formation in Tongwan and Yinzangwan of Huaihua, Western Hunan, and may develop on a large scale. However, the later geologists have made many changes

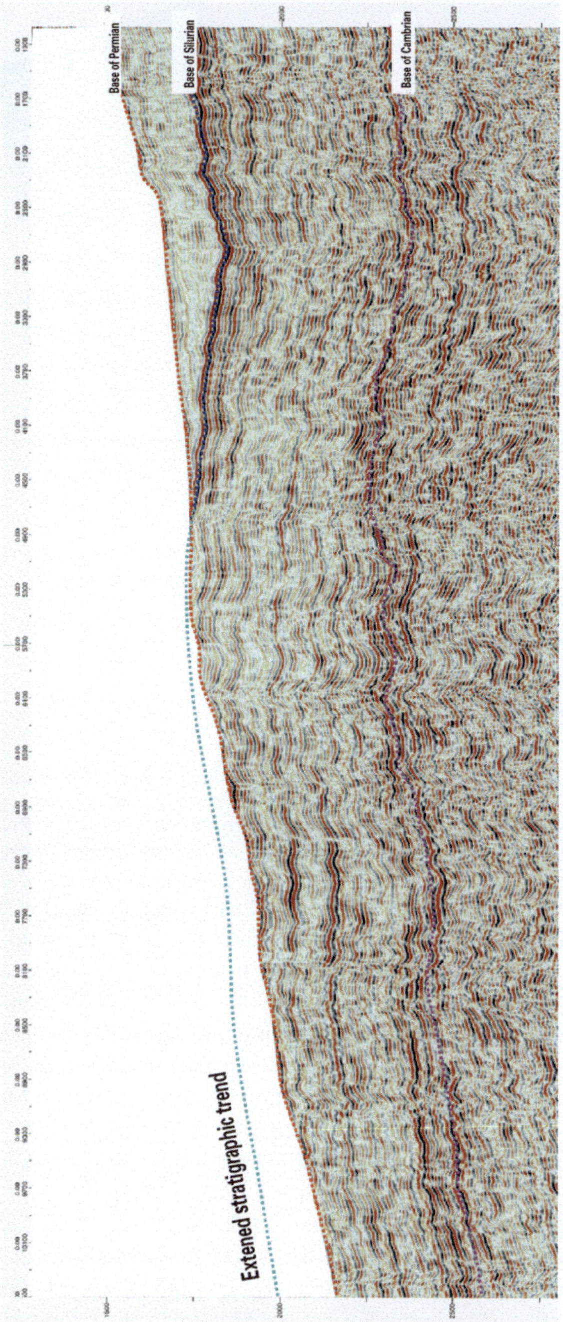

Fig. 19 Stratigraphic structure extension of Upper Cambrian-Ordovician in Leshan-Longnvsi uplift

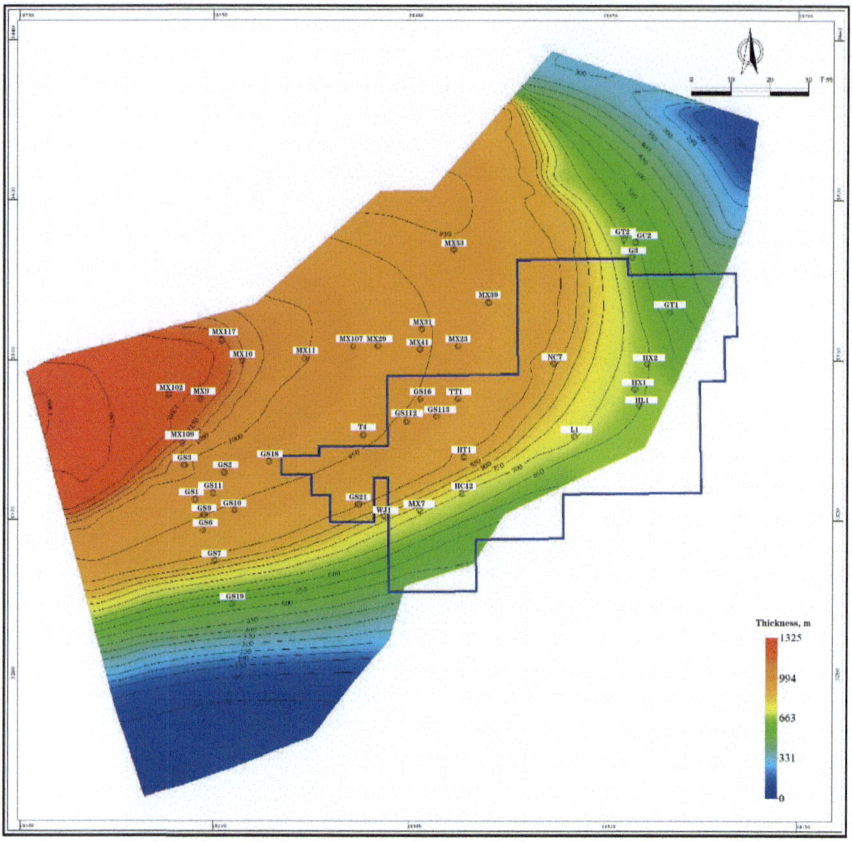

Fig. 20 Erosion recovery of Upper Cambrian-Ordovician in Leshan-Longnvsi uplift

in their understanding of the development period and epoch of the Tongwan movement (Hou et al., 1999; Li et al., 2010, 2013; Xu et al., 2012; Liu et al., 2016). It is believed that the Tongwan tectonic movement is the overall episodic uplift of the Sichuan Basin and its adjacent areas from Late Sinian to Early Cambrian, with local differential rise and fall, which can be divided into three episodes. In the same time, 3 disconformities were developed. One view is that the three episodes of Tongwan movements occurred at the end of the deposition of Deng 2, Deng 4 Member and Maidiping Formation of Early Cambrian, respectively (Wang et al., 2002, 2014; Gu et al., 2014; Wu et al., 2016). Another view is that the Tongwan movement originated between the algae-rich layer and the algae-poor layer in the Deng 2 Member. The second and third episodes occurred at the end of deposition of Deng 2 and Deng 4 Member, respectively (Li et al., 2015). In this study, it is considered that the three episodes were respectively developed between Deng 3 and Deng 2 Member, Deng 4 Member and Qiongzhusi/Maidiping Formation, Maidiping and Qiongzhusi Formation (Fig. 23). Maidiping and Qiongzhusi Formation were developed as the important

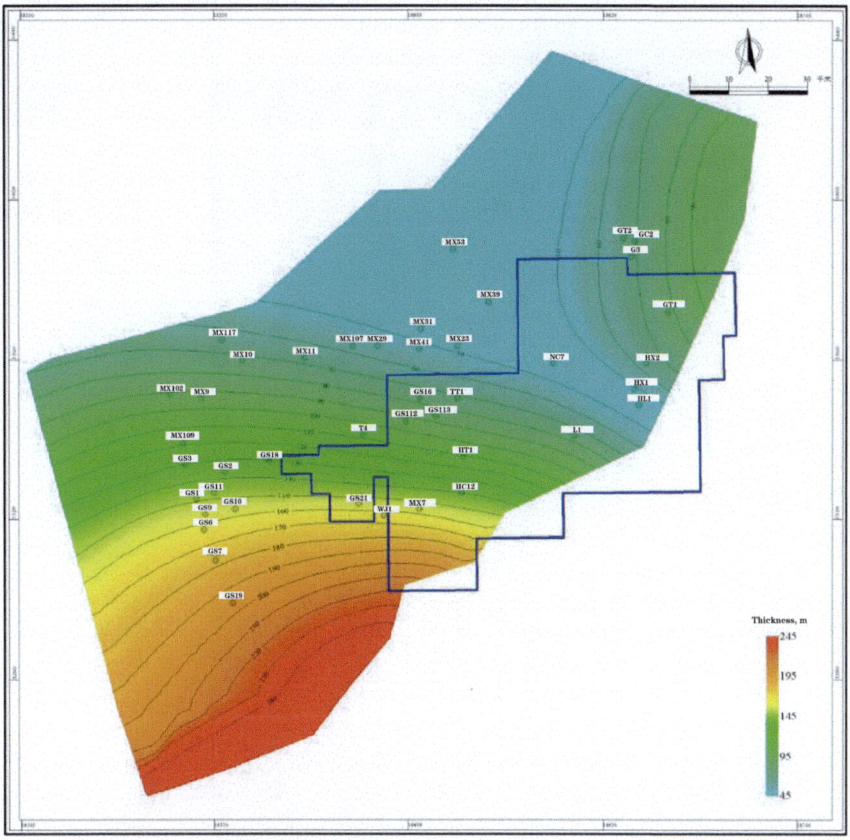

Fig. 21 Thickness of erosion recovery in Leikoupo Formation

hydrocarbon generating centers of Cambrian source rocks. Compared with two sets of weathering crust karst reservoirs in Deng4 and Deng 2 Member of Dengying Formation, large-scale gas field with great exploration potential can be formed.

(1) The First Episode of Tongwan Movement

The first episode of Tongwan movement occurred at the end of the deposition of Deng 2 Member, which showed that the layer between regional clastic rocks of Deng 3 Member and the dolomites of Deng 2 Member was a disconformity.Although the Deng 3 Member is mostly composed of clastic rocks, the lithology vary significantly. In this research, the erosion thickness is not calculated, as the Deng 1 and Deng 2 Member are not divided with few wells drilled in this section of the study area. It is estimated that the erosion thickness is about 100 m.

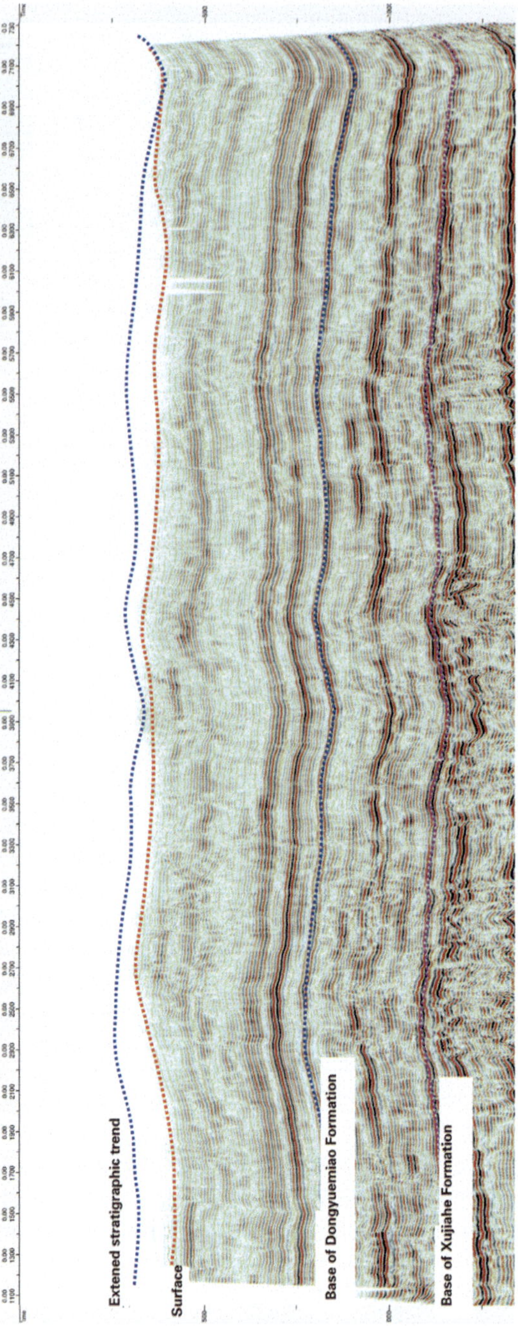

Fig. 22 Stratigraphic structure extension of current Jurassic

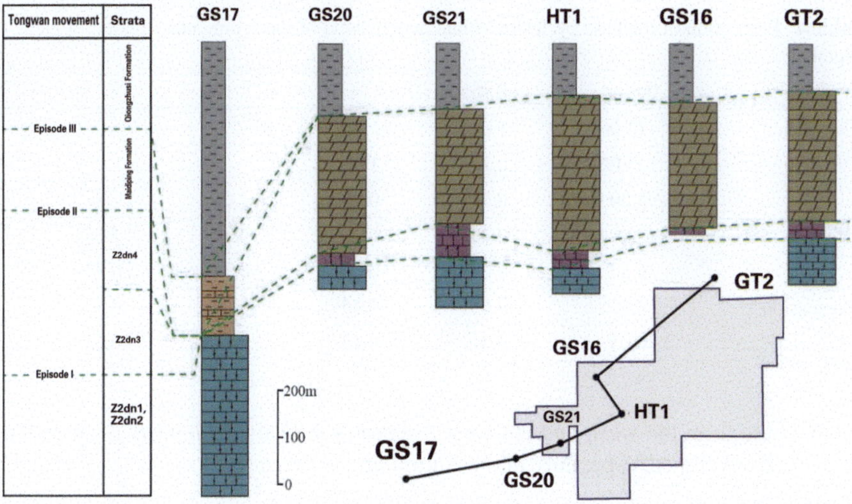

Fig. 23 Periods of Tongwan movement in Central Sichuan

(2) The Second Episode of Tongwan Movement

The second episode of the Tongwan movement occurred at the end of Dengying Formation, which was characterized by the disconformable contact between Maidiping Formation and Dengying Formation in Lower Cambrian. This phenomenon is obviously developed around the study area. The Maidiping Formation is developed with the residual thickness of 0–20 m, and is missing in most wells. The black argillaceous rocks of Qiongzhusi Formation directly cover the dolomites of Deng4 Member. The frequencies are calculated by spectrum analysis with Ln (th/k) or GR logging data in the eight wells drilled through the Deng4 Member. The average cycle thickness and its ratio relationship corresponding to each cycle frequency high point in the spectrum curve are verified and compared repeatedly, as well as the relationship with the natural ratio of the earth orbit cycle in the sedimentary period at that time (Table 4; Fig. 24). As a result, the period can be determined to calculate the formation's average cycle thickness and deposition rate. It is comprehensively considered that the deposition time of Deng4 Member is 2.7 Ma. The denuded thickness of the formation (member) can be estimated by subtracting the thickness of current residual formation from the total deposition thickness.

Finally, the erosion thickness of Deng 4 member can be determined with the stratigraphic extension method (Fig. 25). Affected by Gaoshiti-Moxi palaeo-uplift, the erosion thickness of Deng4 Member in the study area is 140–250 m, and decreases from west to east. The erosion thickness shows that the structures in the study area are greatly affected by the Gaoshiti-Moxi palaeo-uplift formed in the second episode of Tongwan movement, and are located in the central and eastern parts of the palaeo-uplift. In terms of the paleogeomorphology, the structures in the study area are

Table 4 Proportion of the base cycles in Milankovitch cycle of the sedimentary period of Deng4 Member

Cycle	Period	Ratio			
Eccentricity	413.00	1.00			
	125.00	0.30	1.00		
	95.00	0.23	0.76	1.00	
Slope	35.19	0.09	0.28	0.37	1.00
	29.20	0.07	0.23	0.31	0.83
Precession	18.82	0.05	0.15	0.20	0.53
	16.03	0.04	0.13	0.17	0.45

relatively high in the west, and the karst reservoirs in Deng 4 Member is predicted to be well developed, especially in the central and western parts.

(3) The Third Episode of Tongwan Movement

The third episode of Tongwan movement occurred at the end of the Early Cambrian Maidiping Formation, which was characterized by the disconformable contact between the overlying Qiongzhusi Formation and the Lower Cambrian Maidiping Formation. It can be seen that on the east section of Jiangji, South Nanmaji in Central Guizhou, Maidiping Formation is in unconformity contact with the overlying Qiongzhusi Formation, with weathered crust clay layers and limonite layers of 10 cm. The Maidiping Formation is not drilled by the wells in the study area, while it is drilled by Well Gaoshi 17 outside the study area with the thickness of 126 m. It is speculated that the sedimentary thickness of Maidiping Formation in the study area is small, and the erosion caused by tectonic movement in this period is 0–150 m.

2.3 *Erosion Recovery of Caledonian Movement*

At the end of the Middle Silurian, the Indian and Eurasian plates made large-scale subduction and drift to the north and east again. In the weak areas of the Longmenshan fault, the Qinghai-Tibet plate was faulted on the Yangtze block plate, which had a greater impact than any other crustal movements since the Tongwan movement. Not only the Ya'an-Longnvsi palaeo-uplift with the length of 500 km in the Central Sichuan Basin was formed by the crust movement, which is high in the west and low in the east, but also the Luzhou palaeo-uplift was developed in the South Sichuan Basin, which is known as Caledonian orogeny, resulting in serious erosion of the Silurian in the basin. The Sichuan Basin was raised as a whole by the Caledonian movement and denuded for a long time before the Middle Carboniferous transgression. The Caledonian uplift and erosion began at the end of the Middle Silurian. It can be considered that there is no deposition in the Upper Silurian. The erosion thickness in

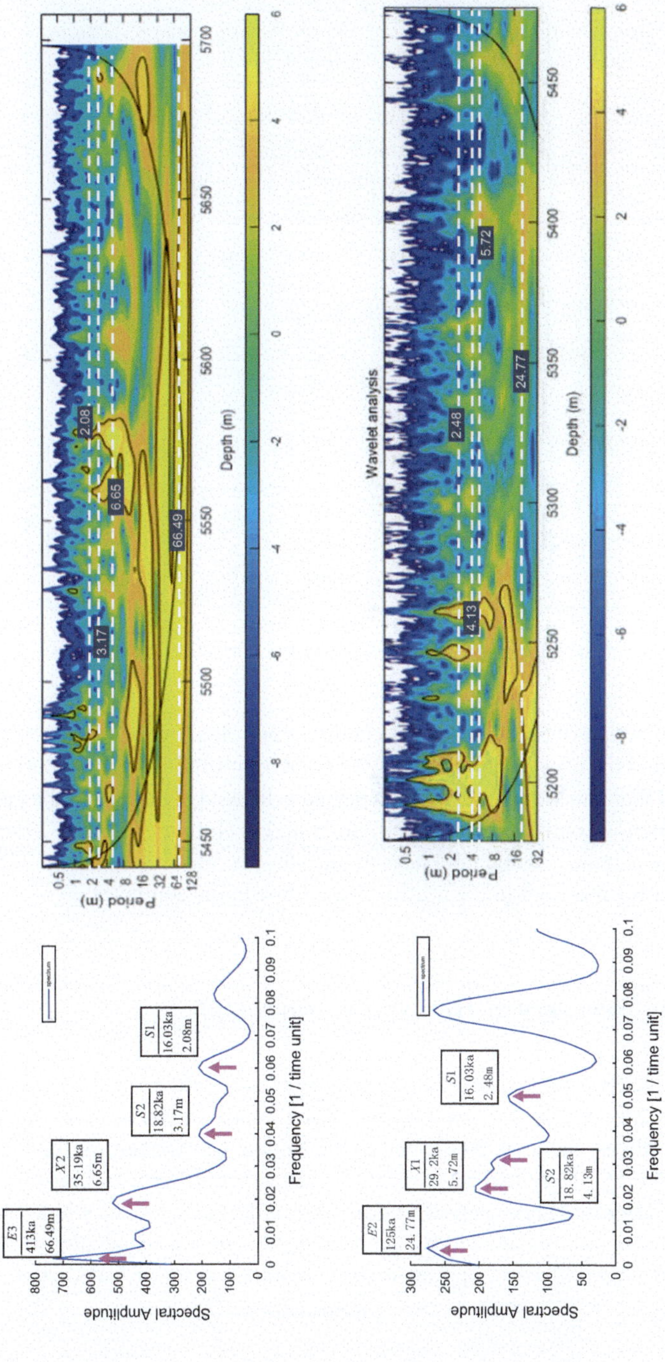

Fig. 24 Spectrum analysis of erosion thickness in Deng4 Member of typical well (Well Hetan 1 in the upper part and Well Gaoshi 20 in the lower part)

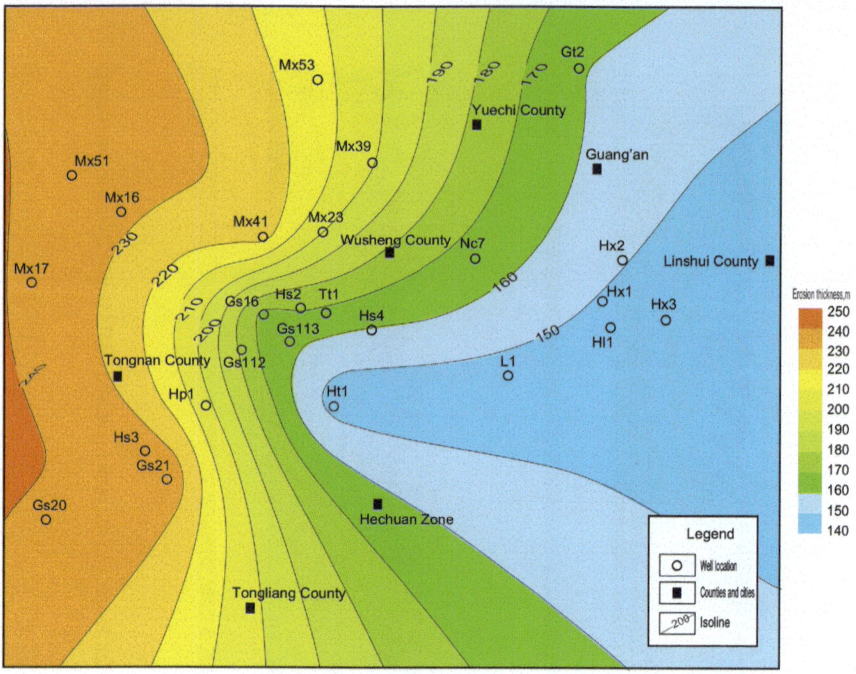

Fig. 25 Contours of erosion thickness of Deng4 Member in Central Sichuan

the Caledonian period of Sichuan Basin is mainly controlled by the Central Sichuan palaeo-uplift, and the erosion thickness in the palaeo-uplift area is 700–1200 m.

The Central Sichuan area of Caledonian period is located in the east of the palaeo-uplift core, which is controlled by the formation and evolution of Leshan-Longnvsi palaeo-uplift. The main erosion thickness in this stage is 500–1000 m (Fig. 26), and the ancient landform is high in the northwest and low in the southeast.

2.4 *Erosion Recovery of Hercynian Movement*

In the Middle Carboniferous, a set of carbonate rocks with small thickness were deposited in the Sichuan Basin by cutting and filling, especially in the low-lying areas of East and Northeast Sichuan. At the end of the Middle Carboniferous, the Sichuan Basin was raised by Hercynian movement. Until the Early Permian transgression, the previously deposited upper part of Middle Carboniferous, even the whole Middle Carboniferous was totally denuded by the large-scale erosion which almost across the whole Late Carboniferous, and even some residual strata of Caledonian were denuded.

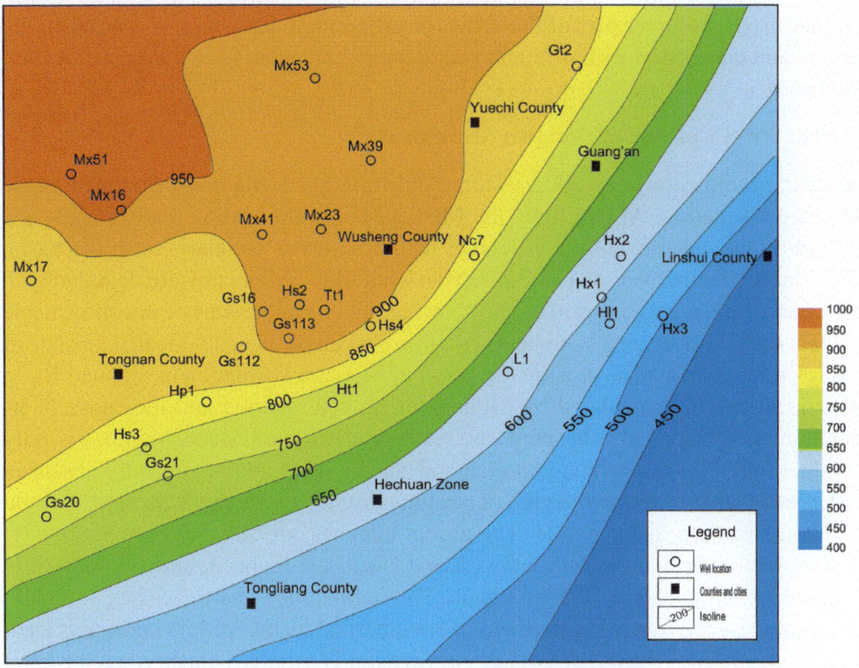

Fig. 26 Erosion thickness of Caledonian strata in Central Sichuan

In the end of Late Carboniferous, the west of Hechuan-Tongnan block was uplifted into land under the influence of Hercynian movement. The Huanglong Formation was not fully preserved due to the strong weathering and erosion. Part or all of the Huanglong Formation on the palaeo-uplift was denuded, forming the ancient karst landform at the top of the Huanglong Formation and the ancient karst system in the formation. With the small sedimentary thickness of Huanglong Formation, it is estimated that the erosion thickness in the study area is about 50 m.

2.5 *Erosion Recovery of Dongwu Movement*

The Dongwu movement occurred in the Permian, which was proposed and named by Li Siguang when he studied the Ningzhen mountains in Nanjing, 1931. The orogeny is generally considered as one of the most important tectonic events in the Late Paleozoic. The tectonic mode, movement nature, dynamic mechanism and impacts of the Dongwu movement on different plates are different. The Dongwu movement occurred between the Middle Permian and the Late Permian in the Yangtze continent where the Sichuan Basin is located. During this period, the crustal tension activities caused by the mantle plume gradually increased, and a large number of basalts were erupted from the inner part of the crust for differential deposition. As a result, the uplifts and depression with different sedimentary characteristics were developed in the basin.

(1) The First Episode of Dongwu Movement

After the sedimentary period, a tectonic uplift of 1–1.5 Ma was developed in the Maokou Formation. At that time, the Maokou Formation exposed the surface and differently denuded in different degrees as the seawater withdrew from the basin. The Mao 4 Member was denuded, the Mao3 and Mao2 Member were denuded with different degrees only in local areas, showing the characteristics of unconformable contact with the overlying Longtan Formation. As can be seen, it is the first episode of Dongwu movement. The erosion thickness of Maokou Formation in Sichuan Basin was quantitatively calculated by Milankovitch cycle method and the cause of its deletion was discussed. The predecessors considered that Mao4 Member is only residual in Yibin-Ya'an-Jiangyou area and Shizhu area, and is generally absent in other areas, with increased degree of stratum absence in Southern Sichuan-Central Sichuan-Northern Sichuan area. The absent thickness of Maokou Formation is 0–200 m, including 0–60 m in Southwest and Northeast Sichuan, and 140–200 m in Southern, Central and Northern Sichuan. The sedimentary characteristics of high in the southwest and low in the northeast were inherited by the Middle Permian's karst landform. In Southwest Sichuan-Central Sichuan-Northern Sichuan area, the karst landform gradually changed from the erosion highland to the upper karst slope and lower karst slope, which is consistent with the sedimentary characteristics of high in the southwest and low in the northeast in Wujiaping period.

The Maokou Formation in Central Sichuan is developed with the residual thickness of 188–227 m, and is roughly in NW–SE banded-distribution. The Moxi 107—Moxi 23—Gaoshi 16—Hetan 1 is developed with a large thickness of 196–213 m, and the northern part of Guangtan 2 is also developed with large thickness.

Based on the GR logging data of 20 wells drilled through Maokou Formation in Central Sichuan, the erosion thickness is calculated by spectrum analysis. The typical results are shown in Fig. 27.

The deposition time of Maokou Formation lasted for as long as 6.29 Ma. Combined with the stratigraphic extension method, the erosion thickness of Maokou Formation is finally figured out. Based on the results of erosion recovery, it can be seen that the erosion thickness of Maokou Formation in Hechuan-Tongnan block is small, and the overall thickness is 45–175 m (Fig. 28). In general, the area controlled by wells shows less erosion in the south and large erosion in the north, reflecting that the ancient landform is relatively high in the north, and the landform of the study area may be greatly affected by the Lujiang palaeo-uplift in the south.

(2) The Second Episode of Dongwu Movement

The regression event at the end of Permian not only lowered the sea level, but also exposed a large area for a long period of time. As a result, the carbonate rocks of Changxing Formation were generally transformed by supergene karst. clearly weathered crust can be seen on the stratigraphic section without reef facies in top

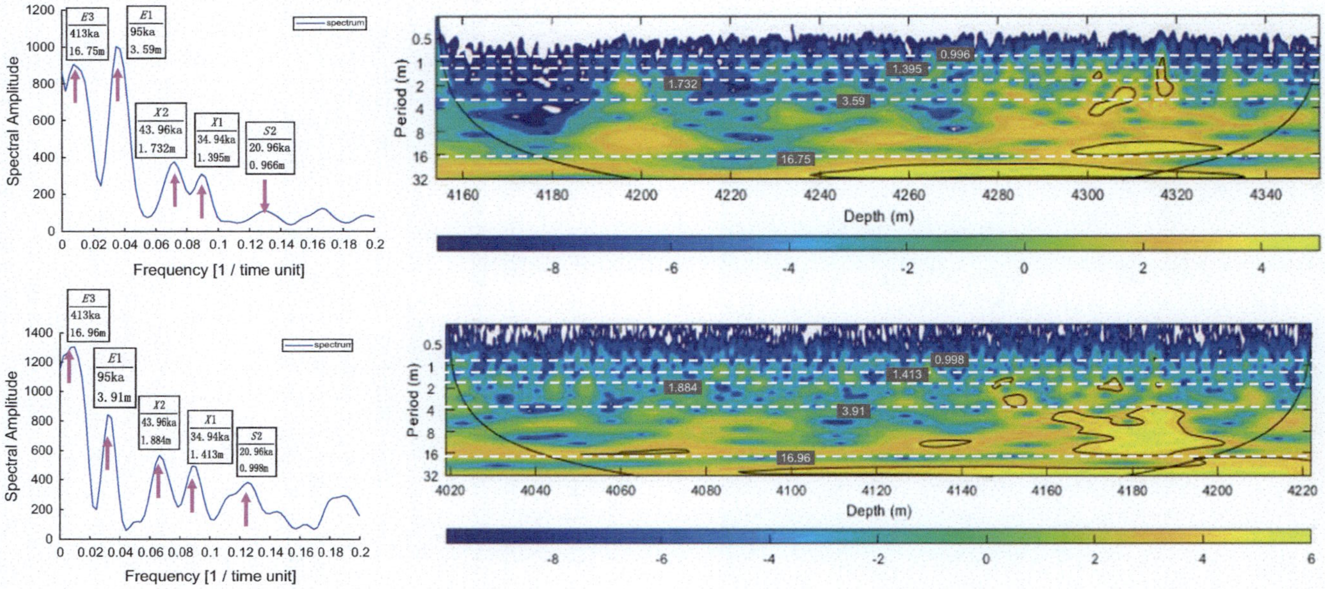

Fig. 27 Spectrum analysis of erosion thickness of typical wells drilled in Maokou Formation of the study area (Well Gaoshi 21 in the upper part and Well Hetan 1 in the lower part)

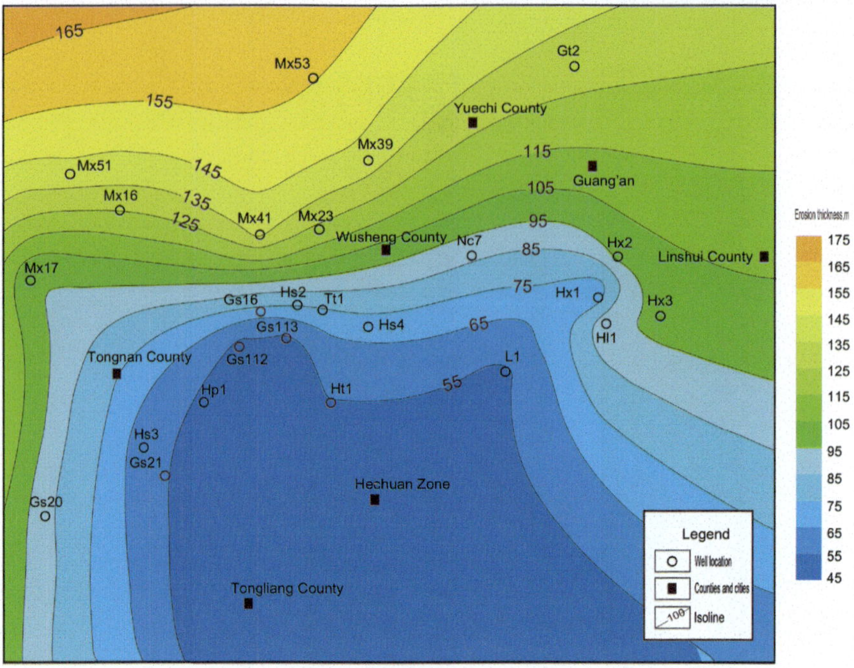

Fig. 28 Erosion thickness of Maokou Formation in the first episode of Dongwu period in Central Sichuan

Changxing Formation of Erlongkou, Fuling area, East Sichuan (Li et al., 2015). At the same time, the exploration confirmed that the high-quality reef-beach reservoirs of Changxing Formation are developed in the middle and upper part (Ma et al., 2005a, b). The regression event at the end of Permian is called the second episode of Dongwu movement.

In the end of Late Permian, the regional paleogeomorphology of Sichuan Basin was in the pattern of "three uplifts and three depressions". The "three depressions" refers to the two deep-water troughs of Kaijiang-Liangping and Chengkou-Western Hubei and depression of Pengxi-Wushengtai. The "three uplifts" refers to the high geomorphic belt of Suining, the platform area from high geomorphic belt of Guang'an to the west side of the Kaijiang-Liangping trough, and the carbonate platform between the Kaijiang-Liangping trough and the Chengkou-Western Hubei trough. The pattern of alternating uplift and depression developed in the early and middle deposition period of Changxing Formation, fell into a pattern in the late period of Changxing Formation, and still existed in the early and middle period of Feixianguan Formation. It is entirely possible for the platform-margin belt and the inner-platform geomorphic high belt to form wave-resistant reefs. The inherited geomorphic uplift characteristics also provide hydrodynamic conditions for the development of overlying oolitic beaches. Among them, the platform-margin belt of Huankaijiang-Liangping trough is a favorable distribution zone for reef of Changxing Formation and oolitic beach of Feixianguan formation, which has been proved by many years of exploration practice. The residual thickness of Changxing Formation in the study area is generally 40–200 m, with a NW–SE thickness increasing belt. The boundary between the inner-platform high belt and the inner-platform depression is obvious, and the inner-platform high zone is a favorable area for the reef-beach bodies.

Based on the GR logging data of wells drilled through the Changxing Formation in Central Sichuan, the erosion thickness is figured out by spectrum analysis. The typical results are shown in Fig. 29.

Combined with the stratigraphic extension method, the erosion thickness of Changxing formation is finally obtained. The results of erosion recovery show that, the erosion thickness of Changxing Formation in Central Sichuan is 30–85 m (Fig. 30), the erosion thickness in the southwest is large and in the northeast is small, reflecting that the erosion thickness is affected by the NE-SW distribution of Luzhou palaeo-uplift in the south.

2.6 Erosion Recovery of Indosinian Movement

After the Triassic, the Sichuan basin was gradually compressed. With the tectonic deformation dominated by compression, large-scale stratigraphic uplift and erosion were developed. In the early episode of Indosinian, the whole Sichuan Basin was raised and denuded, and the marine sedimentary history of the basin was ended. The early episode of Indosinian is an important period for the erosion of the Sichuan Basin again after the Caledonian movement. The late Indosinian movement, which occurred at the end of the deposition of Xujiahe Formation in the Late Triassic, caused obvious uplift and erosion in the study area, resulting in sedimentary discontinuities between the Upper Triassic and the Lower Jurassic, as well as the widespread uplift and erosion at the end of the deposition of the Upper Triassic.

(1) The First Episode of Indosinian Movement

Under the influence of Early Indosinian movement, the whole sea area of Sichuan Basin raised and the sea level dropped. The sedimentary area was exposed as the Leikoupo Formation exposed the surface. The diagenetic environment changed to an epidiagentic environment dominated by weathering and atmospheric fresh water action, and a paleoweathering crust was formed at the top of Leikoupo Formation. In the main area of Sichuan Basin, the erosion surfaces on paleoweathering crust were mainly developed in Lei3 and Lei4 Member. To the east, in the area of Kaijiang-Luzhou palaeo-uplift, the erosion mainly occurs at Lei2 Member and below. For the central Sichuan, the Lei3 and Lei4 Member are mainly denuded. The Lei4 Member of in the northern part of Central Sichuan is developed with a certain residual thickness, while the Lei 3 Member exposed in the south. Therefore, the erosion thickness of early Indosinian period should be the sum thickness of Lei3 and Lei4.

In this research, the Well Nv4 (242.5 m) with the largest sedimentary thickness in the Lei 4 Member area is selected as the benchmark (it is considered not denuded).

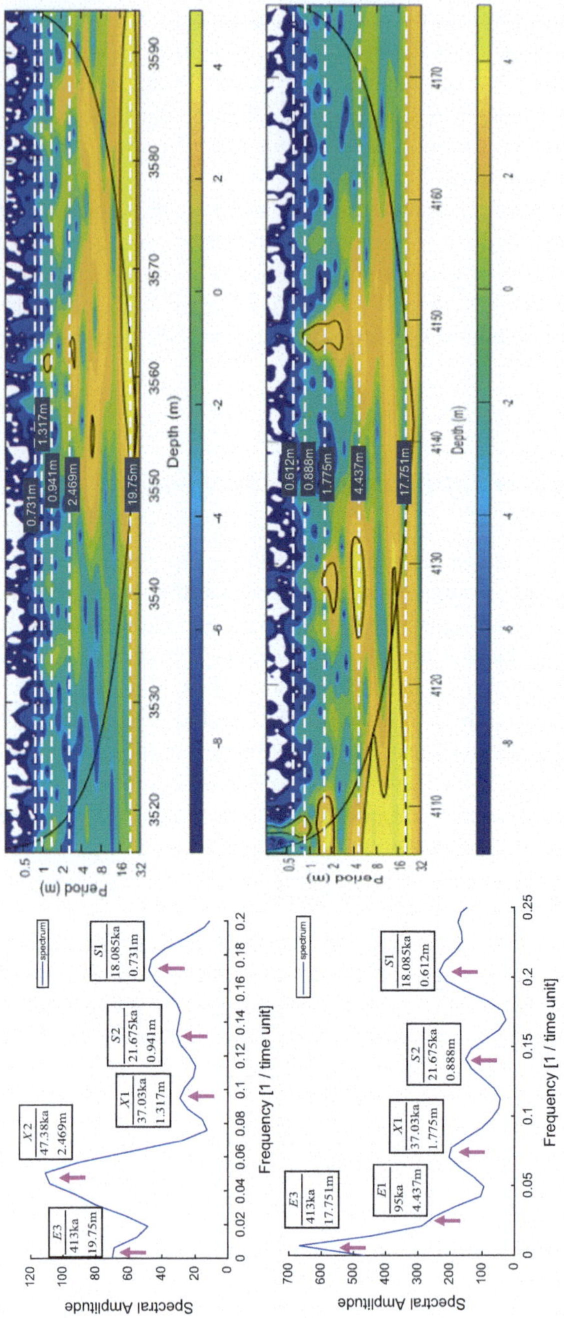

Fig. 29 Spectrum analysis of erosion thickness of typical wells in Changxing Formation (Well Gaoshi 17 in the upper part and Well Nanchong 7 in the lower part)

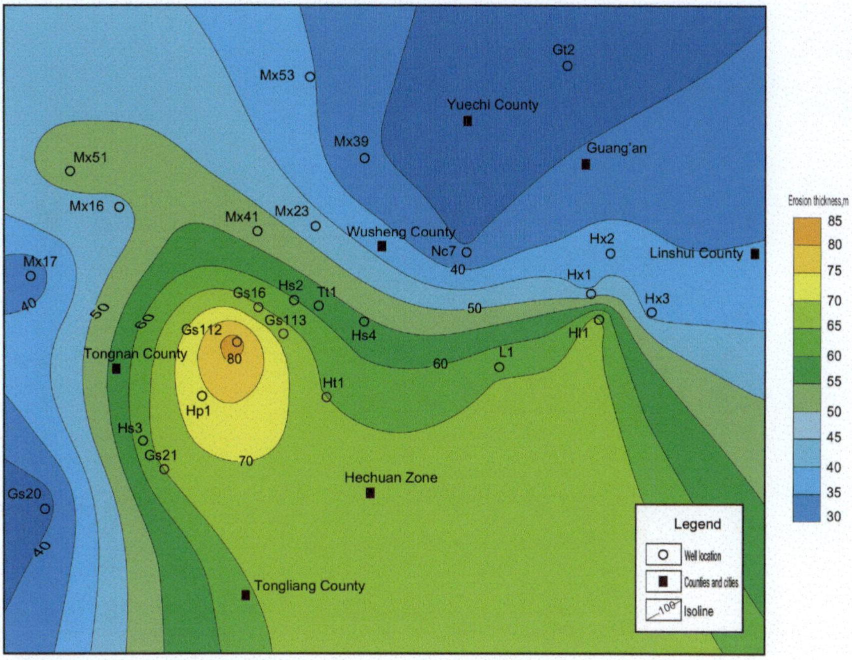

Fig. 30 Contour map of erosion thickness of Changxing Formation in the second episode of Dongwu period in Central Sichuan

The erosion thickness of Lei4 Member is obtained by subtracting the current residual thickness of Lei4 member in the study area. Given the residual thickness of Lei3 Member, and taking pinchout line in Lei4 Member as the boundary, the south area is the erosion area of Lei3 Member, with the thickness of 180 m, and the erosion thickness of Lei3 Member can be calculated by subtracting the current residual thickness. The erosion thickness of Leikoupo Formation is the sum of the erosion thickness of Lei3 and Lei4 Member (Fig. 31). It can be seen that the erosion thickness in the study area is 140–440 m, and the erosion thickness in the north is less than that in the south, reflecting that the structural high point is located in the southern part. The isolines are roughly distributed in NE direction, which are obviously affected by the Lujiang palaeo-uplift.

(2) The Second Episode of Indosinian Movement

The sandstone layers in Xujiahe Formation are the main distribution layers of coal-formed gas in Sichuan Basin. The gas reservoir exploration began in the 1950s. Recently, a series of major breakthroughs have been made, and many large gas fields have been found, such as Qiongxi, Guang'an, Hechuan, Anyue gas fields. The Xujiahe Formation can be divided into six large lithologic segments from bottom to top, namely the first member of Xujiahe Formation (Xu1 Member), Xu2 Member, the third member of Xujiahe Formation (Xu3 Member), Xu4 Member, the fifth member

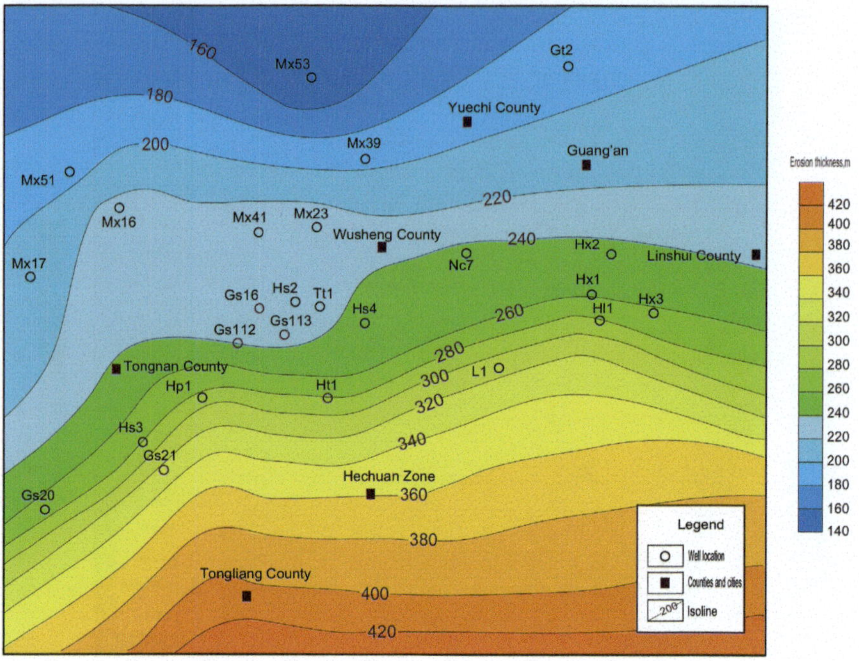

Fig. 31 Contour map of erosion thickness of Leikou Formation in Early Indosinian of Central Sichuan

of Xujiahe Formation (Xu5 Member) and Xu6 Member. Among them, The Xu1, Xu3 and Xu5 Member are mainly developed with mudstones and coal rocks, and the Xu2, Xu4 and Xu6 Member are mainly developed with sandstones. Obvious uplift and erosion were caused by the late Indosinian movement developed at the end of Late Triassic, resulting in certain erosion of Xujiahe Formation.

Based on the erosion recovery of clastic rocks and existing data, the acoustic time difference method is used to recover the erosion of Xujiahe Formation in this research. A linear relationship between the natural logarithm of log acoustic time difference and buried depth is established, and the paleosurface location at the end of the Xujiahe Formation is obtained when extending the relationship between acoustic time difference and depth of the erosion strata to the location with the surface acoustic time difference. The difference between the location of paleosurface and the current surface is the erosion thickness of the formation (Fig. 32).

The results of the erosion recovery of several wells in the study area show that (Fig. 33), the erosion thickness of Xujiahe Formation is 40–250 m, which is large in the southwest and small in the northeast, reflecting the control and influence of Luzhou-Kaijiang palaeo-uplift to the study area.

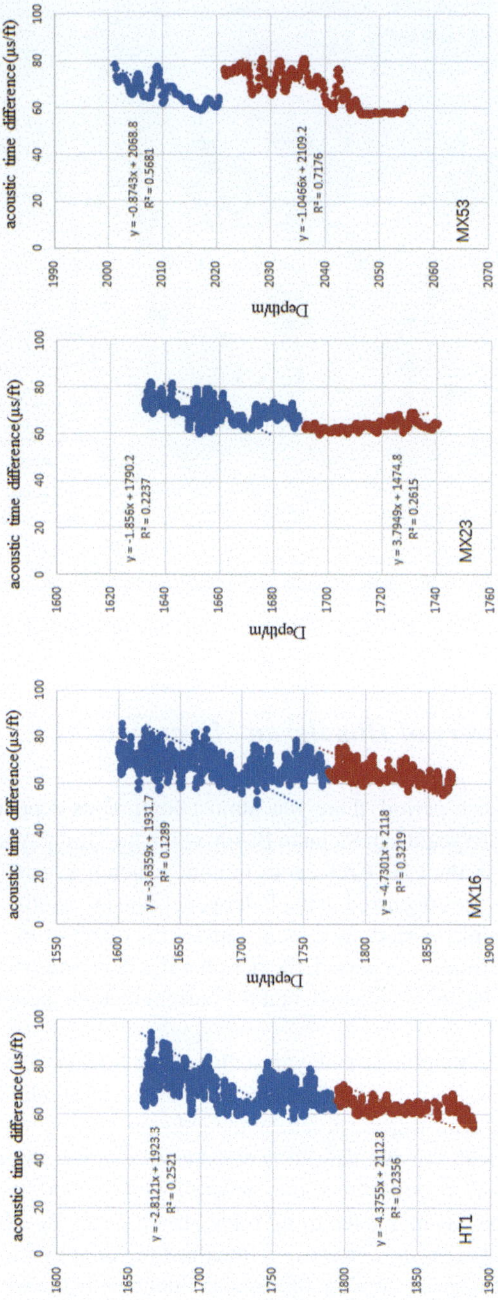

Fig. 32 Erosion recovery of Xujiahe Formation by acoustic time difference method of typical wells in Hechuan-Tongnan block

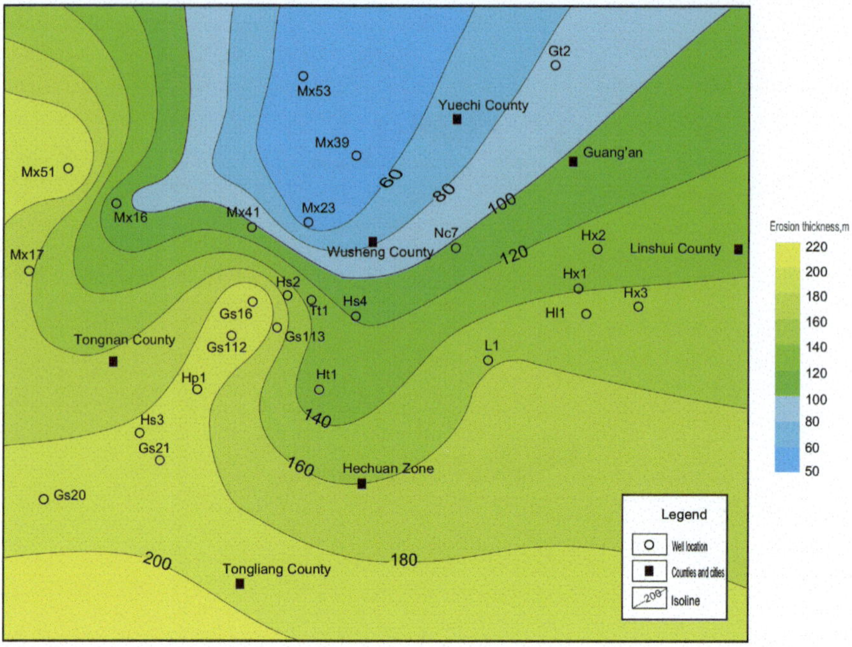

Fig. 33 Contour map of erosion thickness in Xujiahe Formation of Late Indosinian in Central Sichuan

2.7 *Erosion Recovery of Himalayan Movement*

At the end of the Yanshan period, there was almost continuous deposition between the Upper and Lower Cretaceous in Sichuan Basin, while the Himalayan period is the last tectonic period from the Late Cretaceous to now, which has lasted for 97.5 Ma. The main tectonic activities occurred in the Paleogene, causing another fold uplift in Sichuan Basin. After multi-period movements, under the effect of the recent strong tectonic movement in Himalayan period, the peripheral mountain system was raised, the piedmont foreland basin was shaped, and the current tectonic pattern of Sichuan Basin was formed.

Zeng (1988) believed that the Himalayan sediments in Sichuan Basin are mainly composed of sediments from Upper Cretaceous to Paleogene Paleocene, and the erosion strata of Himalayan were restored by the paleogeothermal data. According to the restored Himalayan erosion data, it is considered that the strata in the basin are developed with small erosion thickness, generally 1200–2000 m. Moreover, in the marginal area of the basin, due to strong folds, the strata are developed with the large thickness of about 3000 m. The erosion thickness restored by Zhu et al. (2009) with paleotemperature scale shows that, the erosion thickness in the wells around the basin is large, which is about 2000 m, while the erosion thickness in the wells on the palaeo-uplift formed in the early period is small, which is about 1000 m. Based on

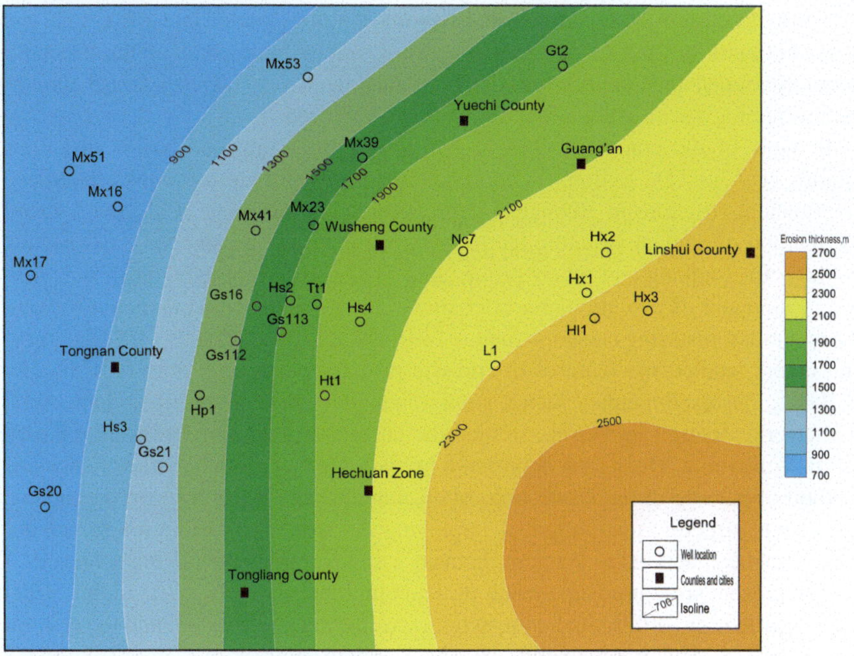

Fig. 34 Contour map of erosion thickness in Himalayan period of Central Sichuan

the previous understanding, the Himalayan erosion thickness of Hechuan-Tongnan block is large, with generally 900–2500 m, as is shown in Fig. 34.

3 Ancient Structure Restoration

In the *Geological Dictionary* compiled by the United States Geological Survey in 1964, the term "Ancient Structure" is interpreted as the previous geological structures or rock sequences of an area in the geological history. Cui Shengqin understood the ancient structure as "the structural condition of one or some development stages (synsedimentary stage and orogenic stage) before the formation of the existing current structure". The prefix "ancient" in the term "ancient structure" has no absolute concept of time, which is relative to present structure. The ancient structure of petroliferous basin controls various elements of hydrocarbon generation and accumulation, and the restoration of ancient structure can provide a reliable basis for comprehensive analysis and simulation of petroliferous basin. Most petroliferous basins in China are developed as superimposed basins, which are formed by complex multicycle evolution and vertical superposition history with multi-stage and multi-type basins. Therefore, the ancient structure restoration is particularly important.

The hydrocarbon prospecting by paleogeomorphology started late in in China, and it was not until the 1970s that importance was given to the research on the structural geomorphology, and attention was drawn to the reflection of oil storage structures in geomorphology, with quite a few results being achieved.

In view of the complex types of multi-cycle superimposed basins in China, Qijiafu, et al., and Du Xudong, et al., conducted exploratory restoration of the Mesozoic ancient structure in Huanghua depression. Mao Xiaoping et al., put forward the method of ancient structure recovery by using physical balanced cross-section method. Jin Jiuqiang and others probed into the Jurassic prototype basin in Northwest China and its evolution characteristics. Zhang Weihua and others put forward the method of restoring ancient structure by use of seismic attributes. Xue Liangqing and others studied the features of Jurassic primary basin in Northwest China by analysing paleosedimentary facies, stratigraphic contact relationship and structural framework. Zhang Liqin et al., studied the ancient structure of Ordovician buried hill in Wumaying, Huanghua depression by the sequence filling characteristics and method of paleolevel reconstruction. Ma Ruhui and others put forward the method for ancient structure restoration by imploying the structural restoration principles, which has been applied to Western Sichuan with favorable result being achieved.

The present structure is the result of the comprehensive effects, including structural deformation, differential compaction, weathering and erosion, and the restoration of ancient structure is to eliminate these effects and restore each stratum to the form at the beginning of deposition in the later stage. In the process of comprehensive analysis of petroleum geology, petroleum geologists hope to get to know the structural characteristics over geologic periods. Usually, structural evolution profile map is prepared to reflect the structural section characteristics in different geological periods, and the plane structural characteristics in different geological periods are reflected by the "pagoda map" and the residual thickness map of stratum compiled by plane structural maps reflecting the present different structural levels.

The theoretical basis of ancient structure mapping is the principle of sedimentary compensation. It is believed that, given relatively constant sedimentation rate and deposition rate, the paleogeographic environment of the basin remains unchanged, and the accumulation thickness of sediments in the basin is roughly equivalent to the amplitude of crustal subsidence. It was the principle of sedimentary compensation that make it possible to reshape the structural development history of the stratum, and it can be inferred that the change of stratum thickness can basically reflect the amplitude and speed of palaeo-uplift.

Currently, there are two most commonly used methods for ancient structure map, one is "thickness method" and the other is "structural restoration method". By the "thickness method", structural layers in different periods are subtracted by each other, and the results can reflect the ancient structure form of the lower stratum when the upper stratum calculated is deposited. This method has great limitations. It is often only applicable to areas with gentle structural fluctuation, and areas with undeveloped folds and faults. While in regions with intense stratigraphic fold deformation and developed faults, due to the influence of fault offset, stratigraphic dip, uneven compaction and uneven erosion, the "ancient structure map" calculated by the "thickness method" often involves huge errors, as a result, complex error correction work is needed.

The "structural recovery method" is a new method for structural evolution analysis. The method involves to gradually restore the current structural form of the target layer to the structural form at the time of deposition, so as to obtain the "ancient structure map" of the target layer in each geological period. The whole process of deformation recovery meets the principles of kinematics and geometry, as well as the requirements of rock deformation stimulation mechanism. During the recovery process, the rock volume remains unchanged. The "ancient structure map" of the target layer made by this method can well avoid the errors caused by faults, inclined strata and even uneven compaction. The structural restoration process is the reverse process of stratigraphic sedimentary deformation. The stratigraphic sedimentary process is usually deposition-compaction-folding-erosion-fracture. Therefore, for a certain target layer, structural restoration is divided into four steps: de-fault, de-fold, de-erosion and de-compaction. Finally, the structural state of the target layer when there is no deformation is obtained.

In the evolution process of the basin, it is the basement subsidence that make the basin form and develop. Since Sleep's study shows that the variation of basement subsidence with time along the Atlantic passive continental margin conforms to the law of exponential function, the analysis of basement subsidence has become an important way to study the genesis of extensional basin along the continental margin and in the plate. In fact, the process of basement subsidence including structural subsidence and load subsidence. The structural subsidence is caused by geodynamic action. Load subsidence is the subsidence passively increased by the further sink of basement made by the weight of sediment itself, when the basin space formed after structural subsidence is filled with sediments. Therefore, the structural settlement is calculated by removing the load subsidence from the basement subsidence.

According to the existing research results, the main mechanisms of sedimentary basin subsidence include equilibrium, deflection and thermal subsidence. The equilibrium model is developed on the base of Archimedes principle, which holds that the rock has no elasticity and each sedimentary column moves independently of each other. So the equilibrium model is also called point compensation or local equilibrium models. The deflection mode is also developed on the base of Archimedes principle, but in this mode, the response of the basement to the load is regarded as an elastic plate stressed and bent in material mechanics. It is considered that the equilibrium compensation occurs not only at the load point, but also in a wide range. The deflection mode is also called the regional equilibrium mode. In the thermal subsidence model, it is believed that the subsidence of the lithosphere is developed by the thermal effect as the lithosphere warms rapidly (such as magma intrusion), while the cooling is much slower, and the density and buoyancy of the cooled rocks are lower than those of the hot rocks. Generally speaking, the subsidence occurred by the thermal mechanism can be divided into two stages: initial rapid subsidence (due to the thinning of the lithosphere) and later rapid subsidence (due to the cooling and contraction of the lithosphere). McKenzie (1978) called the early stage as initial subsidence and the late stage as structural subsidence.

Each of the three subsidence models has its own advantages and disadvantages, which should be selected according to the specific geological conditions of the study area. The equilibrium model is believed to be suitable for areas divided by a large number of high-angle faults, with the load between fault blocks not transmitted to each other (such as the faulted basins in eastern China), and this condition mostly occurs in the primary stage of the development of faulted basins. The deflection model is suitable for large basins with simple structure, such as Ordos Basin. The thermal subsidence model is mostly suitable for extensional basins with distinct lithospheric heating and cooling stages. The simple equilibrium model is adopted in most cases except for special theoretical research, as many parameters and complex problems are involved in the deflection model and thermal subsidence model, and the value of parameters (especially some geophysical parameters) is difficult to determine and often changes with temperature and time. As a result, certain errors are inevitable but minor in general. Taking extensional basins and passive continental margin basins as examples, the post tensional stage or depression stage in the later period is more in line with the deflection model due to the cessation of the fault movement. If the equilibrium model is used, there will be some errors. However, the impact of this error on the structural subsidence is generally only 10–20 m, which will not significantly change the shape of the whole subsidence curve, and the error usually decreases gradually from the mairgin of the basin to the center.

As early as the 1940s and 1950s, petroleum geologists Naiman and Mashkovic of the former Soviet Union developed the ancient structure analysis method of "pagoda map", and systematically analyzed the ancient structure of monocycle platform basins such as Caucasus and West Siberia with this method, with some success being achieved. In 1969, Da-halstrom et al., put forward the concept of balanced section. Roeder and Witherspoon (1978) reconstructed the lithofacies paleogeography of eastern Tennessee by use of the technology of balanced section, and discussed the feasibility of studying the original distribution space of source, reservoir and cap rock assemblages in fold and overthrust structural areas based on the restored section. Supple (1983), Shaw and Suppe (1994) and Rowan & Linares (2000) further developed the idea of ancient structure restoration by balanced section. They not only conducted the quantitative research of deformation process of ancient structure, but also examined the spatial distribution characteristics of local deformation zones in the evolution process of ancient structure traps. Gibbs (1983) systematically applied the balanced section technology derived from the reconstruction of compressed structural area to the ancient structure reconstruction of extensional structural area for the first time. Since then, the study of ancient structure reconstruction of extensional basin has been deepening day by day, which is mainly reflected in the increasing deformation mechanisms (models) of reconstruction.

(1) Ancient structure restoration process

The present structures are developed by the comprehensive action of structural deformation, sedimentary filling, differential compaction, weathering and erosion. In particular, the tectonic movement often leads to the overall change of the basin, which is the biggest influencing factor. The process of ancient structure is to restore the stratum to the structural form at the beginning of deposition by eliminating these comprehensive effects. The paleostructure is a 3D model in a certain geological period which is formed by the restoration of various structural movements developed in the 3D structural model based on the present 3D structural model in the depth domain with the principle of kinematics.

(2) Erosion recovery

Erosion refers to the process in which loose deposits are formed in situ due to the physical and chemical changes in the contact between hard rocks and minerals on or near the surface and the atmosphere, water and organisms, mainly including weathering, mass movement, erosion and transportation. "Restoration erosion" is the reverse process of erosion, which restores the structural form of the target layer to the form before erosion. The results are as described above.

(3) De-fault

Inclined shear is an algorithm suitable for fault with movement. Predecessors jointly created this algorithm by studying the geometry of relevant deformation characteristics of hanging wall. Inclined shear is used to simulate penetrative deformation that is required to be formed in the sliding system at a certain angle along the whole formation of hanging wall, rather than to be formed by the tendency caused by separation along different planes between layers (such as bending and sliding). In the algorithm of inclined shear, a set of parameters selected by the users are used to control the recovery process, which include the moving directions, shear vectors and the distribution of horizontal offset.

In the forward modeling process, the inclined shear algorithm moves the hanging wall of the fault by the specified moving direction and displacement (Fig. 35). In Fig. 35, A represents the initial state (the left side is the amount to be extended), B reveals that an imaginary void space is formed in the blocks between the hanging wall and the footwall during the extension, and the extension area is equal to the void area; C demonstrates that the hanging wall controlled by the shear vector collapses to the fault surface, and the void space disappears. The void between the fault surface and the block of hanging wall is generated by extension. The hanging wall block collapses to the fault surface, and in the process of collapse, the hanging wall is controlled by the moving path specified by the shear vector. The shear vector is either vertical or opposite to the direction of the fault surface. Therefore, it is assumed that the inclined shear deformation is formed in the hanging wall of the fault and moves along a series of parallel lines as shown in Fig. 35, and the movement process cross through the fault surface is developed by the horizontal distance determined by the horizontal

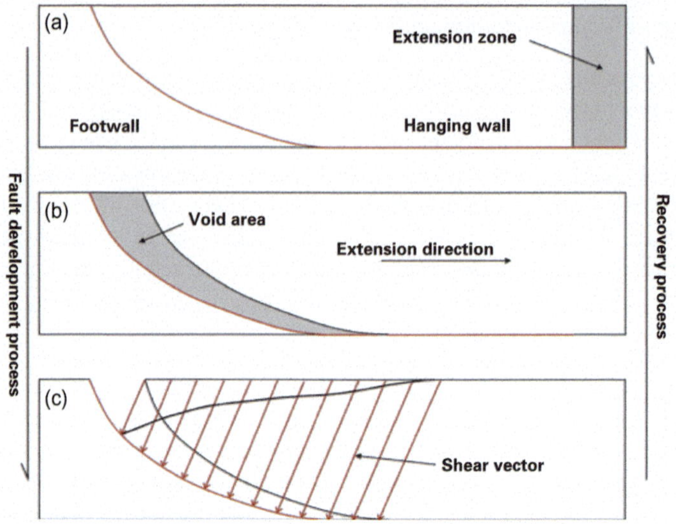

Fig. 35 De-fault by inclined shear

fault distance. The principle of de-fault by inclined shear algorithm: (1) the volume (area in 2D) of hanging wall remains unchanged; (2) the length of the shear vector remains unchanged. Its length refers to the distance between the fault surface and the marker layer of the hanging wall along the direction of shear vector. In addition, according to the same principle, the reverse fault can also be de-faulted to eliminate the formation contraction and thickening caused by the development of reverse fault (Fig. 36).

(4) De-fold

In marine strata, a large number of evaporates were developed by sedimentation. Due to the later tectonic movements, the evaporated layers easily formed detachment faults and salt domes. Especially for the salt domes, with the effect of the salt, the overlying strata were folded in the late stage, and the stratum thickness became larger locally. Therefore, in the process of ancient structure restoration, de-fold should be carried out in order to ensure the authenticity of formation thickness in the process of ancient structure restoration.

The principles of de-fold (Fig. 37):

① In the direction of de-fold, the line length of template stratum remains unchanged;
② In the direction of de-fold, the length of all stratum lines parallel to the template stratum remains unchanged;
③ The surface area of the cylindrical fold producing the fold should remain unchanged;
④ The volume of the fold (area in 2D) remains unchanged;

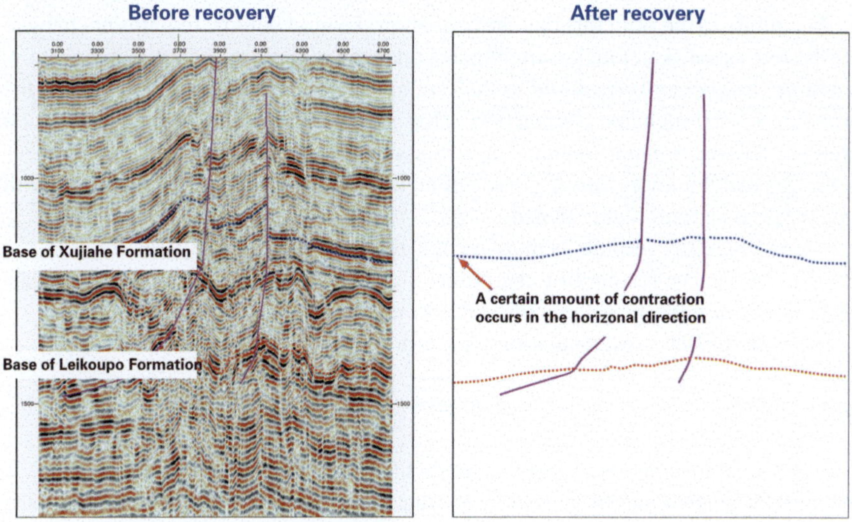

Fig. 36 Stratum thickness increase caused by reverse faults in Central Sichuan

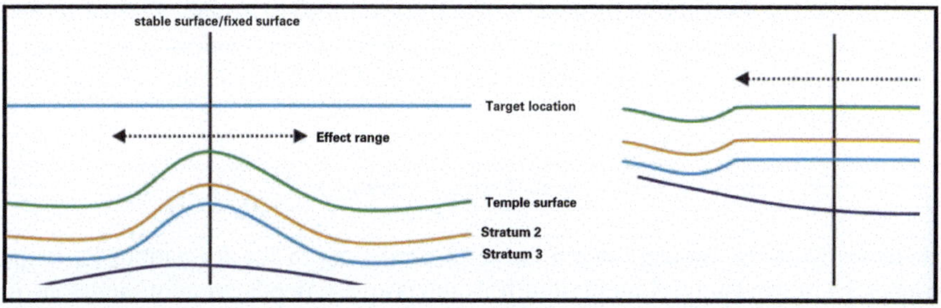

Fig. 37 Schematics of de-fold

⑤ The relative thickness of strata is constant, and the tendency changes by formation separation, which could change the measured thickness at specific points along the template formation.

(5) De-compaction

Compaction is one of the most important diagenesis of sediments, which is the action under heavy load pressure due to the continuous thickening of overlying sediments after deposition. Under the effect of gravity load in later sedimentary strata, the pores between rock particles are smaller and the stratum thickness was thinner, resulting in a certain structural deformation. In order to accurately describe the ancient structure form, it is necessary to restore the strata to the condition before compaction.

According to the principle that the formation skeleton volume remains unchanged (or the formation skeleton density remains unchanged) before and after compaction, the target layers can be restored to the original sedimentary thickness based on the principle of sedimentary compaction. That is, on the premise that the formation skeleton volume remains unchanged, it is assumed that the formation only changes in the vertical thickness during compaction, i.e. vertical thickness gets smaller, with smaller porosity and higher density.

At present, Athy formula is more recognized internationally. By this formula, the compaction can be restored by the porosity curve fitted and regressed by the least square method with the actual core porosity data.

Formula for porosity calculation (Athy formula)

$$\Phi(h) = \Phi_0 e^{-ch} \quad (1)$$

where $\Phi(h)$ is the current porosity at depth h; c is the compaction coefficient; Φ_0 is the initial porosity (surface porosity); h is the depth. The initial porosities of Xujiahe Formation and above in the study area are calculated by the analysis data of sandstone particle size with the calculation formula of initial porosity summarized by Beard & Weyl (1973).

The clastic rocks in the study area are mainly developed in Xujiahe Formation and above. The sedimentary rocks mainly include sandstones and mudstones. The total porosity-depth functional relationship is established according to the percentage of each lithology:

$$\Phi(h) = Ps\Phi s(h) + Pm\Phi m(h) \quad (2)$$

where $\Phi s(h)$ is the porosity-depth curve of sandstone in the formation; Ps is the percentage of sandstone content; $\Phi m(h)$ is the porosity-depth curve of mudstone in the formation; Pm is the percentage of mudstone content; h is the stratum depth; where $P_s + P_m = 1$.

It is impossible to directly restore the strata in the single well with few samples of core porosity data. As a result, in this research, the relationship between measured core porosity and depth is established by the function of porosity and depth with least square statistical regression with the physical property data of cores in Xujiahe Formation.

3.1 Ancient Structure Restoration of the Top Dengying Formation

The principle of "geological balance" in the process of ancient structure restoration is ensured through the operations of de-compaction, erosion recovery, de-fault and de-fold in the overlying strata of Deng4 Member. And then through the backstripping technology, one layer of overlying strata is stripped to restore the paleostructure form of the top Deng4 Member in one period. Finally, the ancient structure evolution map of Top Deng4 Member is drawn. According to the paleostructural map of the top Deng4 Member in different recovery periods (Fig. 38), before the deposition of Longwangmiao Formation, a paleostructural high point along the line of Well Moxi 102—Moxi 11—Moxi 107—Moxi 53 has formed in top Deng4 Member, and Leshan-Longnvsi anticline has initially formed. Before the deposition of Silurian, the paleostructural high point of Deng4 Member is basically consistent with that of Longwangmiao Formation. Due to the strong uplift from the end of Silurian to Carboniferous, the palaeo-uplift of Leshan-Longnvsi was denuded before Permian, but the paleostructural form of top Deng4 Member basically remained unchanged. Permian to Middle Triassic is the stage of stable burial, during this period, Leshan-Longnvsi palaeo-uplift received stable deposition. Therefore, the paleostructural characteristics of top Deng 4 Member in the sedimentary period of Longtan Formation remained unchanged. At the end of the Middle Triassic, tectonic inversion was developed in the Sichuan Basin, with uplift in the southeast and subsidence in the northwest. The Western Sichuan foreland basin developed from the Late Triassic to the Early Jurassic. It inherited the structural pattern at the end of the Middle Triassic and developed foreland basin sedimentation in the Western Sichuan. As a result, the structural form of top Deng4 Member was changed in a certain degree, with the most obvious change of the paleostructural axis migration to the southeast (the paleostructural map before the deposition of Xujiahe Formation in Fig. 38).

After Jurassic, due to the influence of Yanshanian tectonic movement and Himalayan movement, especially the Himalayan tectonic movement, the anticline axis of top Deng4 Member continued to migrate to the southeast and finally stablized.

3.2 Ancient Structure Restoration of the Top Longwangmiao Formation

According to the paleostructural map of the top Longwangmiao Formation in different recovery periods (Fig. 39), before the deposition of Silurian, a paleostructural high point along the line of Well Moxi 102—Moxi 11—Moxi 107—Moxi 53 has been formed in top Longwangmiao Formation, and the stratum at the axis of Leshan-Longnvsi palaeo-uplift is exposed and denuded, so that the Silurian stratum overlaps the underlying Ordovician stratum. Due to the strong uplift from the end of Silurian to Carboniferous, the palaeo-uplift of Leshan-Longnvsi before Permian was denuded, and the Longwangmiao Formation was denuded nearly out in the northwest of the study area. Permian-Middle Triassic was a stage of stable burial, during this period, Leshan-Longnvsi palaeo-uplift received stable deposition. Therefore, the paleostructural characteristics of the top Longwangmiao Formation remained basically unchanged during the deposition of Longtan Formation. At the end of the

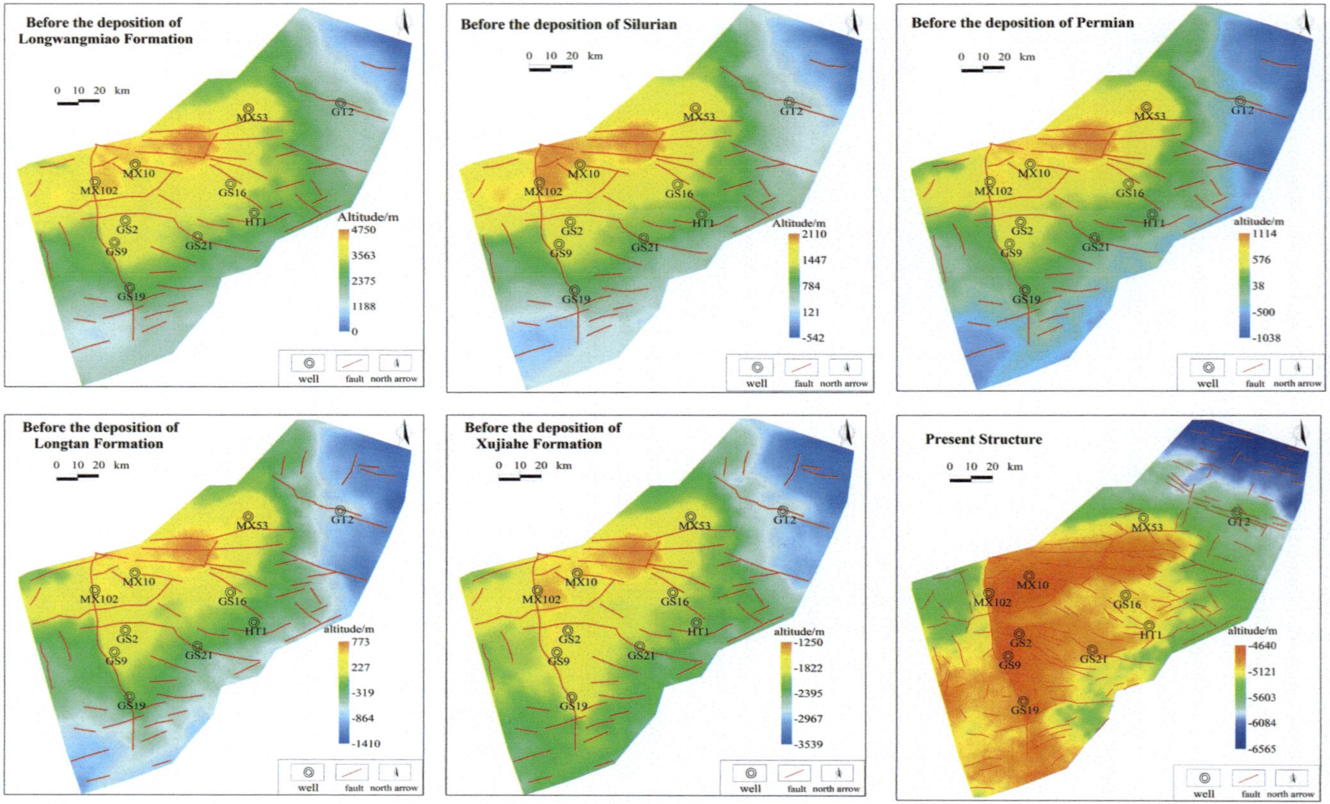

Fig. 38 Ancient structure evolution of top Dengying Formation in Central Sichuan

Middle Triassic, tectonic inversion was developed in the Sichuan Basin, with uplift in the southeast and subsidence in the northwest. The Western Sichuan foreland basin was developed from the Late Triassic to the Early Jurassic, which inherited the structural pattern at the end of the Middle Triassic with foreland basin sedimentation being developed in the Western Sichuan. As a result, the structural form of top Longwangmiao Formation was changed somehow, with the most obvious change of paleostructural axis migration to the southeast (the paleostructural map before the deposition of Xujiahe Formation in Fig. 39). After Jurassic, due to the influence of Yanshanian tectonic movement and Himalayan movement, especially Himalayan tectonic movement, the anticline axis of Top Longwangmiao Formation continued to migrate to the southeast and finally stablized.

3.3 Ancient Structure Restoration of Top Xixiangchi Formation

According to the paleostructural map of the top Xixiangchi Formation in different restoration periods (Fig. 40). Prior to the deposition of Silurian strata, a paleostructural high point along the line of Well Moxi 102—Moxi 11—Moxi 107—Moxi 53 was formed in the top Xixiangchi Formation, and the stratum at the axis of Leshan-Longnvsi palaeo-uplift was exposed and denuded so that the Silurian stratum overlapped the underlying Ordovician. Due to the strong uplift from the end of Silurian to Carboniferous, the palaeo-uplift of Leshan-Longnvsi before Permian was denuded, and the Xixiangchi Formation was denuded nearly out in the northwest of the study area. Permian-Middle Triassic was a stable burial stage, during this period, Leshan-Longnvsi palaeo-uplift received stable deposition. Therefore, the paleostructural characteristics of top Xixiangchi Formation remained basically unchanged before the sedimentary period of Longtan Formation. At the end of the Middle Triassic, the tectonic inversion was developed in the Sichuan Basin, with the uplift in the southeast and the subsidence in the northwest. The Western Sichuan foreland basin was developed from the Late Triassic to the Early Jurassic. It inherited the structural pattern at the end of the Middle Triassic, with foreland basin sedimentation being developed in the Western Sichuan, which caused changes of certain degree in the structural form of top Xixiangchi Formation, with the most obvious change of the paleostructural axis migration to the southeast (the paleostructural map before the deposition of Xujiahe Formation in Fig. 40). After Jurassic, due to the influence of Yanshanian tectonic movement and Himalayan movement, especially the Himalayan tectonic movement, the anticline axis of top Xixiangchi Formation continued to migrate to the southeast and finally stabilized.

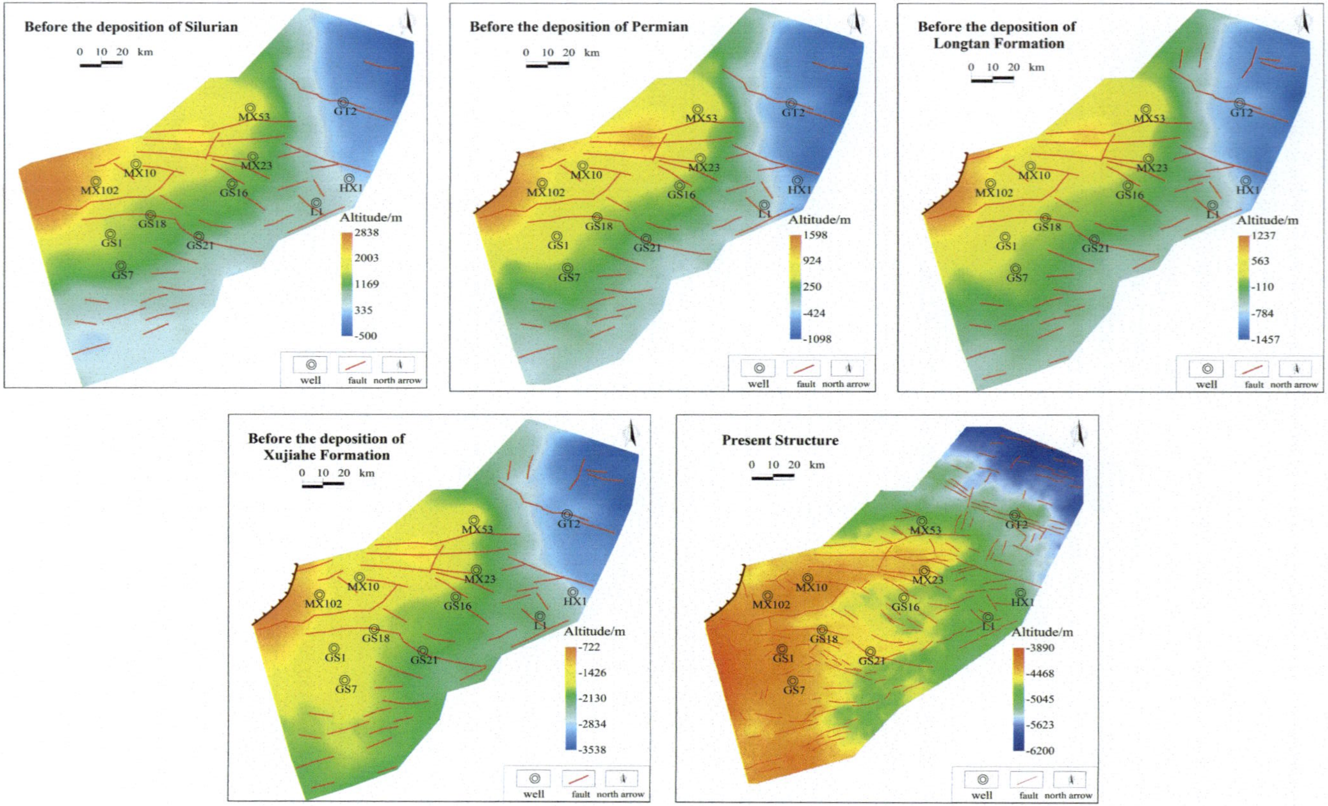

Fig. 39 Ancient structure evolution of top Longwangmiao Formation in Central Sichuan

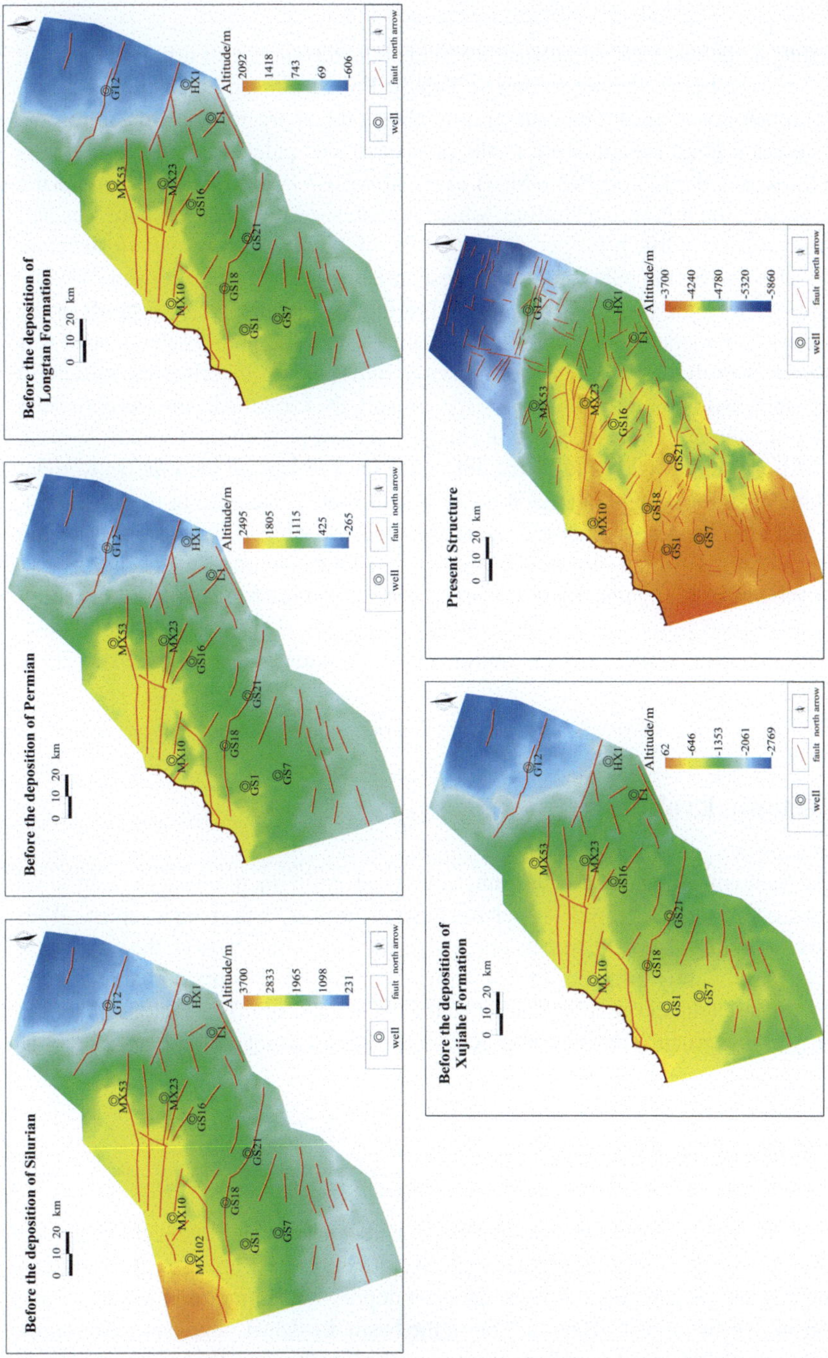

Fig. 40 Ancient structure evolution of top Xixiangchi Formation in Central Sichuan

3.4 Ancient Structure Restoration of Permian Strata

According to the restored paleostructural map of the top surface of different Permian strata (Fig. 41), before the deposition of Xujiahe Formation, the paleostructural map of the top surface of each Permian stratum shows the characteristics of high in the southwest and large buried depth in the northeast. The paleostructural high points along the line of Well Moxi 102—Moxi 117—Moxi 107—Moxi 53 are formed in the top surface of each Permian stratum, which are consistent with the axis of Leshan-Longnvsi palaeo-uplift. It is speculated that the high point of this ancient structure may be affected by the palaeo-uplift of Leshan-Longnvsi. At the end of the Middle Triassic, the tectonic inversion was formed in the Sichuan Basin, with uplift in the southeast and subsidence in the northwest. The Western Sichuan foreland basin was developed from the Late Triassic to the Early Jurassic, and inherited the structural pattern at the end of the Middle Triassic. As a result, the foreland basin was deposited in Western Sichuan. This made certain changes in the top structural form of Permian strata. After the Jurassic, due to the influence of Yanshanian tectonic movement and Himalayan movement, a large number of subsidence occurred in the northwest, uplift and erosion occurred in the southeast, and the ancient structure on the top surface of each Permian stratum showed a monocline inclined to the northwest. Under this influence, paleostructural high points developed in Xujiahe Formation also transformed into hidden high points on the slope. During this period, some nose-shaped bulges developed and inclined to the northwest. Finally, the structures stabilized.

4 Tectonic Evolution

Central Sichuan is located on the inherited local uplift in the eastern section of Leshan-Longnvsi palaeo-uplift, which has been controlled by the palaeo-uplift.. With the development of Lujiang palaeo-uplift after the Indosinian period, the study area has been significantly affected. In this study, the structural evolution characteristics of the study area under the relay control of the two palaeo-uplifts are explored by the recovery of the structural evolution process with the technology of balanced section.

4.1 Paleouplift Characteristics

As a part of Yangtze platform, Sichuan Basin had experienced three tectonic evolution stages since the returning basement consolidation of Jinning movement and Chengjiang movement, including intracraton depression, intracraton rift and foreland basin. As the largest structural unit in the basin, Leshan-Longnvsi palaeo-uplift is a giant uplift with the earliest formation time, the largest scale, the longest duration,

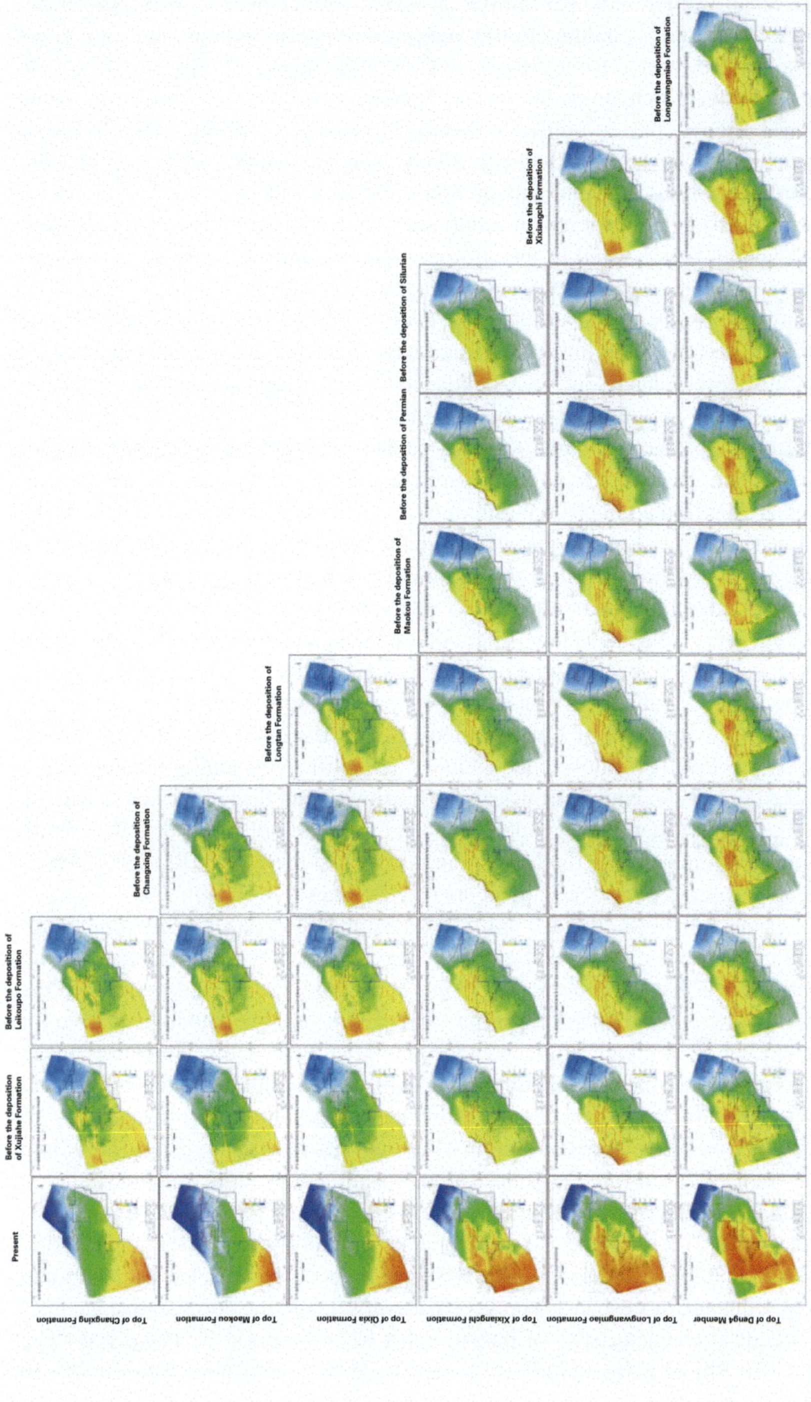

Fig. 41 Ancient structure "pagoda map" of top surface of key layers in Hechuan-Tongnan exploration area

the largest erosion range and the widest coverage in Sichuan Basin. The geological form of Leshan-Longnvsi palaeo-uplift is an asymmetric giant uplift with skirt shape. The core of the uplift is located at the margin of Western Sichuan, and two nose-shaped uplifts are formed from the core to Central Sichuan and Jiangyou. From the margin of Western Sichuan to Southeast, Northeast and South Sichuan, the strata in Leshan-Longnvsi palaeo-uplift are denuded to Lower Cambrian, Middle-Upper Cambrian and Lower Ordovician respectively. The surrounding strata are Middle-Upper Ordovician and Silurian, with an uplift area of about 18×10^4 km^2. Leshan-Longnvsi uplift is characterized by multi-stage or multi-cycle uplift and erosion-uplift. Devonian is deposited in the strong erosion period of the uplift. The shape and scale of pre-Permian palaeo-uplift are the result of the superposition of previous crustal movements, that is, the weathering coverage lasted for as long as 120 Ma. The evolution process after the uplift formation was characterized as uplift in Western Sichuan and depression in Southeast Sichuan in Paleozoic; and depression in Western Sichuan and uplift in Southeast Sichuan in Mesozoic.

Both Luzhou palaeo-uplift and Kaijiang palaeo-uplift belong to Indosinian uplift of Huayingshan. The two palaeo-uplifts are associated with northeast Huayingshan fault, and are distributed in NE-SW direction. They have been shown as early as the deposition period of Jialingjiang Formation in the Early Triassic. After the deposition period of Leikoupo Formation in the Middle Triassic, the rising range of the two palaeo-uplifts gradually increased, and a regional regular uplift was formed with significant erosion in the later stage. The southeast wing of the uplift is steep, while the northwest wing is gentle. The wing of the Luzhou palaeo-uplift is mainly composed of Lei4 Member, and the wing of the Kaijiang palaeo-uplift (West Sichuan syncline area) is mainly composed of Lei5 Member. Before the deposition of Cretaceous, Luzhou palaeo-uplift still developed as inherited uplift. Affected by the deposition of Xujiahe Formation, which is thick in the northwest and thin in the southeast, the slope of the northwest wing is gentle. The Luzhou palaeo-uplift was developed with the largest amplitude during the Indosinian period, and the core of the palaeo-uplift was denuded to Lei2 Member, with an area of less than 8×10^4 km^2.

The Leshan-Longnvsi denuded uplift in Caledonian was developed with particularly significant erosion. The "skylight" in Dengying Formation was developed with large scale, large distribution area and long duration, which is unfavorable to the formation of ancient oil and gas reservoirs and cannot supply enough gas source for late traps. Luzhou-Kaijiang uplift in Indosinian is an inherited sedimentary uplift, which is conducive to the early accumulation of oil and gas into ancient reservoirs and then transform into ancient gas reservoirs by thermal evolution, laying a good foundation for the formation and distribution of present gas reservoirs, that is, providing rich gas sources. Therefore, Luzhou-Kaijiang sedimentary uplift in Indosinian is better than Leshan-Longnvsi denuded uplift in Caledonian.

Xu et al. (2012) divided the evolution process into four stages based on the structural evolution stage of Sichuan Basin, the regional unconformity characteristics and structural sequence in Leshan-Longnvsi palaeo-uplift, and the comparative analysis of the structural characteristics of palaeo-uplift in each period. ① Prototype stage (Sinian). The Sinian palaeo-uplift is characterized by gentle intercraton uplift. In the early stage, it is mainly developed by filling and supplement on the Nanhua volcanic rock basement. In the later stage, the prototype of palaeo-uplift was formed by the tectonic uplift of Tongwan movement. ② Development stage (from Cambrian to pre-Permian). From Cambrian to Ordovician, Leshan-Longnvsi palaeo-uplift was developed as synsedimentary palaeo-uplift. At the end of Ordovician, the palaeo-uplift was strongly folded and deformed, and calmed down before the deposition of Silurian. After the deposition of Silurian, the palaeo-uplift was developed with many overall ups and downs, and was mainly weathered and denuded. Before Permian, the top of the palaeo-uplift was flattened, and the structural pattern was basically formed. ③ Stable burial stage (from Permian to Middle Triassic). With the rise of global sea level in Permian, a set of marine carbonate rocks and clastic rocks with little change in lithology and thickness were deposited in the study area. The Dongwu movement developed at the end of the Early Permian caused the short-term sedimentary discontinuity and regional volcanic eruption between the Upper and Lower Permian in Sichuan Basin. The Permian strata in the basin were developed with little change of thickness. During the deposit period, Leshan-Longnvsi palaeo-uplift showed overall uplift and subsidence, and the form remained basically unchanged. ④ Adjustment and stabilization stage (from Late Triassic to now). In this stage, foreland basins were evolved in Sichuan Basin, and the dynamic environment was mainly compression. The main structures of the basin were the uplift of mountains around the basin and the formation of structural pattern. During this period, the structural pattern of Leshan-Longnvsi palaeo-uplift was adjusted to form the current tectonic pattern.

Huang et al. (2019) analyzed the development of Luzhou palaeo-uplift of Southeast Sichuan during Indosinian, and considered that the palaeo-uplift was controlled by the Indosinian orogeny of the surrounding blocks, and was a frontal uplift formed in the front uplift of the piedmont depression belt during the compression and migration of Jiangnan-Xuefeng orogenic belt from east to west in the southeast mairgin of the Yangtze block. The development of palaeo-uplift was divided into three major evolution stages: ① germination stage (the deposition period of Early Triassic Jialingjiang Formation); ② development period (the deposition period of Middle Triassic Leikoupo Formation); ③ extinction period (the deposition period of Late Triassic Xujiahe Formation).

Central Sichuan is located in the inherited local uplift of the eastern Leshan-Longnvsi palaeo-uplift. Before the Cambrian, a paleokarst highland with a large area and uplift amplitude developed in the eastern Leshan-Longnvsi palaeo-uplift, which covers the current Longnvsi structure, Moxi structure, Anpingdian structure, Gaoshiti structure and Suining ancient structure in the north. So it can be called the Sinian palaeo-uplift in Central Sichuan. At this time, the Gaoshiti structure is located on the upper slope of the western paleokarst highland of the palaeo-uplift in Central Sichuan (Fig. 42). At the same time, the study area is still located in the line of Luzhou-Kaijiang palaeo-uplift, so the regional tectonic evolution is controlled by the relay of the two palaeo-uplifts.

In the NW–SE structural evolution section in the key periods (Fig. 43), the Leshan-Longnvsi palaeo-uplift with near EW direction is obviously developed. Since the deposition of Dengying Formation, the tectonic movements are mainly compression,

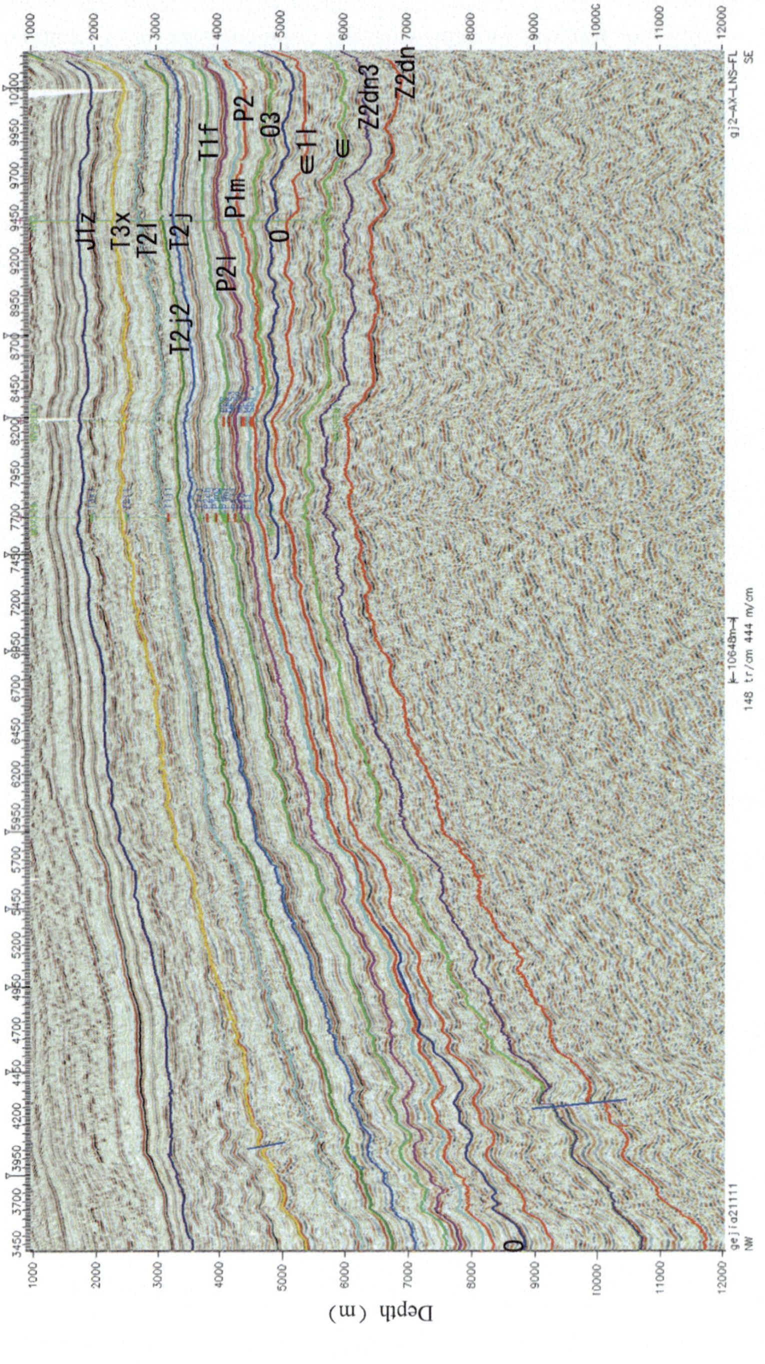

Fig. 42 Typical section of NW–SE direction in Central Sichuan

and the stratum erosion has been caused by the high part of the landform. The Sinian Dengying Formation in the structural core has been in the high part of uplift since the deposition of Sinian, and large-scale traps have been developed. Before the Hercynian movement, the structural high point has been migrating to the southeast. And the structural high point has been migrating to the southeast after the migration to the northwest during the Hercynian movement. The migration of structural highs reflects that the regional tectonic evolution is controlled by the early Leshan-Longnvsi palaeo-uplift and later Luzhou-Kaijiang palaeo-uplift.

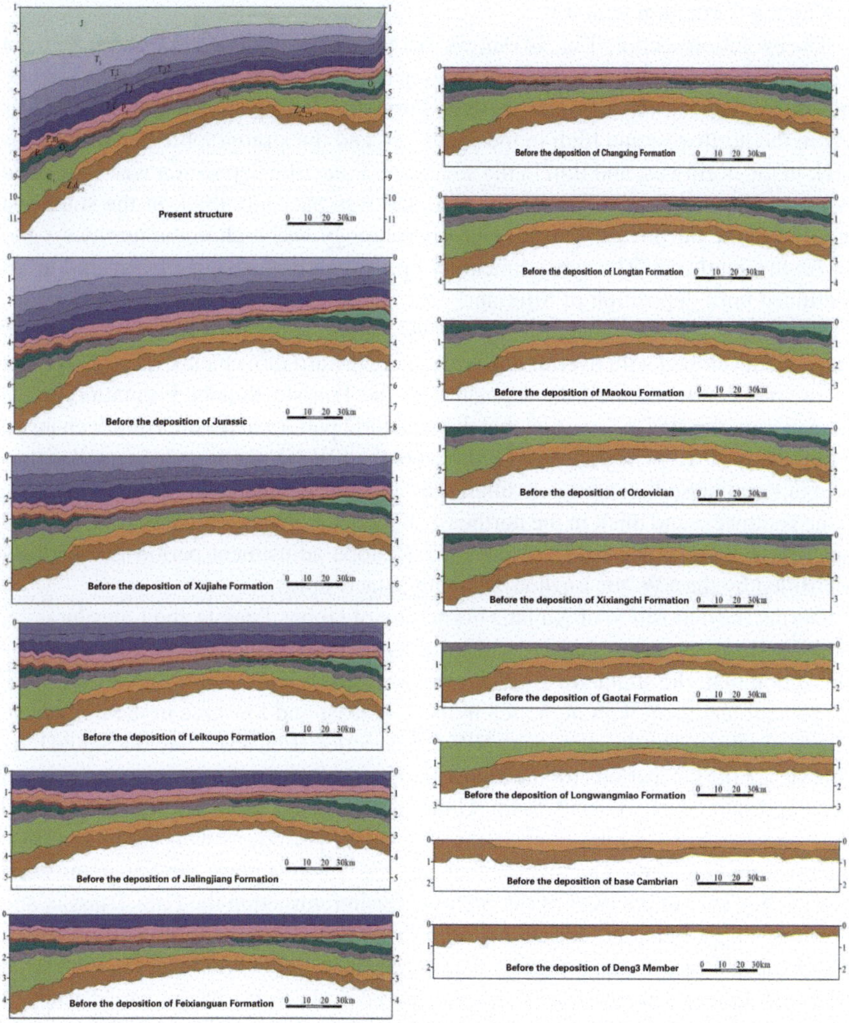

Fig. 43 Geological evolution of NW–SE section in Central Sichuan

4.2 Evolution Characteristics of Regional Sections

Central Sichuan has always been located on the southeast slope of Gaoshiti structure, which is jointly controlled by Leshan-Longnvsi and Luzhou-Kaijiang palaeo-uplift. The structural evolution of Central Sichuan can be divided into three stages: ① the stage of slope formation (controlled by Leshan-Longnvsi palaeo-uplift, before Permian); ② the stage of slope adjustment (controlled by Luzhou-Kaijiang palaeo-uplift, from Permian to the end of Triassic); ③ the stage of foreland transformation (controlled by foreland basins of peripheral orogenic belt, from the deposition period of Xujiahe Formation to now).

The evolution section (Fig. 44) shows that the Cambrian strata in Central Sichuan are well preserved with considerably varying thickness.Before the Yunnan-Guizhou movement (deposition of Longwangmiao Formation), the paleogeomorphology was low in the northwest and high in the southeast, and the stratigraphic deposition was thick in the northwest and thin in the southeast. Later, the deposition was controlled by the palaeo-uplift of Leshan-Longnvsi, and was basically thick in the southeast and thin in the northwest. The paleogeomorphology was high in the northwest and low in the southeast. The synsedimentary uplift with both sedimentation and uplift continued until deposition of Silurian.

Since the Permian, the Leshan-Longnvsi palaeo-uplift with near EW direction was only developed with overall upward and downward movements. This geological feature had continued until the deposition of the Triassic Xujiahe Formation. It can be seen from the NW–SE section that the structural pattern in the area didn't changed significantly (Fig. 44, right). And as shown in the SW-NE section, the overall difference of Permian sedimentation is small. The Triassic sedimentation is obviously thin in the southwest and thick in the northeast, and the landform is high in the southwest and low in the northeast (Fig. 44, left). The slope adjustment period is obviously controlled by the SW-NE Luzhou-Kaijiang palaeo-uplift.

During the deposition of Xujiahe Formation of Upper Triassic and Later Jurassic, the SW-NE section shows the overall subsidence. And it can be seen from the NW–SE section that, the strata are thick in northwest and thin in the southeast, reflecting the ancient landform with high area in the southeast and low area in the northwest, which are affected by the stress with NW–SE direction (Fig. 45). The Middle and Late Triassic are the key periods for the transformation of Sichuan Basin from craton basin to compressional or foreland basin. Under the collision between the South China plate and the North China plate in the Indosinian period, as well as the collision and combination of the Qiangtang block and the Yangtze plate, the Songpan-Ganzi geosyncline fold in the west of the Sichuan Basin returned in the Late Triassic, and the thrust-nappe occurred in Longmenshan from northwest to southeast, forming the Longmenshan thrust-nappe tectonic belt. From then on, the Sichuan Basin entered the development stage of foreland basin. Large-scale thrust nappe was developed in the Longmenshan fold thrust belt, and forced the axis of the top seismic event of the Leshan-Longnvsi palaeo-uplift to move southeast. Therefore, the thickness of Xujiahe Formation was obviously increased from the southeast to the northwest.

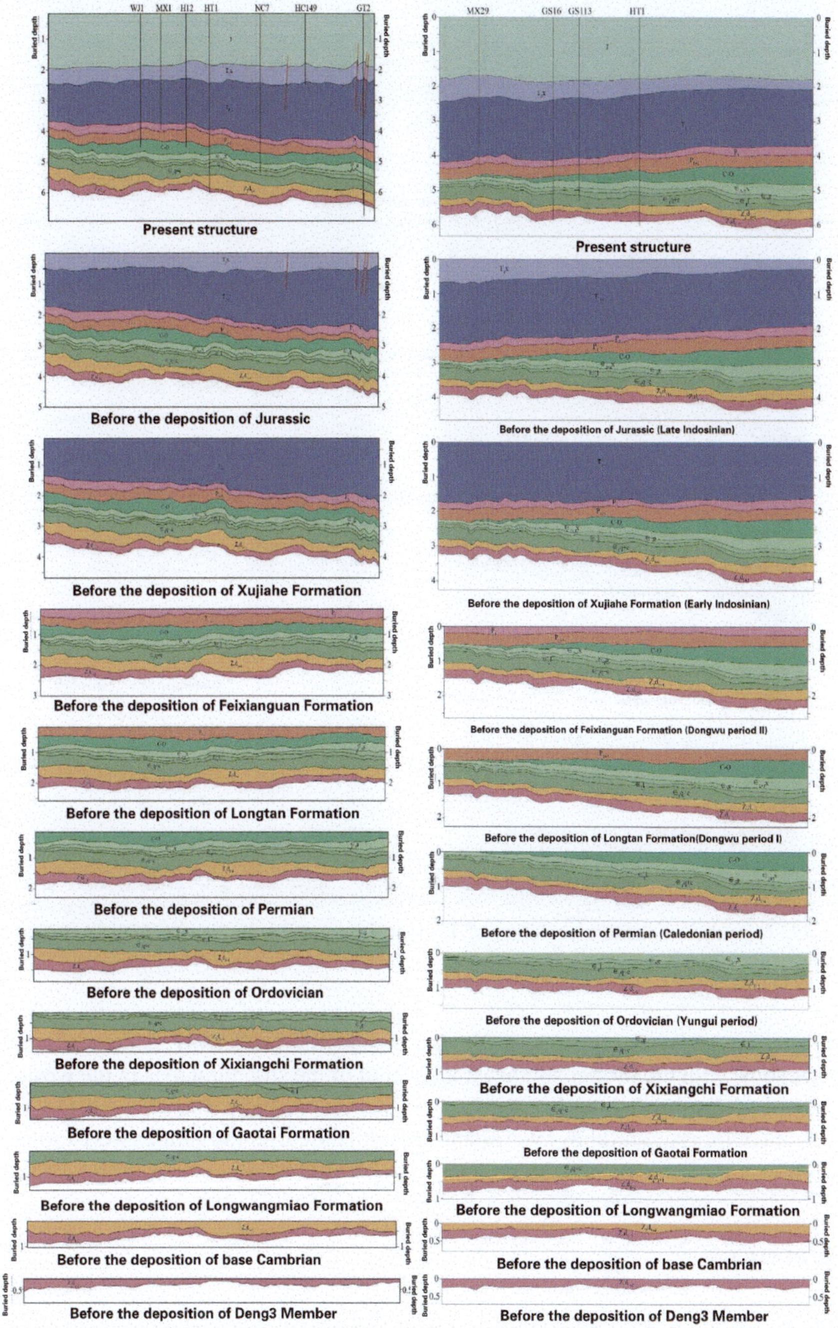

Fig. 44 Evolution of NE section (left) and SE section (right) in Central Sichuan

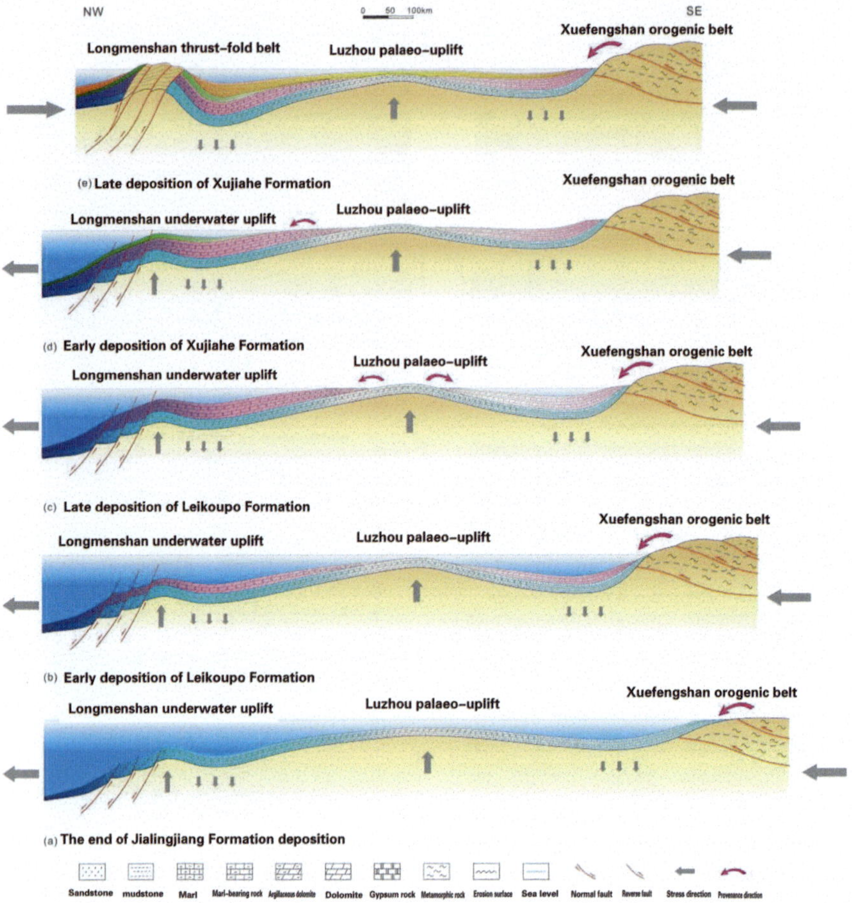

Fig. 45 Tectonic-sedimentary evolution of Luzhou palaeo-uplift during Triassic (according to Huang et al., 2019)

The Jurassic sedimentary strata in the study area were relatively stable. After the Cretaceous, with the further southeast migration of the structural axis in the western Leshan-Longnvsi palaeo-uplift, the Central Sichuan further relatively uplifted, and the strata of Jurassic and above were denuded until the current structures were formed.

References

Beard D C, Weyl P K. 1973. Influence of Texture on Porosity and Permeability of Unconsolidated Sand [J]. AAPG Bulletin, 57(2): 349–369.

Deng B, Liu S G, Liu S, et al. 2009. Restoration of exhumation thickness and its significance in Sichuan Basin, China [J]. Journal of Chengdu University of Technology (Science & Technology Edition), 36(6): 675–686.

Gibbs A D. 1983. Balanced cross-section construction from seismic sections in areas of extensional tectonics [J]. Journal of Structural Geology, 5(2): 153–160.

Gu Z D, Wang Z C. 2014. The discovery of the Neoproterozoic extensional structure in the central Sichuan block, Sichuan Basin and its implications significance of natural gas exploration [J]. Scichina: earth, 44(10): 2210–2220.

Huang H Y, He D F, Li Y Q, et al. 2019. Determination and formation of the Luzhou paleo-uplift in the southeastern Sichuan Basin [J]. Earth Science Frontiers, 26(1): 102–120.

Hou F H, Fang S X, Wang X Z, et al. 1999. Further understandings of the gas-reservoir rocks of Sinian Dengying formation in Sichuan, China [J]. Acta Petrolei Sinica, 06: 16–21.

Jiang Q C, Hu S Y, Wang Z C, et al. 2012. Paleokarst land form of the weathering crust of Middle Permian Maokou Formation in Sichuan Basin and selection of exploration regions [J]. Acta Petrolei Sinica, 33(6): 949–960.

Li Q G, Li K S, Tang H Y. 2010. Unconformity characteristics of Sichuan Basin and its geological implication [J]. Natural Gas Technology, 4(6): 21–25.

Li Q G, Li K S, Zhou Z Z, et al. 2013. Palaeogeomorphology and karst distribution of Tongwan unconformity in Sichuan Basin [J]. Oil & Gas Geology, 34(4): 516–521.

Li H W, Tang H, Su C P, et al. 2015. Characteristics of weathering crust at top of the Upper Permian Changxing Formation in Fuling area, eastern Sichuan Basin and its geological significance [J]. Journal of Palaeogeography, 17(4): 477–492.

Liu S G, Sun W, Zhong Y, et al. 2016. Discussion on the theories of the hydrocarbon formation and distribution of the deep-seated marine carbonates in the Sichuan superimposed basin, China [J]. China Petroleum Exploration, 21(1): 15–27.

Lu Q Z, Hu S B, Guo T L, et al. 2005. The background of the geothermal field for formation of abnormal high pressure in the northeastern Sichuan basin [J]. Chinese Journal of Geophysics, 48(5):1110–1116.

Lu Q Z, Guo T L, Hu S B. 2006. Hydrothermal history and hydrocarbon generation history in Northeastern Sichuan Basin [J]. Xinjiang Petroleum Geology, 42(5): 549–551.

Ma Y S, Cai X Y, Li G X. 2005a. Basic characteristics and concentration of the Puguang Gas Field in the Sichuan Basin [J]. Acta Geologica Sinica, 79(6): 858–865.

Ma Y S, Mou C L, Guo T L, et al. 2005b. Sequence stratigraphy and reservoir distribution of the Changxing Formation in northeastern Sichuan Basin [J]. Earth Science Frontiers, 12(3): 179–185.

McKenzie D. 1978. Some remarks on the development of sedimentary basins [J]. Earth and Planetary Science Letters, 40(1): 25–32.

Roeder D, Witherspon W. 1978. Palinspastic map of East Tennessee [J]. American Journal of Science, 278: 543–550.

Rowan M G, Linares R. 2000. Fold-evolution matrices and axia-l surface analysis of faul-t bend folds: Application to the Medina anticline eastern Cordillera, Columbia [J]. AAPG Bull, 84(6): 741–764.

Suppe J. 1983. Geometry and kinematics of fault-bend folding [J]. American Journal of Science, 283: 684–721.

Shaw J H,Suppe J. 1994. Structural trend analysis by axial surface mapping [J]. AAPG Bull, 78(5): 700–721.

Wang Z C, Zhao W Z, Peng H Y. 2002. Characteristics of multi-source petroleum systems in Sichuan Basin [J]. Petroleum Exploration and Development, 29(2): 26–28.

Wang G Z, Liu S G, Liu W, et al. 2014. Process of hydrocarbon accumulation of Sinian Dengying Formation in Gaoshiti structure, Central Sichuan, China [J]. Journal of Chengdu University of Technology (Science & Technology Edition), 41(6): 684–693.

Wu S J, Wei G Q, Yang W, et al. 2016. Tongwan Movement and its geologic significances in Sichuan Basin [J]. Natural Gas Geoscience, 27(1): 60–70.

Xu H L, Wei G Q, Jia C Z, et al. 2012. Tectonic evolution of the Leshan-Longnüsi paleo-uplift and its control on gas accumulation in the Sinian strata, Sichuan Basin [J]. Petroleum Exploration and Development, 39(4): 406–416.

Yuan Y S, Sun D S, Li S J, et al. 2013. Caledonian erosion thickness reconstruction in the Sichuan Basin [J]. Chinese Journal of Geology, 20(3): 581–591.
Yuan Y S, Yang S C, Hu S B, et al. 2008. Tectonic subsidence of Qiongdongnan Basin and its main control factors [J]. Chinese Journal of Geophysics, 51(2): 376–383.
Zeng D F. 1988. A preliminary study on the restoration for the various denuded sequences of Sichuan Basin [J]. Petroleum Geology & Experiment, 10(2):134–141.
Zhang Y B, Wang G H, Yu Z W, et al. 2013. Milankovitch cycles and high-frequency sequences of the Middle Permian Maokou Formation in Sichuan Basin [J]. Journal of Palaegeography, 15(6): 777–786.
Zhu C Q, Xu M, Shan J N, et al. 2009. Quantifying the denudations of major tectonic events in Sichuan basin:Constrained by the paleothermal records [J]. Geology in China, 26(6): 1268–1277.

Chapter 3
Lithofacies Paleogeography and Sedimentary Evolution

In the research of carbonate lithofacies paleogeography, the method with single-factor analysis and multi-factor graphs is used. During the study, well data and outcrop data are widely used, while seismic data are rarely applied due to the poor precision. In this research, the accuracy of lithofacies paleogeography map is greatly improved by the comprehensive application of geological and seismic data with large seismic data. The primary objective is to establish a chart showing the relationship between microfacies and their petrological characteristics by the description of color, lithology, sedimentary structure, sedimentary rhythm and reservoir pore types based on core and outcrop data. On this basis, the logging lithofacies model is established. The characteristics of single-well sedimentary facies are then analyzed with the reconstruction of drilling lithology characteristics based on the results obtained from logging interpretation, thin section data, gas test data and logging data. The sedimentary correlation section is completed, and then the spatial distribution of sedimentary elements (thickness of formation and high-energy facies belt) is implemented. At the same time, the sedimentary characteristics in each well are analyzed by combining sedimentary interpretation through seismic response and a large number of seismic profiles. Several seismic technologies are used to analyze the paleogeomorphology background of each stratum during the sedimentary period. Finally, fine lithofacies paleogeography maps are compiled by Well-Seismic Combination.

1 Lithofacies Paleogeography of the Sinian Dengying Formation

Sinian is characterized by the development of the oldest microbial algal mounds in geological history. Algal mound is mainly built by microorganisms (cyanobacteria and other micro algae, bacteria, etc.), or is mainly built by microorganisms, with

Q. Li and T. Wang, *Paleogeography and Gas Exploration of Marine Multiple Formations in Central Sichuan Basin*, https://doi.org/10.1007/978-981-99-2528-5_3

the participation of other phyla (corals, bryozoas, stromatoporoids, sponges, echinoderms, cephalopods, brachiopods, phyla calcium algae, etc.), and it is a carbonate buildup mainly composed of marl and in a curved shape.

Microbialites have been studied for more than 200 years, but the terminology was defined very late. Burne (1987) first proposed the term "Microbialite", and then Riding (1991) changed it to "Microbolite", but "Microbialite" is still used at up to today. Microbialites are sediments (rocks) formed in situ by benthic microbial communities, which are formed by the capture and coherency of clastic sediments, and the inorganic or organic induction mineralization related to microbial activities. The rock types of microbialites have been expanded from carbonate rocks to phosphorite, iron rock, manganese rock and organic shale. The formation environment ranges from ocean (coastal to deep sea), lake, river and fountain to cave, weathered crust and desert. The types of rock-building microorganisms are diverse, including photosynthetic prokaryotes (cyanobacteria), eukaryotic microalgae (such as brown algae, red algae, diatoms, etc.), chemical autotrophs or heterotrophic microorganisms (such as sulfur bacteria, etc.), some metazoan (such as ostracod and crustacea, etc.) (Riding et al., 2001) and viruses. Epiphyton, Frutexites, Girvanella, Rothpletzella, Hedstroemia, Izhella, Renalcis, Wetheredella, Nuia (Huang and Bian 1983). Epiphyton, Frutexites, Girvanella, Rothpletzella, Hedstroemia, Izhella, Renalcis, Wetheredella, and Nuia ("an algae with unknown genetic relationship", Huang and Bian 1983).

The classification scheme and related terms of microbialite are still in the process of continuous supplement and improvement. Luo et al. (2013) believe that the big differences in the classification scheme are mainly related to the classification basis adopted by various researchers. Currently, the classification of microbial carbonate rocks is mainly based on microstructure, macrostructure and megastructure. Microstructure is often difficult to preserve and is rarely used for classification. Macrostructures, such as lamination, clot, micro finger, and lump are easy to observe and thus serves as the main basis for classification. Megastructures, such as the whole microbialite buildup and bioherms/reefs, are rarely used as the basis for classification. On the basis of Riding (2000), Mei et al. (2007) classified oncolites and laminites into microbial carbonate rocks. Han (2009) believes that the microbial framework rocks, built by epiphytons and renalcis and largely developed in Cambrian, do not have the typical structure of microbialite divided by Riding (2000) and Mei et al. (2007). Therefore, most experts of microbialite believe that microbial carbonate rocks should be described and classified according to the actual situation. In general, common microbial mounds contain many sedimentary structures, such as stromatolites, clots, micro-finger shape, lumps, nuclear stones, laminites, dendrolities and spongiostromata. In this research, the petrology nomenclature refers to the rules of sedimentary structure and lithology.

1.1 *Sedimentary Facies*

In this research, the petrological characteristics and microfacies are clarified by describing cores and thin sections based on the coring wells and peripheral outcrops. Deng2 and Deng4 Member in Gaomo gas field are mainly deposited as platform margin, while the eastern parts are mainly deposited as restricted platforms dominated by dolomite, and the microfacies of the microbial algal mound are mainly developed in the restricted platform.

1.1.1 Petrological Characteristics

Based on core observation analysis, the petrological characteristics of Dengying Formation can be identified and summarized as microbial dolomite, granular dolomite, argillaceous silty dolomite and breccia/dissolved-deposited dolomite.

Microbial Dolomite

(1) **Algae-bacteria dolomite**

The algae-bacteria dolomite is microbial dolomite formed by the dolomite sedimentation of cyanobacteria (bacteria) or chondrites microorganisms. It is mainly composed of muddy-silty dolomite and sand debris, showing the characteristics of bonding (Fig. 1). Some of these dolomites are locally rich in bacteria and bond with plaster to form dark lumps, some grow in dendritic bifurcation to form pseudotree-root structure or layered bird-eye structure and some grow competitively with cements. Bacterial lump dolomite is composed of lumps of different sizes and shapes. The dark lumps are rich in bacteria. They have irregular bird-eye pores between the lumps, and possess distinctive dissolution boundaries. The spotted algal dolomite is formed in an extremely shallow subtidal and intertidal environment. The plexiform algal dolomite may be formed in the deep-water subtidal environment, and the bacteria related to the cement may be related to the activities of atmospheric fresh water.

(2) **Clot dolomite**

Clot dolomite is mainly formed by the secretion of bacteria and algae and the bond of microcrystal dolomite, including clot dolomite and algal clot dolomite (Fig. 2) which are both formed by bacteria and algae microorganisms. The traces of bacteria and algae in algal clot dolomite are obvious and the color is uneven. The difference may be related to the different enrichment of bacteria and algae.

This kind of dolomite is generally fine, with the color of grayish-white, porcelain-white, porcelain-gray and cyan-gray, and it is uniform and fine texture with porcelain fracture. Generally, it is relatively tight with few dissolution pores. This kind of dolomite is the most important rock of carbonate platform facies in shallow water

Fig. 1 Algae-bacteria dolomite. **a** Gray algae-bacteria dolomite, Well Gaoshi 16, 5459.1 m, Deng4 Member. Algal bacteria are mainly composed of dark lumps and branches. The size and shape of bacterial lumps are uneven, irregular bird-eye pores are developed between the lumps, and some are dissolved. The algae bacteria are filled with muddy-silty dolomites and sandy dolomites, showing the bonding characteristics overall. **b** Gray algae-bacteria dolomite, Well Gaoshi 21, 5262.9 m, Deng4 Member. Algal bacteria are mainly composed of dark lumps and branches. Bacterial lumps are uneven in size and different in shape. Irregular bird-eye pores are developed among the lumps, and corrosion pores are developed locally

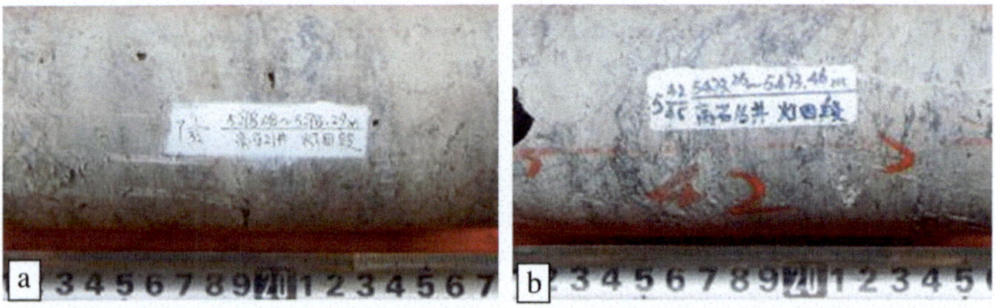

Fig. 2 Clot dolomite. **a** Gray clot dolomite, Well Gaoshi 21, 5278.1 m, Deng4 Member. The dark part is the microcrystal dolomite secreted or bonded by algae and bacteria, which performs as isolated and irregular spot. **b** Gray clot algal dolomite, Well Gaoshi 16, 5473.3 m, Deng4 Member. The dark part is microcrystal dolomite secreted or bonded by algae and bacteria, showing flocculent and irregular porphyritic

of Dengying Formation. Its color is gray, light color and light grayish-white. The dolomite deposited in deep-water subtidal is dark-gray or cyan-gray. The single-layer dolomites are mainly developed as medium layer and thin layer, and thick lump locally. Some dolomites are developed as mound shape, which can form huge microcrystal clot lime-mud mound or dolomite-mud mound.

(3) **Laminite dolomite**

The laminite dolomite is essentially a dolomite rich in bacteria-algae mat traces (Fig. 3). In general, laminate dolomite can be divided into several types according to the thickness, density and structural characteristics of the algal lamina. Those with sparse algal lamina and horizontal or wavy occurrence are called algal laminite dolomite (Fig. 3a, d). Algal stromatolite dolomite (Fig. 3b, c) is characterized by dense algal lamina and occurrence in wall, column and mound. Algal framework dolomite (Fig. 3e) is characterized by dense and tangled algal lamina, with wave resistant structure and framework pores. The laminite dolomite is mainly developed in two environments, one at the top of shallow tidal flats and the other at the marl mounds. There are sheet-like pores and window pores between the laminations, which are filled with microcrystal dolomites. The other is in subtidal deep water, and the laminas are generally formed with mound-shaped arcs, which can form large marl mound, such as the Deng1 Member in Jinkouhe, Ebian (Fig. 3f).

(4) **Thrombolite dolomite**

Thrombolite has been translated into Chinese term as "凝块叠层石". In the classification system of microbial carbonate rocks, the concept of "thrombolite" was initially interpreted by its founder Aitkeno as "cryptoalgal structure related to stromatolite, lack of lamina and characterized by macro thrombolite fabric". With the development of research, many new types of thrombolites have been found. Up to now, there are two structural types in the thrombolites. One is formed by bacteria and algae bonding various particles and marl (dolomitic mud) as network, dendritic, porous or lump thrombolites (Fig. 4). The previous groups or aggregate particles such as algae lumps and grapevine can be divided into thrombolites. The other is the lump thrombolites formed by bacteria and algae bonding microcrystal dolomites (calcite) or marls. The outer margin of the thrombolite has a coating composed of micritic dolomite. Between the lumps are marl or dolomitic mud, or recrystallized microcrystal, microsparry dolomites (Fig. 5), or ctenoid structure formed after dissolution. In the process of diagenesis, the internal dissolution and recrystallization inner lump may leave a dark envelope. The thrombolite must be formed under a certain hydrodynamic condition. As a result, it is mainly distributed in shallow-water carbonate platforms, and can form large-scale marl mound or dolomite muddy mound. The thrombolites can be developed as favorable reservoir rocks with dissolution pores easily formed inside or at the margin of the structure. In the Gulf of Mexico, a large amount of oil and gas have been found in the thrombolite marl mound.

The development degree of thrombolite dolomite in the eastern part of Central Sichuan is relatively low, and only developed in Well Gaoshi 16 (Fig. 4). While large-scale thrombolite dolomites are developed in the western platform margin, with many dissolution pores (Fig. 5).

Granular Dolomite

The particle composition of granular dolomite are mainly algae particles and sand debris, with uniform distributed particles. Beddings composed of dark algae particles were observed with a size of about 1 mm, and could be visible by the naked eyes, mainly composed of sparry cementation and needle-like corrosion pores (Fig. 6). Granular dolomites are the deposited in high-energy beach Belt, and widely distributed in each member of Dengying Formation.

Fig. 3 Laminite dolomite. **a** Algal laminite dolomite, Well Gaoshi 16, 5471.9 m, Deng4 Member. The laminae are sparse and the occurrence is horizontal, with bedding dissolution pores. **b** Algal stromatolite dolomite, Well Gaoshi 21, 5286.2 m, Deng4 Member. The dark algae laminae are dense, and some occurrences are developed as wall and mound. **c** Algal stromatolite dolomite, Well Gaoshi 21, 5266.9 m, Deng4 Member. The dark algae laminae are dense, and some occurrences are developed as wall and mound. **d** Algal laminite dolomite, Well Gaoshi 16, 5311.2 m, Deng4 Member. The laminae are sparse and horizontal. **e** Algal framework dolomite, Well Gaoshi 16, 5482.9 m, Deng4 Member. The algal laminae are dense and tangled, with wave resistant structure and framework pores. **f** Laminite marl mound, Ebian Jinkouhe profile, Deng1 Member, Zhang (2017), with an exposure height of about 2 m

Muddy-Silty Dolomite

The muddy-silty dolomite is composed of dolomite less than 0.01 mm. The color of muddy-silty dolomite is light gray, and those containing organic matters are dark gray. It is developed as medium and thin layers (Fig. 7). The muddy-silty dolomites are widely developed in the research area of Hechuan-Tongnan. In addition, the muddy-silty dolomites are usually produced simultaneously with silicification, which shows

Fig. 4 Thrombolite dolomite. **a** Network, dendritic, porous or lump thrombolites formed by bacteria and algae bonding with various particles and marls (dolomitic mud), with high development degree of dissolution pore, Well Gaoshi 16, 5464.7 m. **b** Lump thrombolite, developed with a high degree of dissolution pore, Well Gaoshi 16, 5468.4 m

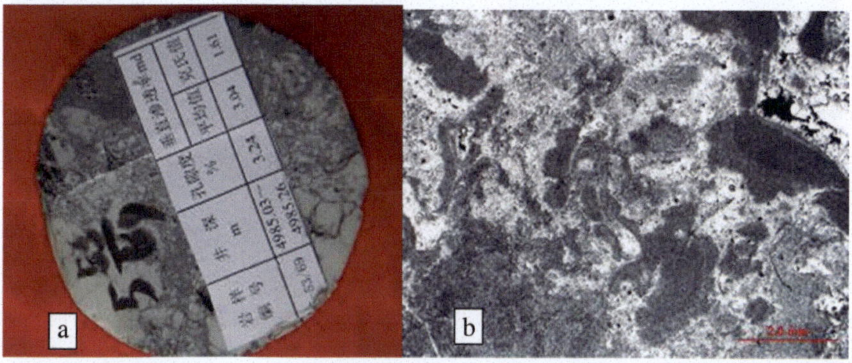

Fig. 5 Thrombolite dolomite. **a** Thrombolite dolomite, Well Gaoshi 1, 4985.1 m. Dark spotted thrombolite coexists with light-gray sand debris. **b** Thrombolite dolomite, Well Gaoshi 1, 4985.1 m. With dark irregular thrombolite and sparry cementation

that the dolomites are generally intercalated with black-gray and gray-white layered silicalites, homogeneous lump and banded silicalites (commonly known as "black silicon" and "white silicon") or interbedded with them (Fig. 8).

Breccia Dolomite

The breccia dolomite in carbonate strata are mainly composed of karst breccia dolomite and collapse breccia dolomite. The composition of karst breccia dolomite is similar to that of the upper and lower dolomites and the breccias are of different sizes, disorderly accumulation and no bedding. The matrix between the breccias is muddy micritic dolomite (Fig. 9), which is produced after the filling or collapse of

Fig. 6 Granular dolomite. **a** Algal granular dolomite, Well Gaoshi 21, 5249.3 m. The dark part is algae particles with beddings, and the particles are clear and visible to the naked eye. The light colored part is sand debris, and needle-like corrosion pores are developed. **b** Granular dolomite, Well Gaoshi 21, 5291.8 m. A large number of needle-like corrosion pores are developed. **c** Granular dolomite, Well Gaoshi 16, 5450.8 m. The particles are visible to the naked eye, and the dark part is algae particles. Corrosion pores are developed with a few bedding corrosion pores. **d** Algal granular dolomite, Well Gaoshi 16, 5470.2 m. The dark part is algae particles with bedding, and the particles are clear and visible to the naked eye. The light colored part is sand debris. Needle-like dissolution pores and bedding dissolution pores are developed

the dissolution caves. Slump breccia dolomite is formed by collapse and redeposition after plastic crumple fracture. The breccia of slump breccia dolomite is the same as that of the surrounding rocks, and it is a sedimentary marker rock of deep-water carbonate slope facies. The genesis of slump breccia dolomites here has nothing to do with the primary sedimentation, and are mostly the products of weathering crust and karst.

Fig. 7 Muddy-silty dolomite. **a** Dark-gray muddy-silty dolomite, horizontal bedding, Well Gaoshi 16, 5445.9 m. **b** Gray muddy-silty dolomite, with a small amount of spotted algal laminae, Well Gaoshi 21, 5301.5 m

Fig. 8 Silicified muddy-silty dolomite. Gray-white silicified strips, lamina and dark micrite can be seen inside the siliceous, Well Gaoshi 21, 5318.9 m

1.1.2 Combination Types and Characteristics of Sedimentary Facies

In Moxi-Gaoshiti area, the sedimentary facies combination types of dolomite in Dengying Formation is prominently represented as mound-beach complex in both vertical evolution sequence and plane differentiation pattern. Hechuan-Tongnan area is located in the east of Gaoshiti area, in which carbonate platform deposits of Dengying Formation are widely distributed, mainly developing intra-platform algal mounds, granular beaches and sea between mound and bank related to microbial rocks (Table 1).

Fig. 9 Breccia dolomite. **a** Breccia dolomite, Well Gaoshi 16, 5464 m. The breccia is composed of single micritic dolomite, with different sizes, disorderly accumulation and angular shape. Between the breccias are dark micritic matrix. **b** Breccia dolomite, which can also be recognized as cracked dolomite, Well Gaoshi 21, 5281.3 m. The original rock is gray algae-bacteria dolomite, with irregular high-angle cracks filled by dark micritic material, and the top is dark micritic material. The whole rock shows brecciated or cracked

Table 1 Sedimentary facies types and petrological characteristics of Dengying Formation

Open (restricted) platform		
Microbial mound	Granular beach	Sea between mound and bank
Algae-bacteria dolomite, clot dolomite, laminated dolomite, thrombolite dolomite, microorganism, dissolution pore. Carbonate buildup formed under the action of cyanobacteria and bacteria	Algal granular dolomite and doloarenite, needle-like dissolution pore	Dark-gray micritic silty dolomite and algal micritic silty dolomite
Deng2 and Deng4 Member		

Intra-platform Algal Mound

Algal mound is created by microorganisms (cyanobacteria and other microalgae, bacteria, etc.), or mainly formed by microorganisms, with the participation of other phyla (corals, bryozoas, stromatoporoids, sponges, echinoderms, cephalopods, brachiopods, phyla calcium algae, etc.), which is a carbonate buildup mainly composed of marl and with curved characteristics. A large number of microorganisms and microorganism related algae can be clearly seen from the cores of Well Gaoshi 16 and Gaoshi 21. The sedimentary features are as follows. (1) The lithology of algal mound is complex. The algal mound is mainly composed of algae-bacteria dolomite, clot dolomite, laminated dolomite and thrombolite dolomite. Among them, laminated dolomite, algal laminated dolomite, algal laminated dolomite and algal framework dolomite can be further identified, and laminated dolomite is the most developed one. (2) Different from the Phanerozoic large skeleton reefs, the main constructors of algal mound in Dengying Formation are microorganisms, macroalgae, algal secretions and micropaleophytes. In the process of frame-building and poreforming by bacteria, algae and their biochemical actions, framework rocks with different algae-bacteria traces and occurrences (also be called microbial framework rock) are developed, such as algae bacteria, thrombolite, spongiostromata, stromatolite, laminated rock and snowflake rock, which are different from the rock characteristics of Phanerozoic. The research on the vertical microfacies combination and lateral superposition of algal mound is still under exploration. According to the convention point of view on biological build-up framework rock, mound base (characterized by gravel debris and granular dolomite), mound core (characterized by various microcrystalline clot dolomite, microbial framework rock and thrombolite framework rock), mound top (characterized by stromatolite and laminated dolomite), mound wing (characterized by gravel debris dolomite), and intra-mound depression can be identified in the algal mound. Moreover, there should be differences in microfacies superposition pattern and scale between the algal mounds on the margin of the platform and inside the platform.

Intra-platform Granular Beach

There is no obvious difference in the formation of Sinian intra-platform granular beach and Phanerozoic granular beach. The intra-platform granular beach is mainly composed of algal granular dolomite and doloarenite, which is the product of high-energy environment. Based on the development of granular beach in Well Gaoshi 16 and Well Gaoshi 21, granular beach is generally symbiotic with algal mound, but the occurrence frequency and cumulative thickness of granular beach are both less than that of the intra-platform mound.

Mound Interbank Sea

A relatively low negative stratigraphic unit is developed in the mound-beach body with lateral discontinuities. Relatively fine-grained carbonate sediments can be deposited in this stratigraphic unit, such as micritic silty dolomites. Due to collapse or storm, a small amount of mound-beach sediments can migrate to fine-grained sediments between mounds and beaches to form micritic silty dolomite containing algae. Moreover, due to the high-frequency rise of sea level (transgression), dark-gray thin-layered micritic dolomite can be formed, usually with horizontal beddings.

In short, the sea between mound and beach is the product of low-energy conditions, and it can form a variety of combination and superposition patterns with high-energy mound-beach in space and vertical direction. In addition, the sea between mound and beach related to transgression can also be used as an interface for high-frequency sequence identification and correlation.

1.2 Sedimentary Characteristics of Backbone Section

1.2.1 Sedimentary Characteristics of Well Gaoshi 16

The drilling depth of Well Gaoshi 16 is 5720 m, and Deng3 Member is drilled. There are six coring samples at the upper part of the Deng4 Member, with the coring length of about 30 m.

The sedimentary characteristics of the coring well section from bottom to top are as follows (Fig. 10). (1) Eight high-frequency cycles are developed in the lower part (from the 6th to the 3rd coring), composed of sea between mound and beach, and microbial mound. The bottom of each high-frequency cycle is thin-layered micritic silty dolomite, and the upper part is gray thick-layered laminated dolomite. The cycle thickness gradually thickens upward. Reservoir spaces such as dissolution pore holes are developed. Dissolved breccia can be seen in the third coring. (2) In the upper part (from the 2nd to the 1st coring), there are five high-frequency cycles, composed of the sea between mound and beach, and the intra-platform granular beach. The lower part of each high-frequency cycle is micritic silty dolomite, and the upper part is gray granular dolomite. The upward thickness change law of the cycle is not obvious. In the first coring, it can be seen that the hydrothermal minerals are filled in the dissolution pores. Dissolution pores and holes are also developed.

1.2.2 Sedimentary Characteristics of Well Gaoshi 21

The completed drilling depth of Well Gaoshi 21 is 5670 m, and Deng2 Member is drilled in this well. There are 13 coring at the upper part of the Deng4 Member, and the coring length is about 85 m.

The sedimentary characteristics of the coring profile from bottom to top are as follows (Fig. 11). (1) The lower part (from the 13th to the 11th coring) is gray laminated dolomite, the lamina is mainly composed of algal lamina structure, and the whole rock is deposited by microbial mound. It is intercalated with 3 sets of dark gray thin-layer micritic silty dolomites, and is deposited as intermound sea. A high-frequency cycle is formed by micritic silty dolomite and laminated dolomite, with the thickness tends to increase gradually upward. Dissolution pores and holes can be seen in the cycle. (2) The middle part is mainly gray algal dolomite (from the 10th to the 8th coring), which is mainly deposited as intra-platform beach. Three dark-gray thin-layered micritic dolomite layers are developed between algal dolomites, which form a high-frequency cycle with algal dolomite, and the cycle thickness tends to thicken upward. (3) The upper part (from the 7th to the 1st core) is assembled by micritic dolomite, algal granular dolomite and laminated dolomite as a whole, which is deposited as the combination of mound and beach. Three sets of combinations can be identified as a whole. Multiple high-frequency cycles of thin-layered micritic dolomites and algal dolomites-laminated dolomites can be further identified within each combination, and the upward thickening trend isn't very clear. Needle-like

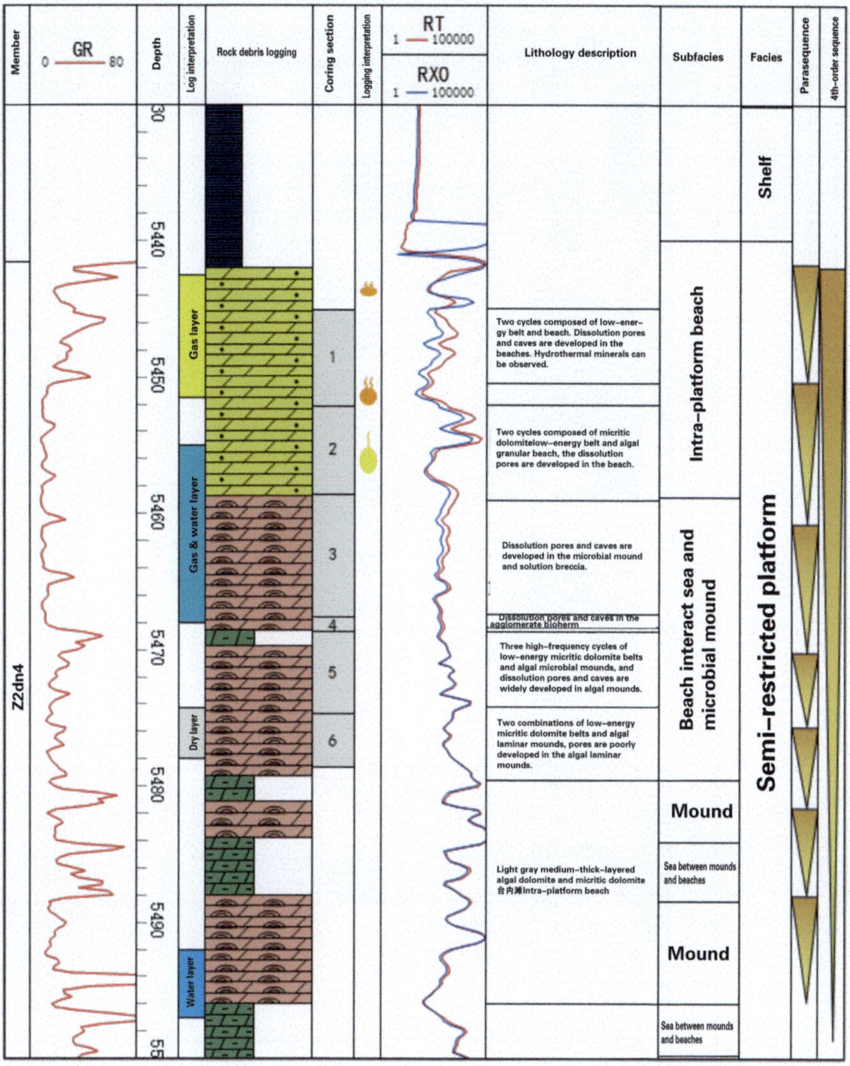

Fig. 10 Column map of sedimentary characteristics in cores of Deng4 Member in Well GS 16

dissolution pores are mostly developed in algal dolomite, dissolution pores are mostly developed in biological mound, and upward dissolution pores are increased.

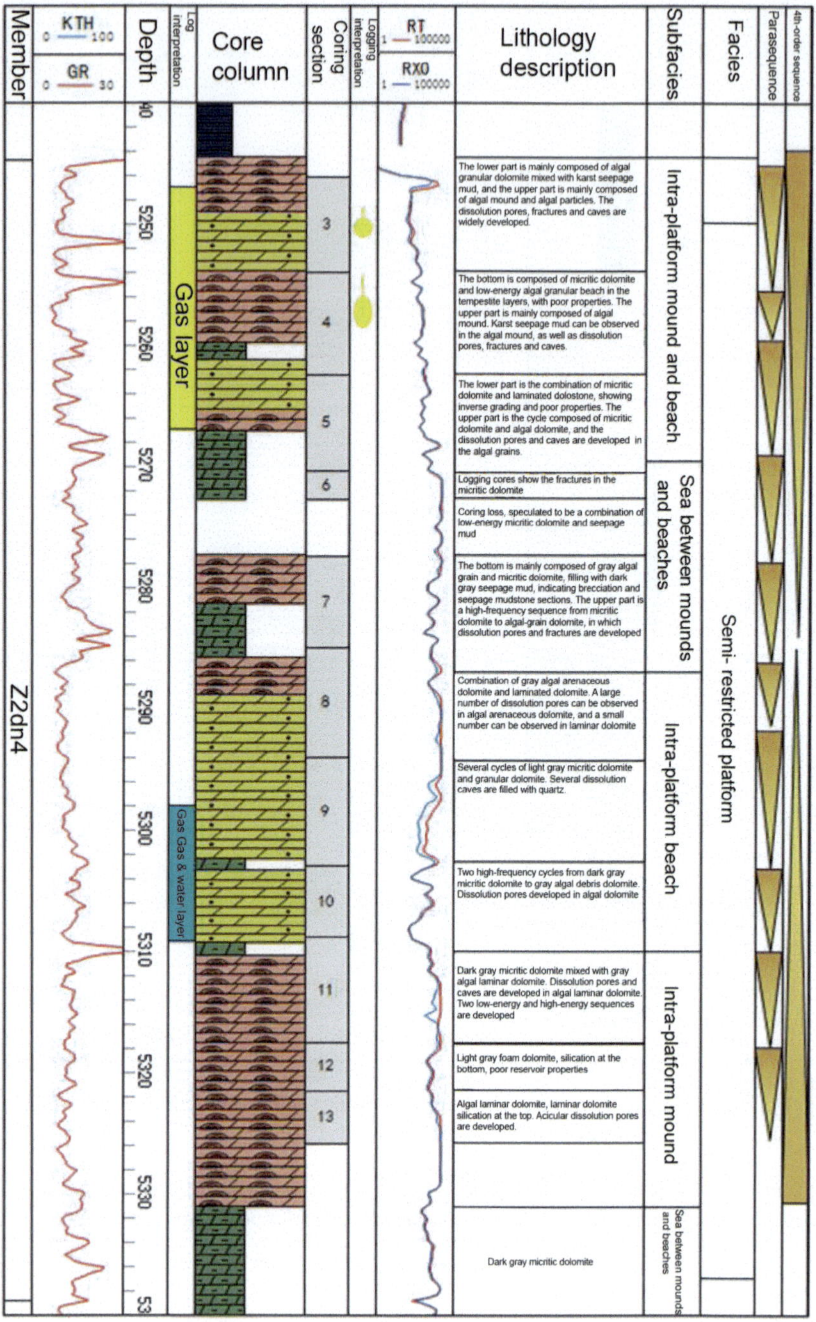

Fig. 11 Column map of sedimentary characteristics in cores of Deng4 Member in Well GS 21

1.2.3 Sedimentary Characteristics of Outcrop Profile in Dengying Formation, Xianfeng

Xianfeng profile is located at the southwest margin of the basin, with a stratigraphic thickness of about 800 m. It can be divided into two third-order sequences (Sq1 and Sq2). The sedimentary facies are mainly platform margin, shelf and restricted platform (Fig. 12).

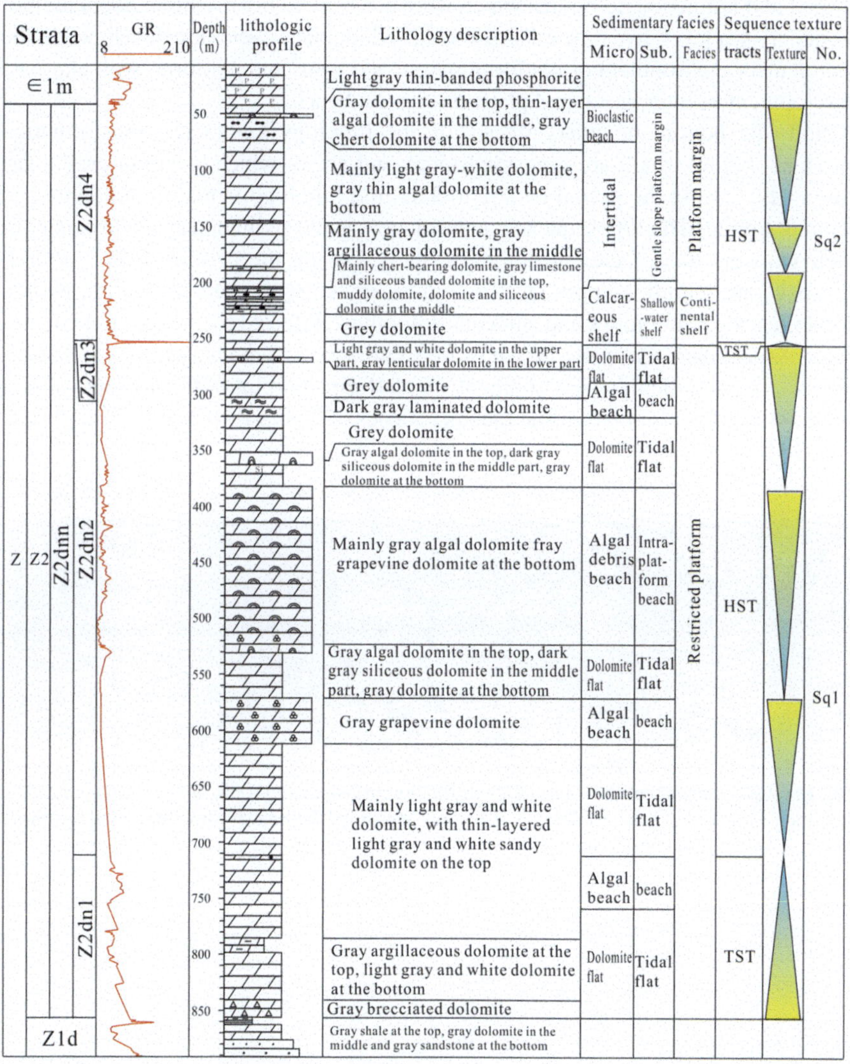

Fig. 12 Column map of Sinian sedimentary facies and sequence stratigraphy of Xianfeng profile

Sq1: the bottom boundary of Sq1 is the lithology-lithofacies transformation interface, below which are the gypsum dolomites of Doushantuo Formation, above which are the muddy dolomites, and the water body suddenly deepens. During the period of Sq1-TST, the lithology is mainly dolomite mixed with muddy limestone, gravelly dolomite and shale, and the bottom is gray sandstone. It is deposited as the sedimentary facies of restricted platforms, divided into the subfacies of tidal flat and the microfacies of dolomite flat. During the period of Sq1 to HST, the rocks were mainly composed of algal dolomites, grape dolomites mixed with dolomites, bean-shaped and laminated dolomites which were affected by the rhythmic fluctuation of secondary sea level, five upward shallowing sedimentary cycles are developed as the sedimentary environment of platform margin and restricted platform, with alternate development of inter beach and platform-margin beach.

Sq2: the bottom boundary of Sq2 is the lithology-lithofacies transformation interface, below which are gray mudstones and above which are light-gray algae dolomites. During the period of Sq2 to TST, the lithology is mainly dolomite and beam dolomite. The Sq2 sequence is developed as the sedimentary environment of shelf with the microfacies of dolomitic shelf. During the period of Sq2 to HST, the rocks are mainly composed of dolomites mixed with chert dolomites, muddy dolomites, algal dolomites and siliceous dolomites. Affected by the rhythmic fluctuation of secondary sea level, three upward shallowing sedimentary cycles are developed as the sedimentary environment of platform margin, with the subfacies of gentle-slope platform margin and the microfacies of bioclastic beach and inter beach (Fig. 13).

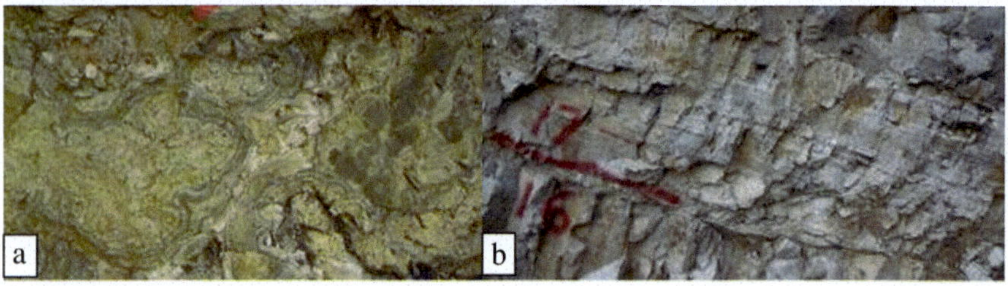

Fig. 13 Rock characteristics of Dengying Formation in Xianfeng profile. **a** Gray snowflake algal dolomite, with rich "grape and lace" structures and algal laminated structures, half-restricted platform, Deng2 Member. **b** Siliceous banded dolomite of Deng4 Member, open platform

1.3 Analysis of Sedimentary Facies

1.3.1 Logging Facies

Based on the sedimentary analysis of cores in Well Gaoshi 16 and Gaoshi 21, core location is carried out, and the natural gamma-ray logs with high lithology sensitivity are selected as the main series in logging facies analysis. The logging facies characteristics of microfacies combination in Dengying Formation are summarized as follows: (1) Sea between mound and beach—biological mound—granular beach. There are two characteristics in this combination. One is a relatively thin-layered structure, with natural gamma-ray logs from high amplitude to low toothed amplitude, showing a multiple funnel-shaped combination (Fig. 14). This structure may indicate the law of high-frequency sea-level change and may develop in the early stage of highstand system. The other is the structure of thin-layered inter-beach sea and thick-layered mound beach. The natural gamma-ray logs show the characteristics from thin-layered high amplitude to thick-layered micro-toothed box (Fig. 15). This structure may indicate the continuous decline of sea level, which is mainly developed in the late stage of the highstand system. (1) Combination of microbial mound and sea between mound and beach. This facies combination is formed by thick layers and similar logging response between thin-layered inter-beach sea and thick-layered mound beach. The natural gamma-ray logs show the characteristics of thin-layered high amplitude to thick-layered toothed box (Fig. 16), with relatively low gamma value. The structure of the facies combination may also be related to the continuous decline of sea level.

1.3.2 Single-Well Analysis of Sedimentary Facies

Due to the limited data of coring wells and the limitation of cuttings logging description, it is difficult to reflect more abundant sedimentary connotation. Therefore, the base of the research work is the reasonable reconstruction of comprehensive histogram of sedimentary facies in single wells. The idea of comprehensive histogram reconstruction in this research is as follows (Fig. 17). (1) Sedimentary facies analysis shall be completed for all drilled wells as far as possible. (2) Taking the sedimentary knowledge of outcrop, sequence and core as a guide for sedimentary facies reconstruction, For example, three microfacies are mainly developed in Deng4 Member, so they should also be developed in the profiles of uncored wells. (3) Generally speaking, based on the idea of marine sequence, the high-energy facies belt is mainly developed in the upper part of the sequence/subsequence set. Based on this principle, the lithology and lithofacies of profiles in the uncored wells can be judged. (4) Sedimentary facies identification by the establishment of logging facies model. It can be divided into three steps. (a) Clarify the characteristics of sedimentary facies (subfacies) in coring wells. (b) Convert the coring depth into logging depth by core location. (c) Based on the basic principle that the logging curve characteristics can

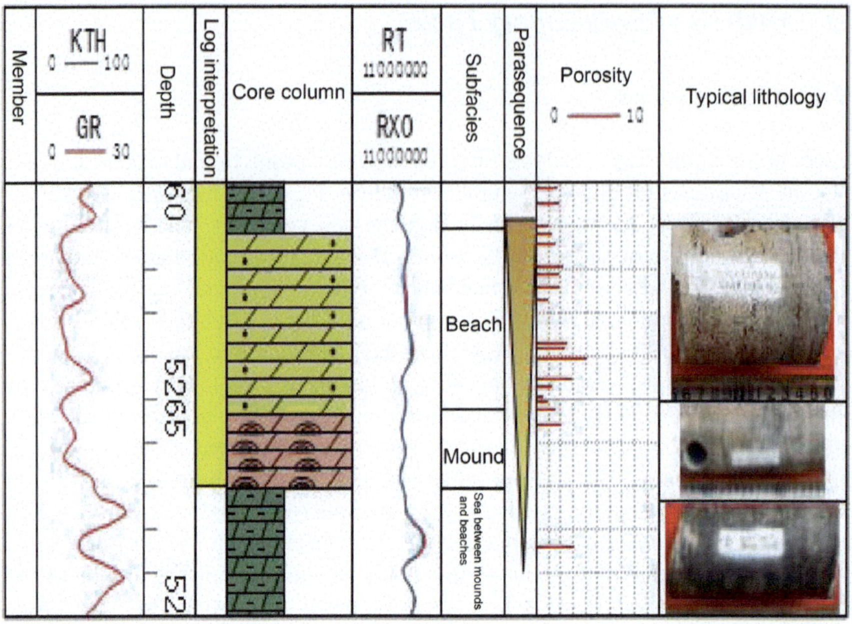

Fig. 14 Logging facies characteristics of microfacies combination of intra-beach sea—mound—beach in Well Gaoshi 21

reflect the characteristics of sedimentary water body and the basic logging facies model summarized by predecessors, the logging response models of typical sedimentary facies (subfacies) in the research area are established. (5) The data of cuttings logging, logging interpretation, cuttings thin section and gas test results shall be collected completely for the guidance of lithology and lithofacies identification, for example, first the basic lithology knowledge of cuttings logging data should be respected. Second, logging curves interpreted as gas reservoir or reservoir generally can be understood as high-energy facies. Thirdly, the identification results of sporadic cuttings thin sections can be used to support the understanding of lithology.

Based on the above five constraints, the sedimentary interpretation of uncored wells can be realized. Taking Longwangmiao Formation of Well Hetan 1 as an example, the following describes the process of sedimentary interpretation. (1) There is only a small amount of coring data in Longwangmiao Formation of Well Hetan 1, with original logging cutting description of light gray dolomite, which lacks sedimentary connotation. (2) The conventional logging curves, cuttings logging description, logging interpretation results and a small amount of cuttings thin section description results can be collected. (3) There is a small amount of coring at the upper part of Deng4 Member, and the sedimentary connotation of the coring profile can be directly loaded on the cutting profile. (4) There are three points in the rock debris thin section. The sparry residual sandy dolomite can be identified as granular beach and the silty dolomite can be identified as inter-beach sea. (5) The well section interpreted as

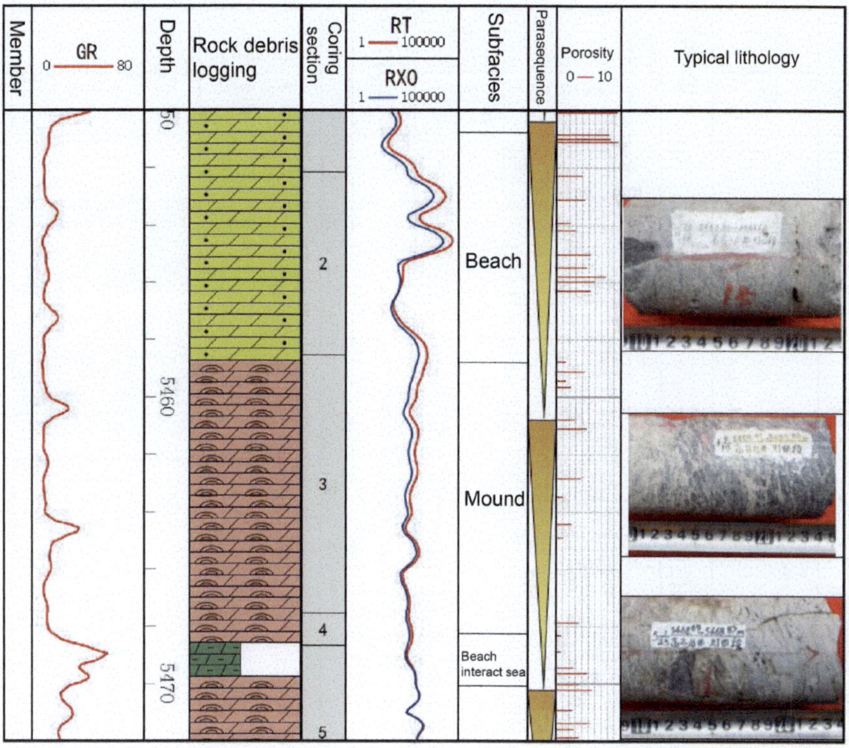

Fig. 15 Logging facies characteristics of microfacies combination of intra-beach sea—mound—beach in Well Gaoshi 16

gas reservoir or poor gas reservoir by the loggings can be understood as high-energy facies. (6) Based on the shape and value of natural gamma-ray, the logging facies relationship between granular beach and inter-beach sea of Longwangmiao Formation is established (generally speaking, the shape of natural gamma-ray from inter-beach sea to granular beach shows a toothed funnel type, in which the granular beach has low gamma value and the inter-beach sea has high gamma value). Based on this idea, most lithology/lithofacies identification can be completed. Conversely, the interpretation of logging facies can be constrained by the understanding of core data and cuttings thin sections. It can be seen that the sparry doloarenite section in the cutting profile corresponds to low natural gamma-ray, while the silty dolomite section corresponds to the high natural gamma-ray. This further proves the rationality of logging geological analysis. Thee comprehensive column of single-well sedimentary facies in Well Hetan 1 is determined and completed through this understanding and analysis process.

The comprehensive column of single-well sedimentary facies is completed based on the above ideas, for example, a typical intra-platform beach is deposited in Well

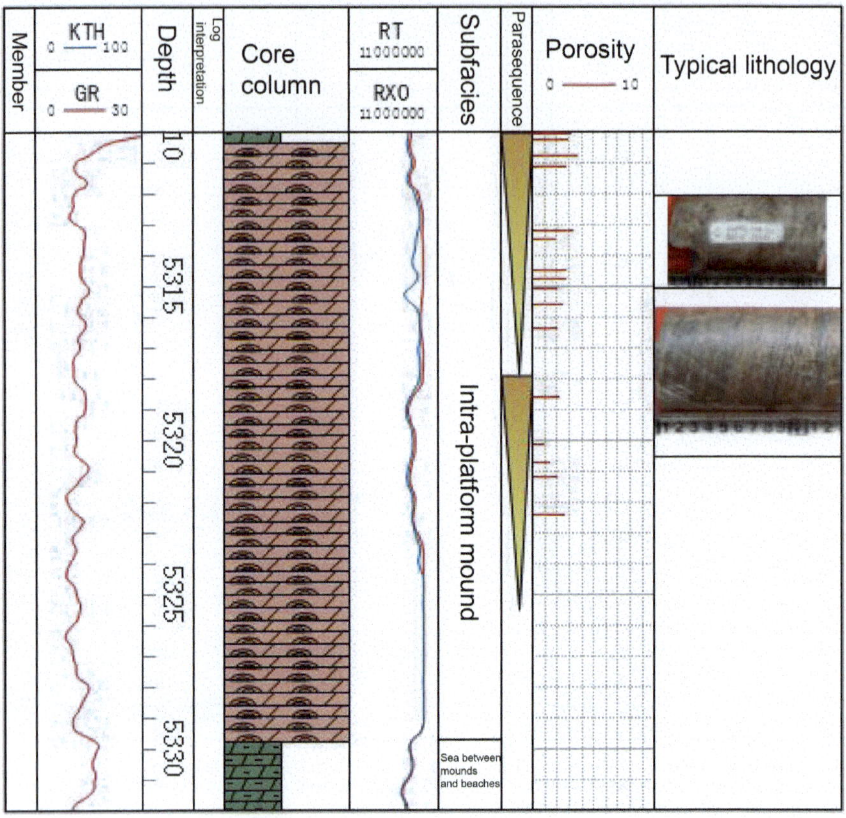

Fig. 16 Logging facies characteristics of intra-beach sea and mound in Well Gaoshi 21

Gaoshi 16. This well reveals Deng4 and Deng3 Member. The Deng4 Member is developed in 5440–5704 m, with a thickness of 264 m and the Deng3 Member is not drilled through, with a thickness of about 20 m. The Deng3 Member is mainly composed of dark-gray to blue-gray muddy deposits, which is a typical shelf deposition. In the early stage, the Deng4 Member was mainly deposited by carbonate platform, with a large set of gray-dark to gray thin-layered micritic dolomite, which was shown as a typical subfacies of inter-beach sea. With the change of sea level, light gray middle-layered intra-platform granular beach was deposited, established as a subfacies of the intra-platform beach. In the middle and late stage of the highstand system, the high-frequency sea-level decreased slowly, and the microbial mound grew rapidly, forming the combination of low-energy belt and microbial mound, and the combination of low-energy belt-algal debris beach-mound. The combination of microbial mound and beach thickens upward. The maximum thickness of beach is 30 m, and the maximum thickness of microbial mound is 16 m. Well Gaoshi 16 is identified as restricted platform facies based on the combination lithologic characteristics.

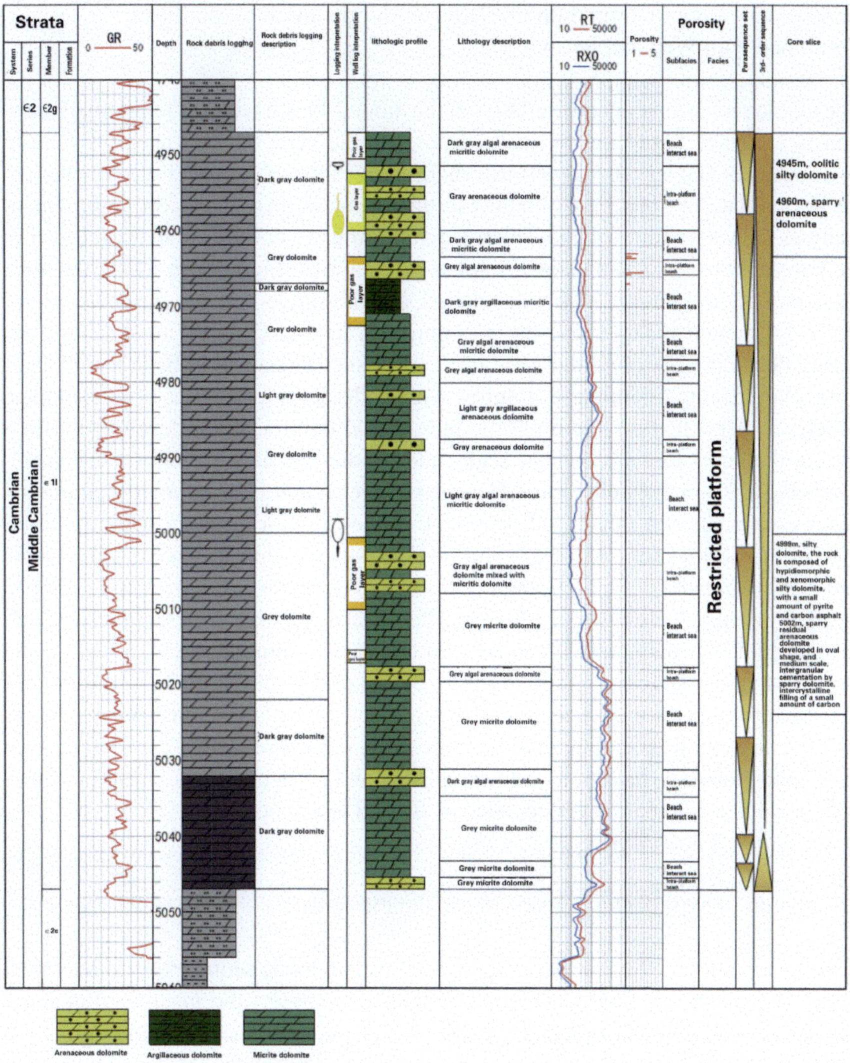

Fig. 17 Construction process schema of sedimentary facies in single well (uncored well) (Well Hetan 1)

1.3.3 Sedimentary Facies Correlation

Based on the detailed analysis of sedimentary facies in single wells, the correlation profile is constructed to analyze the spatial distribution of sedimentary facies.

Sedimentary Facies Correlation of Platform Margin and Intra-platform Combined with Well and Seismic Data

The macro sedimentary characteristics of the mound-beach body in Dengying Formation is analyzed based on the well-seismic profile across Well Gaoshi 1–Gaoshi 2–Gaoshi 18–Gaoshi 16 of platform margin and intra-platform (Fig. 18). Firstly, the coring data of Sinian are obtained in these four wells and the core sedimentary analysis shows as follows. (1) Thick-layered laminate dolomites are developed in well area of Gaoshi 1 and Gaoshi 2 (platform margin) (Fig. 19), in which the algal framework dolomites are highly developed, and the single-layer thickness of microbial algal mound is large, and the dissolution pores of microbial algal mound are highly developed. Around the well area of Gaoshi 18 (transition belt, east of platform margin), algal laminar dolomites with underdeveloped pores are developed, which are deposited as microbial mounds. Algal dolomites are developed with a certain thickness and a particular scale, and needle-like corrosion pores are developed (Fig. 20). In the well area of Gaoshi 16, a relatively isolated microbial mound beach combination is developed inside the platform, with a certain scale of cumulative vertical thickness. Most dissolution pores are developed in algal frameworks and algal laminated dolomites, and needle-like dissolution pores are also developed in algal granular beaches. (2) Based on the idea of seismic facies, the cross-well seismic profile is dissected. It can be seen that the stratum thickness of Deng2 and Deng4 Member decreases gradually from the margin of the platform to the interior of the platform in the east. The mound beach body in the platform margin presents as large-scale mound reflection, and the large-scale mound reflection can also be seen inside the platform (there may be multiple secondary depressions in the platform, and the depression margin is deposited as mound beach body).

Correlation of Sedimentary Facies Inside the Platform

Inside the platform, the correlation profile of well-seismic sedimentary across Well Moxi 41, Gaoshi 16 and Hetan 1 is established to dissect the macro sedimentary characteristics of the intra-platform mound-beach bodies in the Dengying Formation (Fig. 21). It can be seen from the profile that: (1) the stratum thickness of Deng4 and Deng2 Member does not change obviously; (2) the seismic events are dominated by parallel sheet reflection, and several mound reflections can be seen, which have a certain scale and horizontal continuity. It is interpreted as intra-platform mound-beach complex and sea between mound and beach (restricted-platform to lagoon).

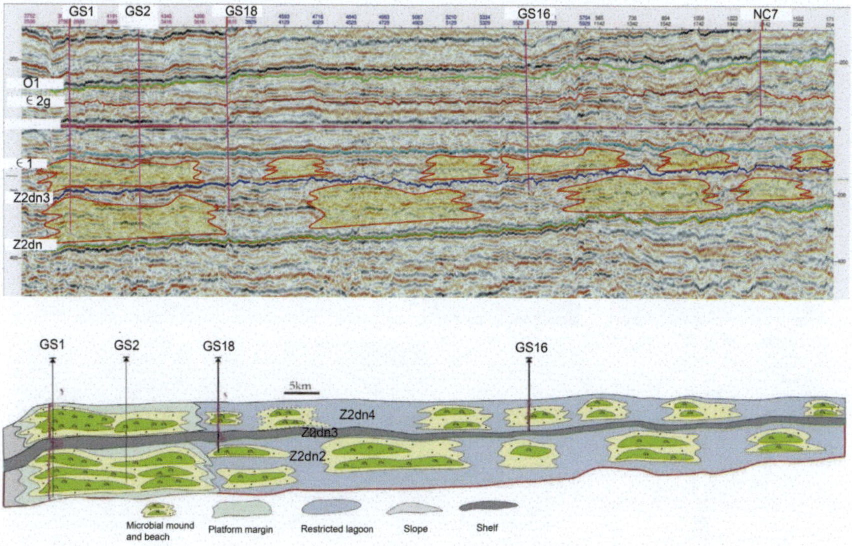

Fig. 18 Well-seismic sedimentary facies correlation across Well Gaoshi 1–Gaoshi 2–Gaoshi 18–Gaoshi 16

1.4 Seismic Facies Characteristics

1.4.1 Distribution Characteristics of Karst Paleogeomorphology

The distribution of karst paleogeomorphology in carbonate weathering crust has an important influence on the development degree of karst reservoirs in carbonate strata. Different karst geomorphic units often lead to different development degrees of karst reservoirs, and control the distribution of oil and gas in the later stage to a certain extent. The restoration of weathering crust karst paleogeomorphology is an important means to predict the distribution of karst reservoirs. Therefore, it is very important to study the development law of carbonate weathering crust karst reservoirs by the characteristics of weathering crust karst paleogeomorphology.

The residual thickness method is commonly used in karst paleogeomorphology restoration. Under the similar denudation effect, the area with thicker stratum thickness reflects that the landform is higher at that time, while the area with thinner stratum is the low-lying place during that time. Therefore, a special profile under the interface to be restored (generally the isochronous surface in the research area) is selected as the datum and flattened, then the residual thickness above this datum represents the shape of ancient landform. Characteristics: it is used with limited application and harsh condition, and is only used when the denudation effect is similar, that is, no structural extrusion uplift. It is applied when the sedimentary water body is retreated and exposed in a large scale in the stable platform sedimentary area, and the whole area is evenly exposed and denuded. The contact style between the

Fig. 19 Petrological characteristics of mounds in platform marginal. **a** Algal framework dolomite, Well Gaoshi 1, 4967.3 m, algal laminae is irregularly built, and framework pores expanded by corrosion can be seen. **b** Algal laminite (spongiostromata) dolomite, Well Gaoshi 1, 4981.5 m. Algal lamina is built horizontally or as irregular wavy, and the degree of pore development is low. **c** Spongiostromata dolomite, ordinary thin section, Well Gaoshi 1, 4951.6 m. **d** Grayish-brown algal laminite dolomite, with a small amount of dissolution pores and fractures, Well Gaoshi 2, 5015.2 m. **e** Algal laminite dolomite, with dissolution pores and fractures, cast thin section, Well Gaoshi 2, 5015 m

Fig. 20 Petrological characteristics of transition belt (east of platform margin). **a** Algal laminite dolomite, dark algal laminar intercalated with light-gray sandy laminar, horizontal-wavy, microbial mound, Well Gaoshi 18, 5184.8 m. **b** Algal dolomite with beddings, needle-like dissolution pores, granular beach, Gaoshi 18, 5177 m. **c** Dark-gray micrite limestone, sea between mound and beach, Well Gaoshi 18, 5179.8 m

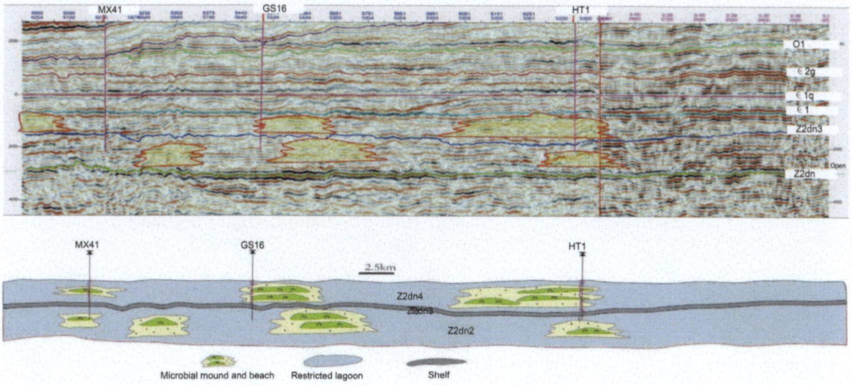

Fig. 21 Well-seismic sedimentary facies correlation across Well Moxi41–Gaoshi16–Hetan1

Upper Dengying Formation and the Lower Cambrian is an overlapping contact and disconformity, so this method is suitable for the karst model in the near horizontal karst strata.

On the eastern platform margin belt, the overall thickness of the Dengying Formation is large in the west and small in the east gradually decreasing from 1100 to about 400 m. The Dengyin Formation is narrow along the western platform boundary, and the strata inside the platform are thick in the east and thin in the west (thickness less than 400 m). The stratum surrounding Well Gaoshi 21, Gaoshi 16, and Hetan 1 is roughly 700–800 m thick (Fig. 22).

Deng2 Member is thick in the west and thin in the east as a whole with a thickness of 150–730 m. The Gaoshiti-Moxi platform margin belt is developed with the largest thickness. The intra-platform strata are developed with the thickness of about 200–500 m, and the intra-platform strata of Well Gaoshi 18–Gaoshi 16–Hetan 1 is developed with the largest thickness of about 500 m. A thick belt of more than

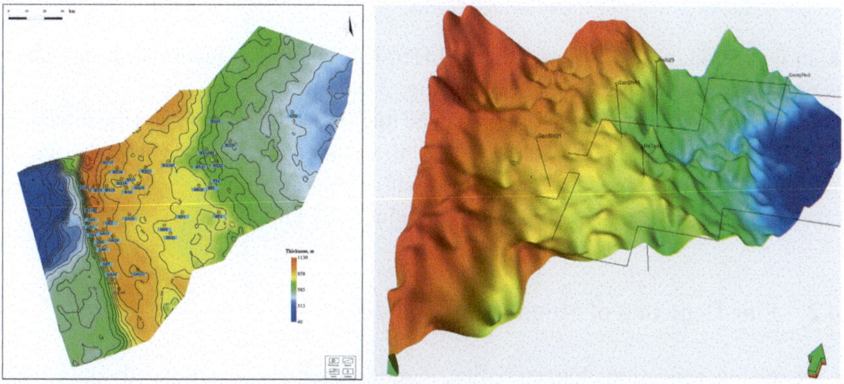

Fig. 22 Thickness and paleogeomorphology map of Dengying Formation in Central Sichuan

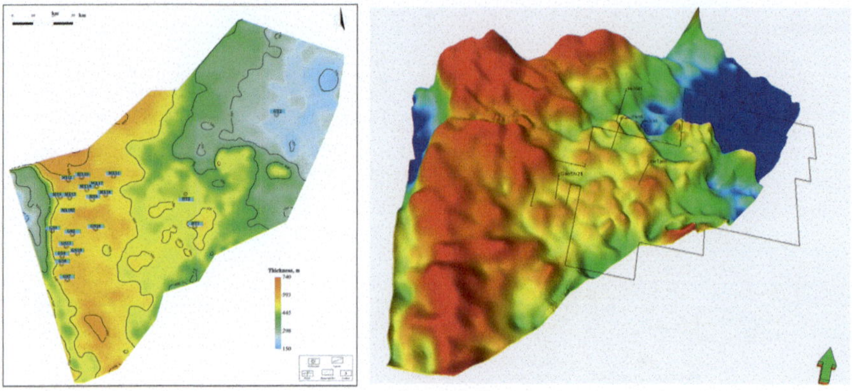

Fig. 23 Thickness and paleogeomorphology map of Deng1 and Deng2 Member of Dengying Formation in Central Sichuan

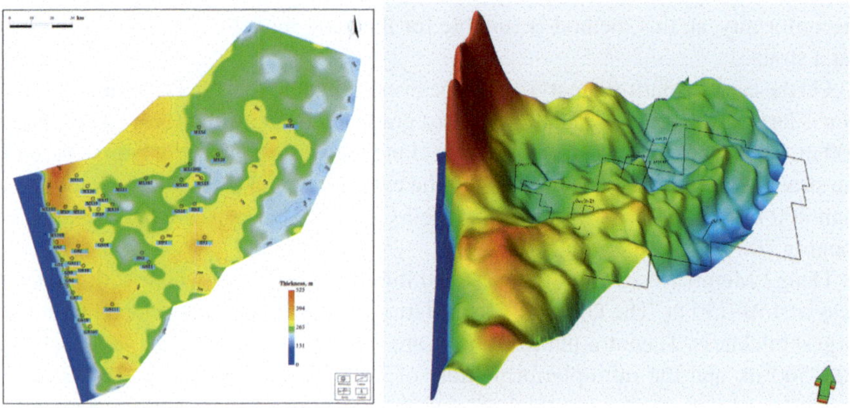

Fig. 24 Thickness and paleogeomorphology map of Deng4 Member in Central Sichuan

400 m is formed in the northeast of Well Hetan 1, which is a favorable zone for the development of the mound-beach body (Fig. 23).

The Deng4 Member is developed with the thickness of 0–525 m, and the thickness of platform margin belt is the largest. The intra-platform area is developed with the thickness of 240–400 m, and is developed with a paleogeomorphic high belt formed as the mound-beach body with an area of about 2700 km^2 (Fig. 24).

1.4.2 Fine Depiction of Mound-Beach Bodies

As for seismic reflection characteristics, mound-beach bodies are mostly developed as mound-shaped random reflections with weak amplitude. The inter-beach seas are mostly developed with relatively good continuity and relatively strong amplitude. In addition, the regional stratum thickness of the mound-beach body is slightly larger than that of inter-beach sea. Therefore, by drawing the thickness map of Deng4 and Deng2 Member, the geomorphic height of Deng2 and Deng4 Member can be roughly reflected. In addition, since Deng2 and Deng4 Member were exposed and denuded after deposition, the current thickness can only represent the residual thickness. However, since the sediments of mound beach are more resistant to the erosion of chemical weathering than the sediments of inter-beach sea, the residual stratum thickness can basically reflect the ancient landform at that time.

According to the drilling results of Well Moxi 9 and Gaoshi 18 in Gaoshi-Moxi area, the Deng4 Member in Well Moxi 9 is developed with the daily gas production of 10×10^4 m^3, and the sedimentary microfacies of mound-beach body. The Deng4 Member of Well Gaoshi 18 is developed with the daily gas production of 24.07×10^4 m^3, and the sedimentary microfacies of mound-beach body. Microbial mound-beach body have the same sedimentary construction, structure, and lithologic characteristics as those in the platform margin belt of the continental margin.

On the seismic profile, microbial mound-beach bodies are characterized by mound-like disorder and weak reflections, the stratum thickness increases (Fig. 25), and the area of a single mound-beach complex is large. Seismic attribute analysis is a comprehensive analysis technology for extracting reflected wave attributes along the target layer, that is, purposeful processing of post stack seismic data based on the accurate calibration of direct physical potential and 3D layer interpretation obtained by spectrum imaging technology through the quantitative relationship between the characteristics of seismic reflected wave-layer and amplitude spectrum and phase spectrum. Then the mound-beach body is predicted by the comprehensive analysis of multi parameters, which are the 3D physical information of the reservoir extracted along the layer, including instantaneous amplitude phase (Fig. 26), similarity (Fig. 27), sweet spot attribute (Fig. 28) and wave impedance profile (Fig. 29).

Intra-platform mound-beach bodies are widely developed in Dengying Formation, east of Gaoshiti-Moxi platform margin belt. Coring, conventional logging data and imaging logging data show that reservoirs of mound-beach body are encountered in Deng4 Member of Well Gaoshi 16.

The coring footage of Deng4 Member is 33.56 m, the core length is 28.5 m, and the core recovery is 84.92%. It is composed of brownish-gray and gray dolomite of 20.84 m, brownish gray and gray dissolved dolomite of 2.87 m and limestone of 4.79 m. Fractures and holes are relatively developed in the middle part of the rock, but are underdeveloped in the upper and lower parts. There are 170 small holes, 35 middle holes, 13 large holes, 35 fractures and 42 gas emitting pores.

Conventional logging data of 34# reservoir in Deng4 Member indicate that pores are developed, array acoustic energy attenuation is obvious, the imaging data indicate that fractures and dissolved pores are developed, nuclear magnetic resonance data show that this profile is mainly composed of small and medium pores, logging data show "gas invasion" at the depth of 5443.00–5444.00 and 5450.00–5452.00 m, which can be comprehensively interpreted as gas reservoir.

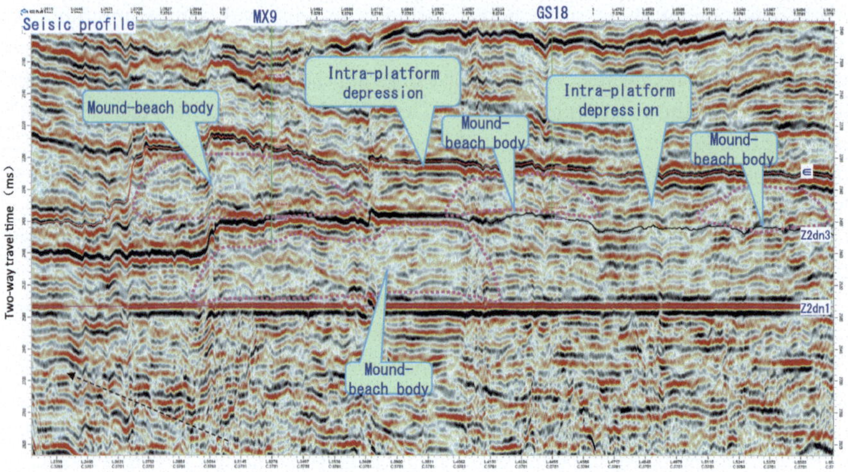

Fig. 25 Seismic profile of mound-beach body in Dengying Formation of Well Moxi 9–Gaoshi 18

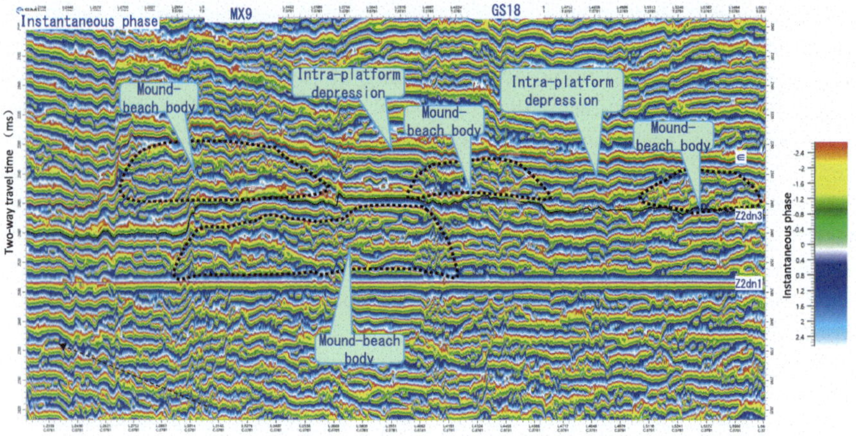

Fig. 26 Instantaneous phase profile of mound-beach body in Dengying Formation of Well Moxi 9–Gaoshi 18

Conventional logging data of 35# reservoir in Deng4 Member indicate that pores are developed, array acoustic energy attenuation is obvious, and imaging data suggest that dissolved pores are formed. NMR data shows that this profile is developed with mainly small and medium-sized pores, and the discrimination result of NMR fluid is gas–water characteristics. Logging shows "gas logging anomaly" at 5454.00–5457.00 m of well section, which is comprehensively interpreted as gas water layer.

Intra-platform mound-beach bodies are widely developed in Dengying Formation, east of the Gaoshiti-Moxi platform margin belt, which are different from the typical mound-beach bodies with mound-like reflections in the platform margin belt. The

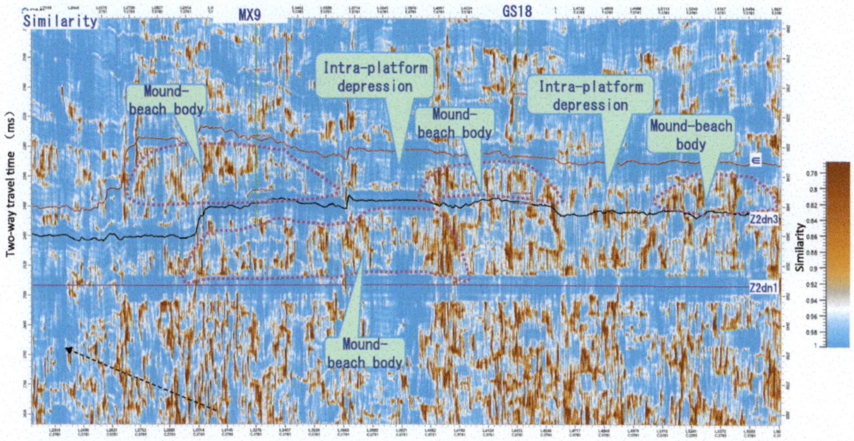

Fig. 27 Similarity profile of mound-beach body in Dengying Formation of Well Moxi 9–Gaoshi 18

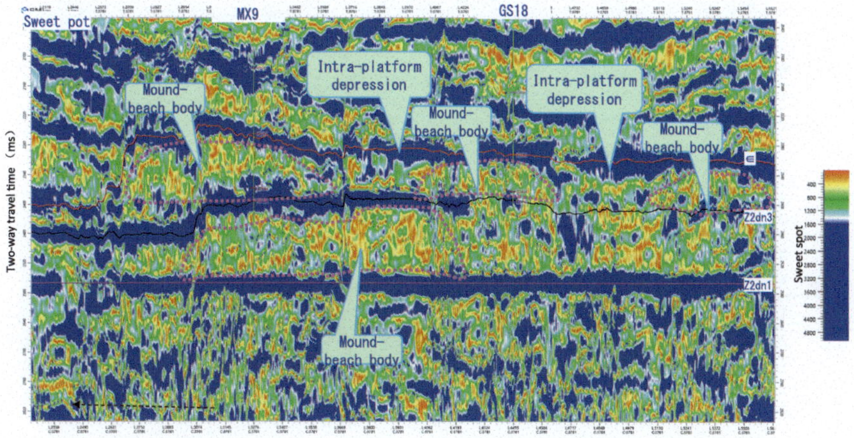

Fig. 28 Sweet spot profile of mound-beach body in Dengying Formation of Well Moxi 9–Gaoshi 18

intra-platform mound-beach body is characterized by atypical mound-like reflection and superimposed weak-amplitude chaotic reflections with wide trough and weak reflection (Fig. 30). The characteristics of mound-beach body can be more clearly described by various instantaneous phase of seismic attributes across Well Gaoshi 16 (Fig. 31), such as similarity (Fig. 32), sweet spot attribute (Fig. 33) and wave impedance profile (Fig. 34).

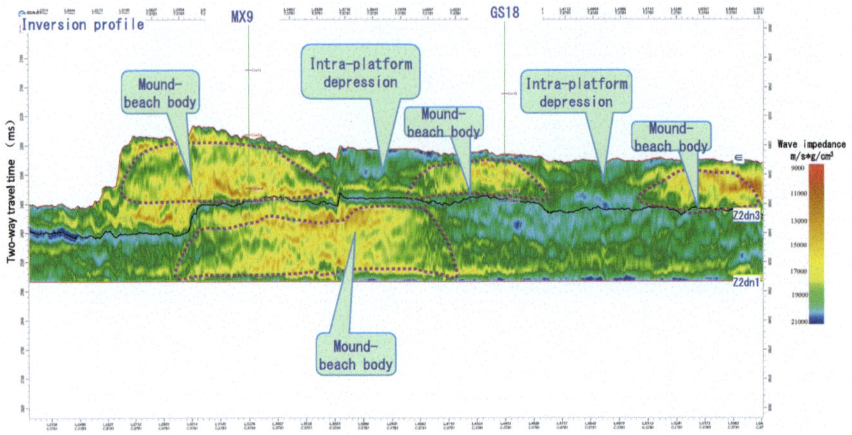

Fig. 29 Wave impedance profile of mound-beach body in Dengying Formation of Well Moxi 9–Gaoshi 18

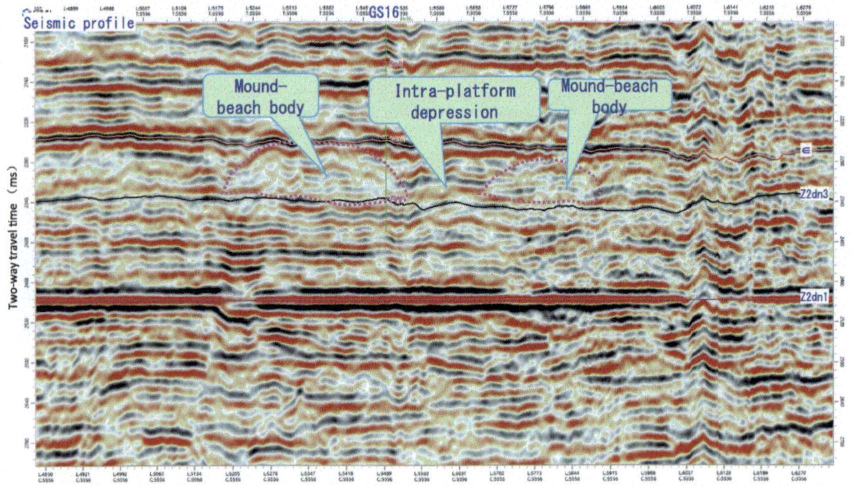

Fig. 30 Seismic profile of mound-beach body in Deng4 Member of Well Gaoshi 16

The distribution characteristics of mound-beach body can be more clearly described by the attribute extraction from 2 and 3D seismic data, such as amplitude (Fig. 35) and the number of wave crests (Fig. 36), based on the mound-like weak-amplitude chaotic reflection of mound-beach body in Dengying Formation.

Combined with the logging identification of mound-beach reservoir, seismic profile interpretation, fine characterization of seismic attribute plane and profile, the thickness map of mound-beach body is further extracted with the plane distribution through the interpretation of top mound-beach body in Deng2 and Deng4 Member. In the intra-platform area, 8 intra-platform mound beaches are depicted in

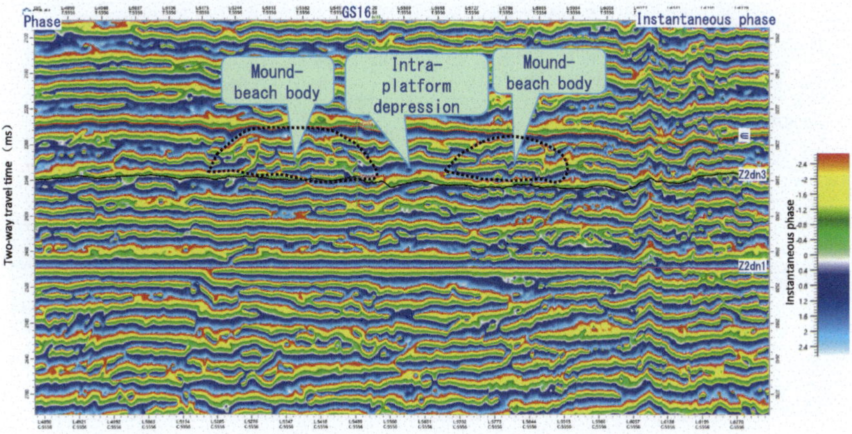

Fig. 31 Instantaneous phase profile of mound-beach body in Deng4 Member of Well Gaoshi 16

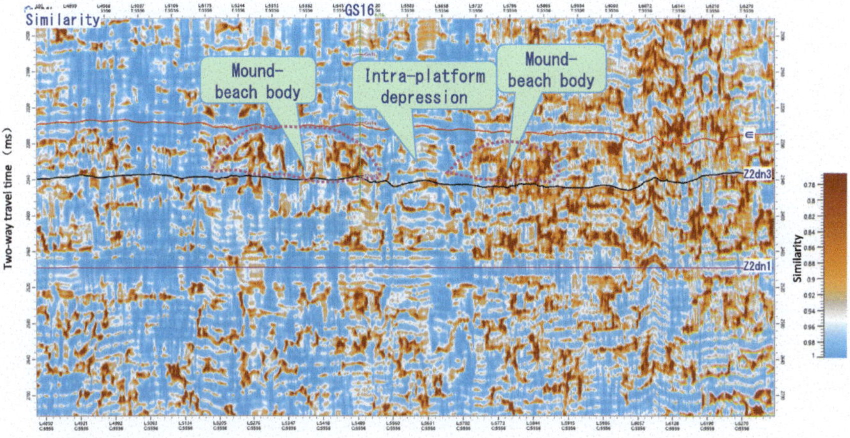

Fig. 32 Similarity profile of mound-beach body in Deng4 Member of Well Gaoshi 16

Deng2 Member, with a total area of about 980 km^2 (Fig. 37). And 8 intra-platform mound beaches are depicted in Deng4 Member, with a total area of about 781 km^2 (Fig. 38). In general, the mound beach body is the most favorable exploration target in Dengying Formation, which provides reference and basis for lithofacies paleogeography description and exploration well deployment.

The seismic facies analysis shows that the seismic facies of mound beach body are characterized by mound-like shape with high amplitude, disordered reflection structure, weak amplitude, and poor continuity. The seismic facies of trough is characterized by strong amplitude, strong continuity and parallel reflection, as shown in Fig. 39. The warm-tone area represents the seismic facies of mound-beach body, and the cold-tone area represents the seismic facies of inter-beach sea.

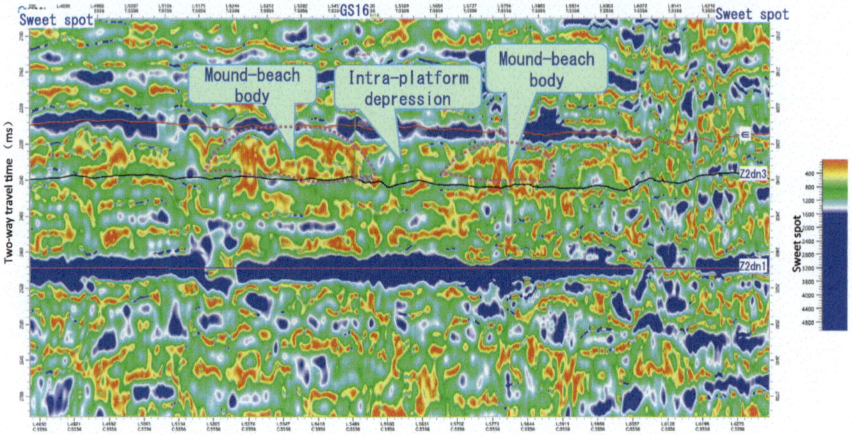

Fig. 33 Sweet spot profile of mound-beach body in Deng4 Member of Well Gaoshi 16

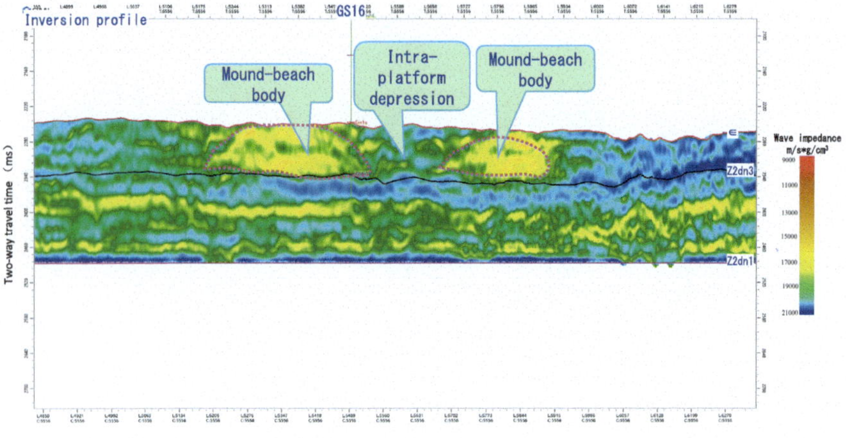

Fig. 34 Porosity profile of mound-beach body in Deng4 Member of Well Gaoshi 16

1.5 Lithofacies Paleogeographic Pattern

The logging facies interpretation model is established based on the sedimentary geological analysis of outcrops and cores. According to the detailed interpretation of sedimentary facies in single wells, across-well facies analysis and seismic facies analysis, the lithofacies paleogeographic maps of Deng2 and Deng4 Member are drawn with the seismic reflection characteristics of mound-beach body and inter-beach sea based on the single factor of formation thickness of Deng2 and Deng4 Member.

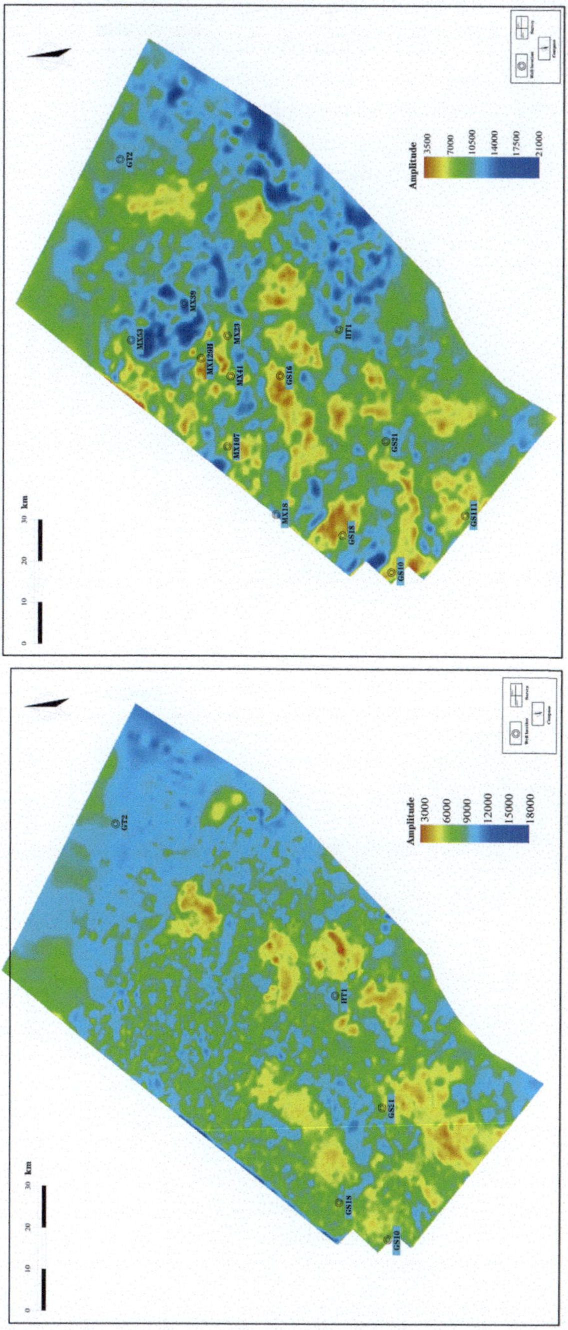

Fig. 35 Amplitude attribute plan of Deng2 Member (left) and Deng4 Member (right) in Central Sichuan

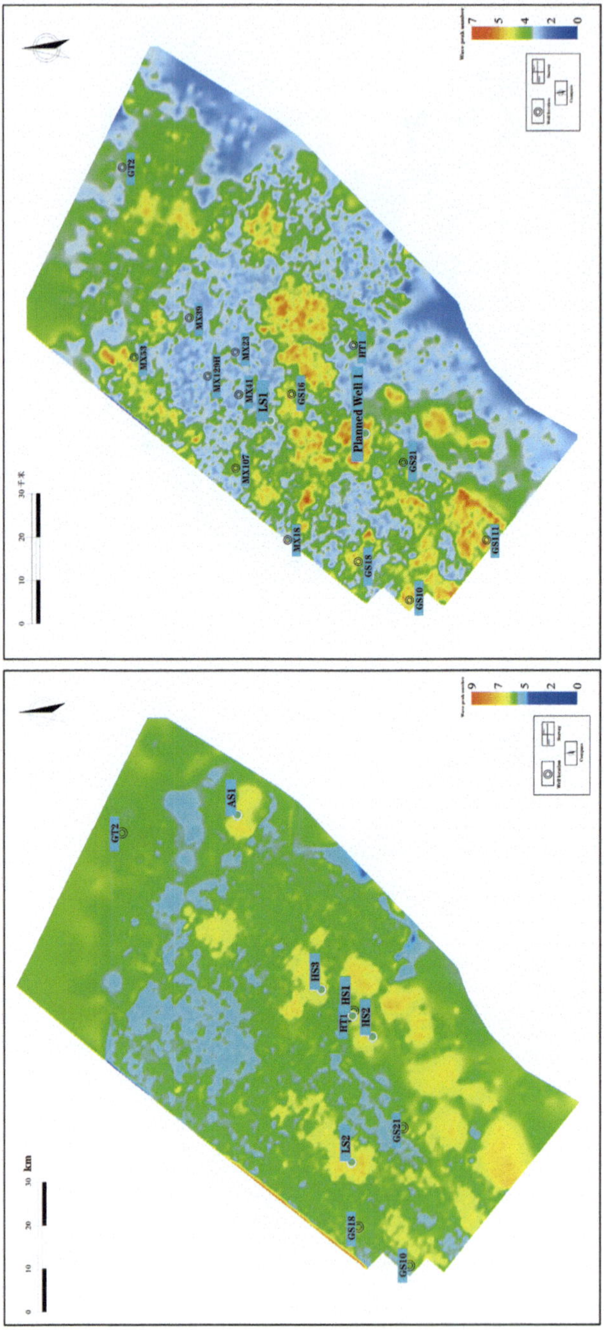

Fig. 36 Attribute plan of wave crest number of Deng2 Member (left) and Deng4 Member (right) in Central Sichuan

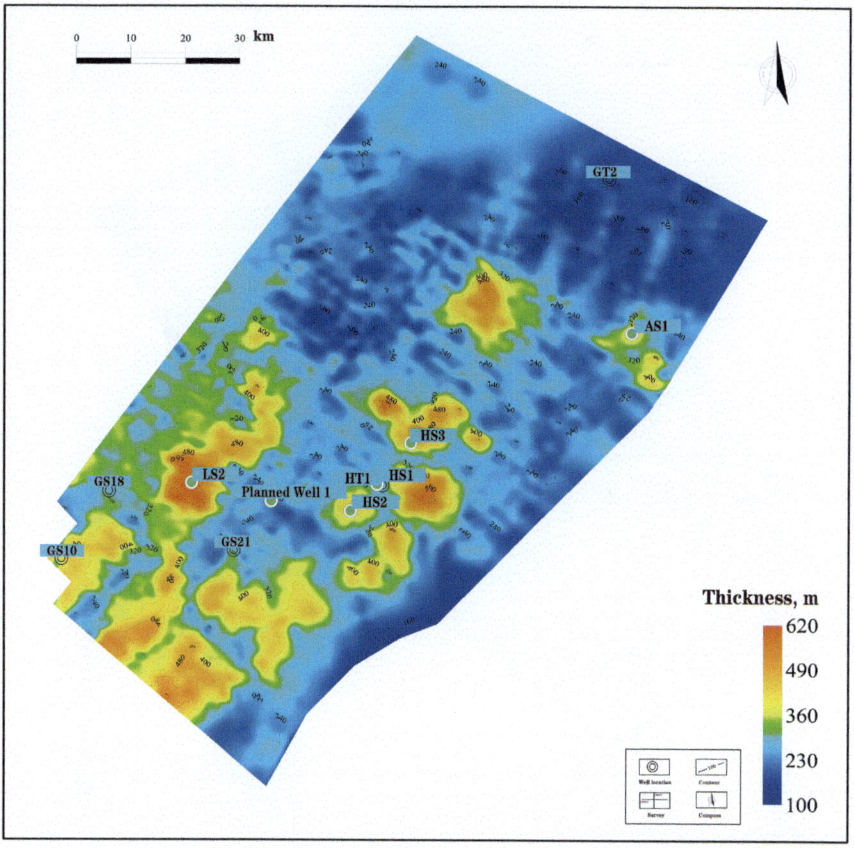

Fig. 37 Thickness map of intra-platform mound and beach of Deng2 Member in Central Sichuan

1.5.1 Lithofacies Paleogeographic Pattern

A nearly NS platform margin belt is developed in the west of Hechuan-Tongnan exploration area during the deposition period of Deng2 Member. The slope belt is developed in the west of the platform margin belt. The platform margin belt comprises algal mound and platform margin beach with large scale and contiguous distribution (Fig. 40). An intra-platform mound beach of NE direction is developed in Deng2 Member with some small scale of the mound-beach body compared to the platform margin belt.

The lithofacies paleogeographic pattern of Deng4 Member is basically consistent with that of Deng2 Member. A nearly NS platform margin belt is developed in the west, and the west of the platform margin belt is a slope belt. The platform margin belt is mainly composed of algal mound and platform margin beach with large scale and continuous distribution (Fig. 41). An intra-platform mound beach of NE direction is developed in Deng4 Member with small scale of mound-beach body. However,

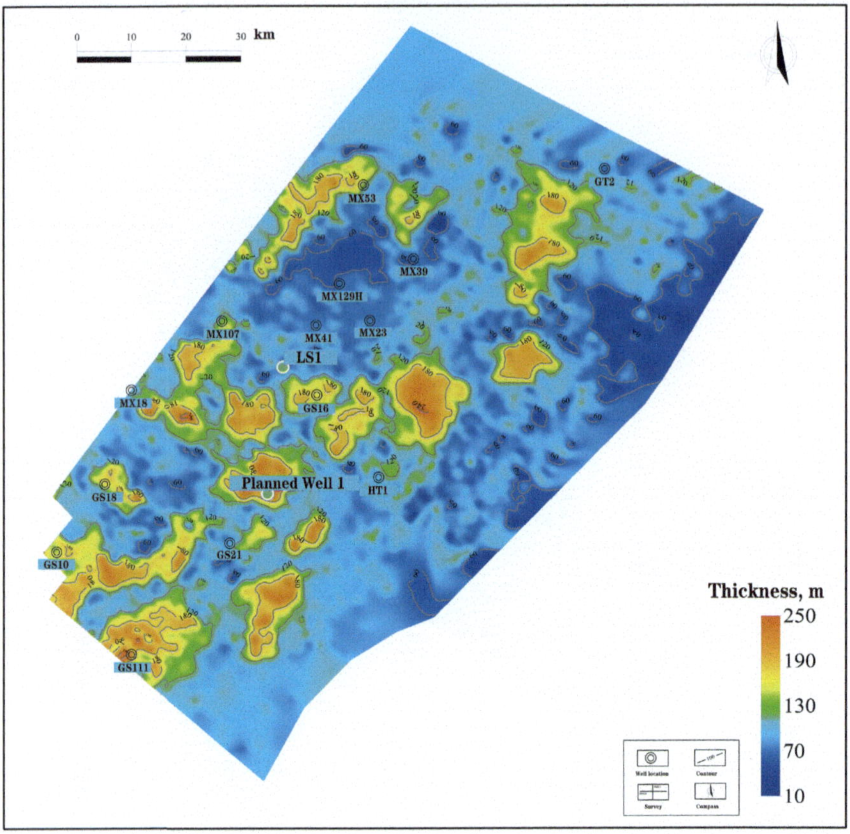

Fig. 38 Thickness map of intra-platform mound and beach of Deng4 Member in Central Sichuan

compared with Deng2 Member, the scale of intra-platform mound beach in Deng4 Member is relatively large.

1.5.2 Sedimentary Model

The Upper Yangtze platform is a half-isolated rimmed carbonate platform during the Dengying period (Zhang 2017). The platform deposition are characterized by algae-bacteria dolomitic marl mounds, inter-mound depressions and granular beaches, or mound-beach complexes. After several relatively uniform ascending and descending movements, the process of platform evolution, carbonate platforms with different characteristics are deposited from Deng1 to Deng4 Member. Among them, carbonate platforms are deposited in Deng1 and Deng2 period. Rift trough is developed in the west of Gaoshi-Moxi gas field. In Hechuan-Tongnan research area, east of the

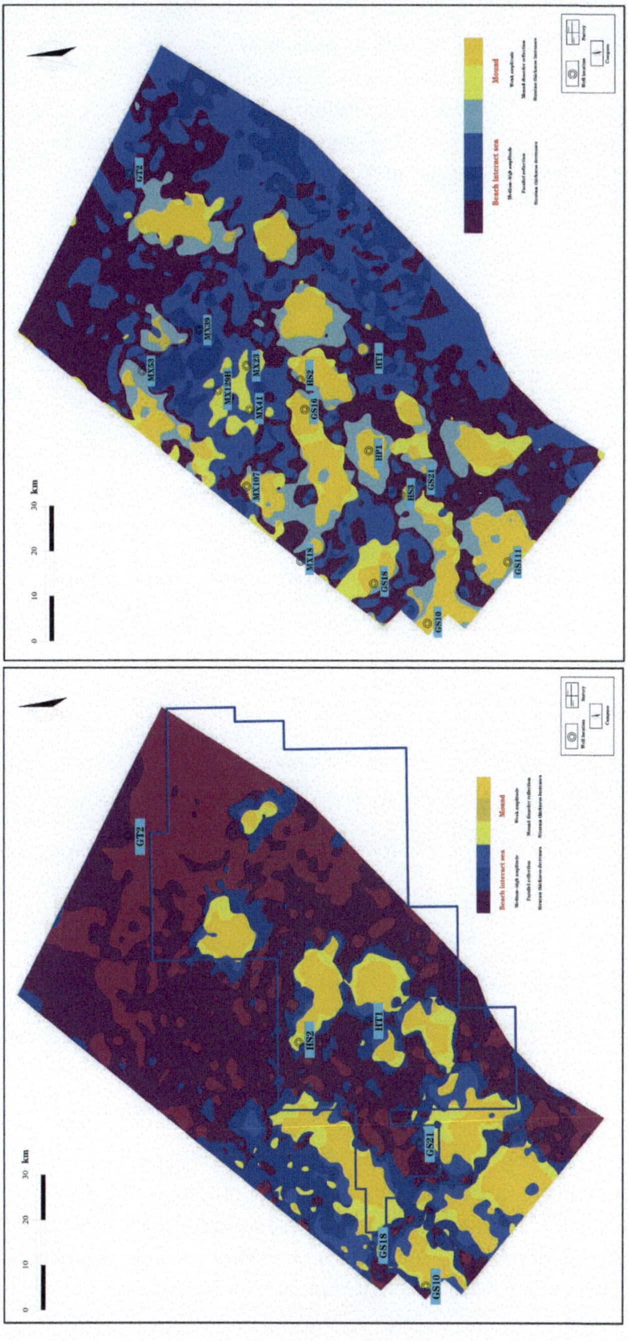

Fig. 39 Seismic facies of Deng2 Member (left) and Deng4 Member (right) in Central Sichuan

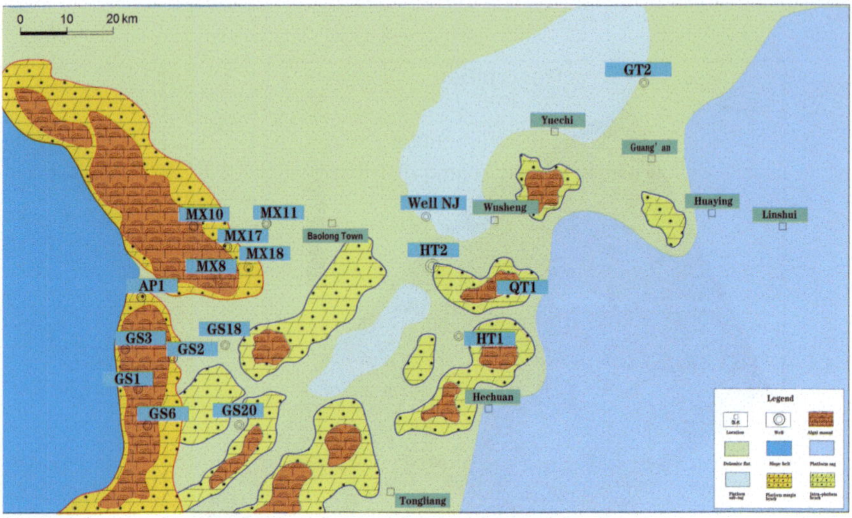

Fig. 40 Lithofacies paleogeography map of Deng2 Member in Central Sichuan

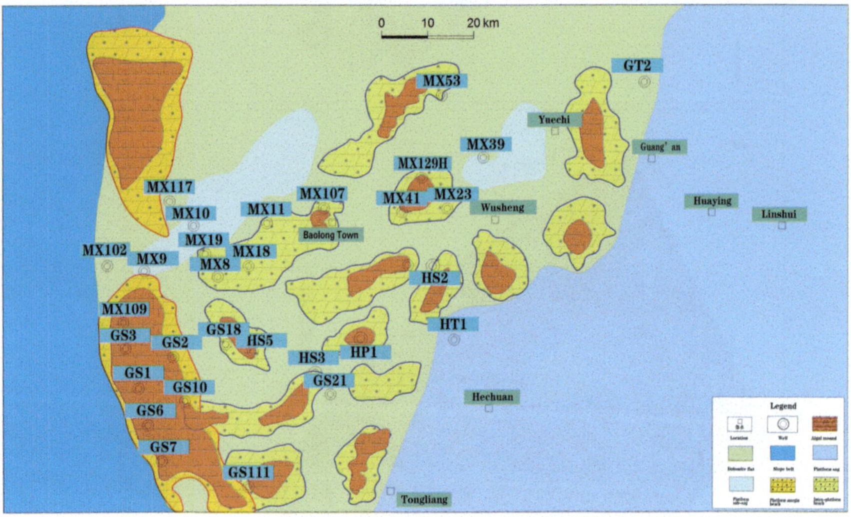

Fig. 41 Lithofacies paleogeography map of Deng4 Member in Central Sichuan

Gaoshi-Moxi gas field, multiple secondary small depressions are developed, with low sedimentary energy. Intra-platform microbial mound-beach bodies are developed around the depressions. The sea level rose during the Deng3 period, and the Sichuan Basin was formed as a shelf environment. Dark-gray and blue-gray muddy clastic rocks were developed in the platform. In the Deng4 period, the carbonate platform continued to grow, and the intra-platform depressions and biological mound-beach bodies may continue to grow based on Deng2 Member, showing the extensive development of inter-platform mound-beach body in Deng4 Member. The multiple movements of the platform also resulted in different degrees of differentiation within the platform, and a nearly NS extension rift trough was formed in the west of the platform. The formation of rift trough changed the sedimentary pattern of the platform and affected the distribution of sedimentary facies belt on the platform.

It should be pointed out that the sedimentary model of Dengying Formation in Sichuan Basin is different from the classical sedimentary model of rimmed platform. In the sedimentary model of Dengying Formation, the platform margin of Gaoshi-Moxi gas field is similar to Wilson rimmed platform margin, while inside the platform, there may be several small-scale depressions, which may be formed by the difference of faults or micro paleogeomorphology. There is relatively high paleotopography around the depression, where microbial mound-beach bodies can rapidly grow to a large scale and become a favorable exploration target (Fig. 42).

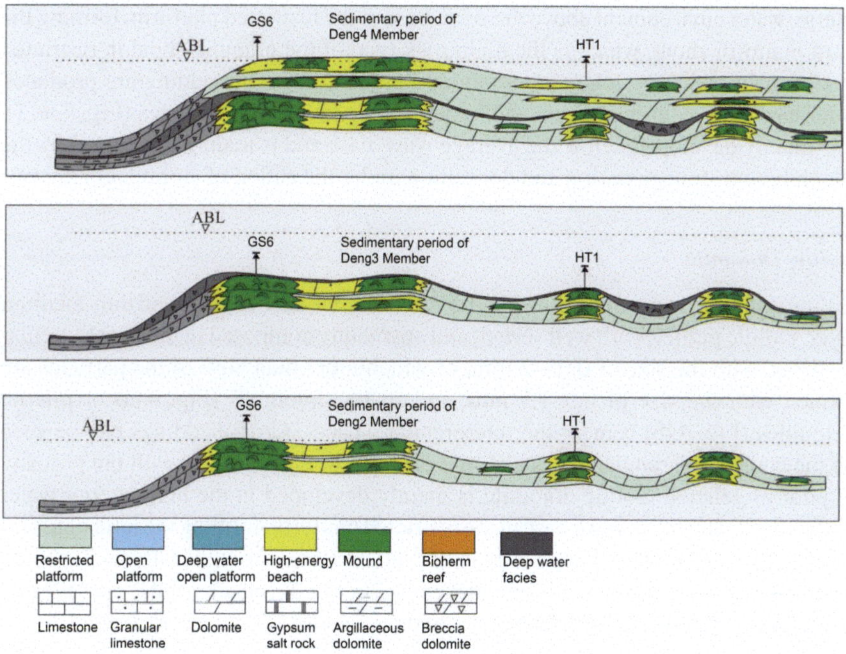

Fig. 42 Sedimentary model of Dengying Formation

2 Lithofacies Paleogeography and Sedimentary Evolution of the Longwangmiao Formation

2.1 *Sedimentary Facies Type*

2.1.1 Petrological Characteristics

Granular Dolomite

Doloarenite

The doloarenite is mainly formed by the color of gray-dark to gray and medium-thick layers, and the particles are mainly sand debris, occasionally a small amount of oolitic, bioclastic and gravel debris. The particle size of sand debris is 0.3–1.5 mm, and sparry dolomites cement the intergranular. Dense needle-like dissolution pores and dissolution holes are commonly developed in the core, as a result, the porosity of the core is general to good (Fig. 43). Doloarenite is mainly developed in the high-energy water environment above the wave base in the restricted platform, forming the intra-platform shoal, which is the main rock type of the granular shoal in restricted platform. In addition, doloarenite, muddy lamina and storm bedding are produced simultaneously with thin layers and underdeveloped dissolution pores (Fig. 43e, f). It is mainly developed below the average wave base and is mainly influenced by the allochthonous transportation and deposition under the action of storm and gravity.

Oolitic Dolomite

Oolitic dolomite is mainly formed in the color of gray-dark to gray, and thin-medium layer. Oolitic particles are well sorted, and are mainly composed of elliptical-circular particles, with the size of 0.05–2 mm, of which more than 80% of the particles are formed with the size of 0.5–1.5 mm, and sparry dolomites (Fig. 44a) or micrite dolomites (Fig. 44b) cement the intergranular spaces. Cross beddings can be seen on the core. Intergranular pores and pinholes can be seen locally, with the porosity of poor to general. Oolitic dolomite is mainly developed in the high-energy water environment above the wave base in the restricted platform, forming the intra platform shoal, and is the main rock type of granular shoal in the restricted platform. It should be noted that oolitic dolomite is only developed in Moxi well area and is not revealed in the coring profiles of Well Tongtan 1 and Guangtan 2 in the research area.

Cristal Dolomite

Crystal dolomite usually refers to the type of dolomite in crystalline form, and they lack any residual sedimentary structure. In the crystal dolomite, dark-gray dark spots are mainly composed of salty crystals, and light-gray bright spots are mainly

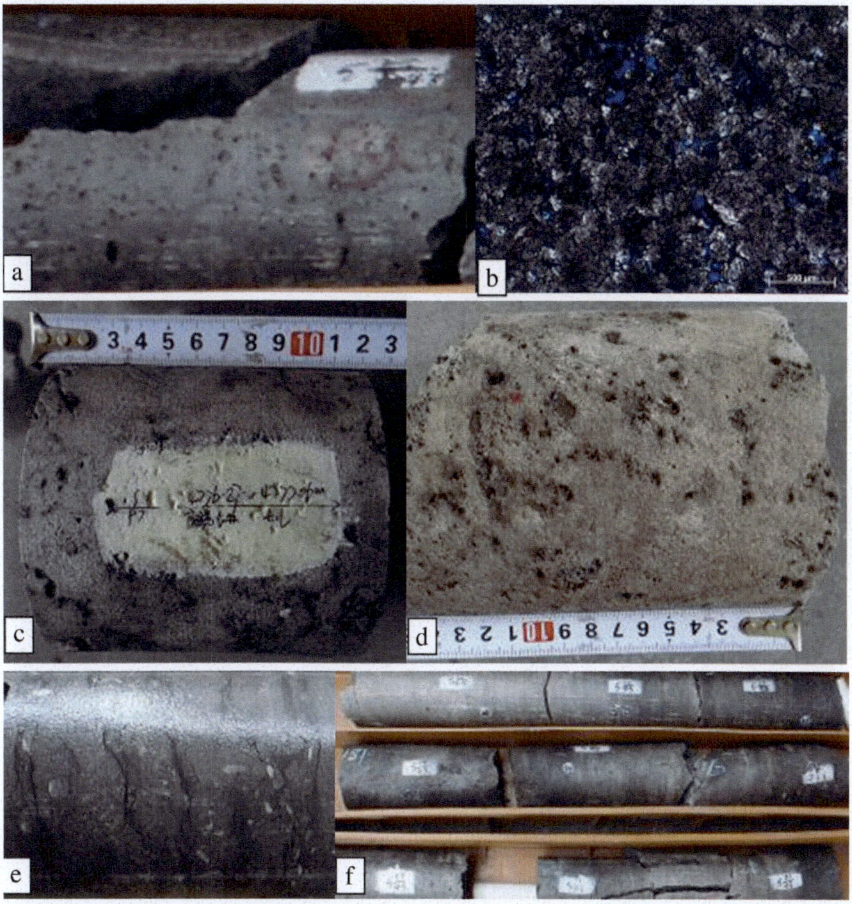

Fig. 43 Petrological characteristics of doloarenite. **a** Doloarenite with dissolution pores, Well Tongtan 1, 4998.65 m. **b** Residual doloarenite, intergranular dissolution pores, Well Tongtan 1, 4998.65 m. **c** Gray doloarenite with dissolution pores, Well Moxi 16, 4776.3 m. **d** Doloarenite with dissolution pores, Well Moxi 20, 4612.2 m. **e** Gravelly micrite doloarenite, with storm bedding composed of irregular muddy laminae, Well Guangtan 2. **f** Cycle of dark gray micrite dolomite and gray micrite doloarenite. Gravel debris and muddy lamina can be seen in the doloarenite with reverse rhythm, Well Tongtan 1

composed of micrites. Some crystal dolomites are formed by "spots" with light and dark interaction with unclear beddings. A few pinholes and dissolved pores are developed in the core. The origin of crystal dolomite is complex, fine-grained dolomite, for example, may be related to a low-energy sedimentary environment (Fig. 45a, b), whereas coarse-grained dolomite may be related to complete dolomitization (Fig. 45c).

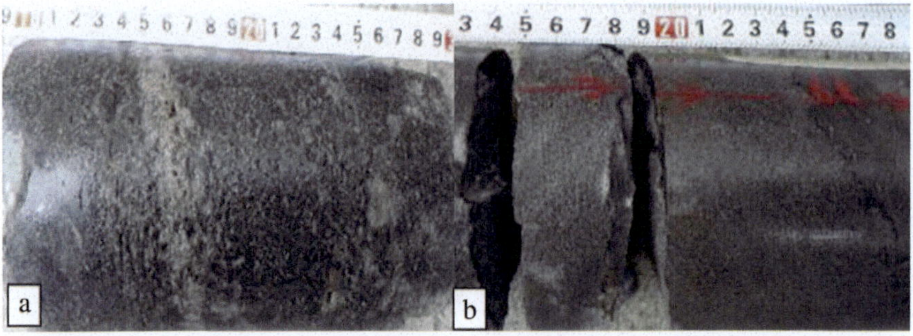

Fig. 44 Petrological characteristics of oolitic dolomite. **a** Dark-gray oolitic dolomite, cross beddings, clear grain margin, sparry cementation, Well Moxi 203, 4770.2 m. **b** Oolitic dolomite, with unclear grain margin and mainly micrite cementation, Well Moxi 203, 4787.2 m

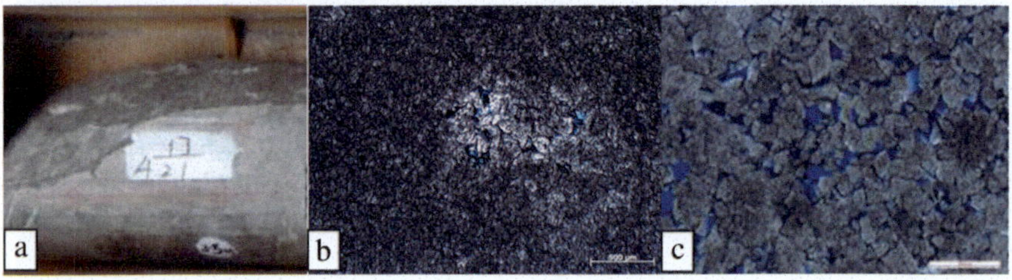

Fig. 45 Petrological characteristics of crystal dolomite. **a** Light-gray fine-grained dolomite, Well Tongtan 1, 4992.55 m. **b** Fine-grained dolomite, no obvious residual structure, a small number of dissolution pores, Well Tongtan 1, 4992.55 m. **c** Coarse-grained dolomite, dissolution pores, Well Gaoshi 113, 4990 m

Micritic Dolomite/Muddy Micritic Dolomite

In practical work, as the products of low-energy sedimentary environment, micritic dolomite and muddy micritic dolomite are difficult to distinguish, and can be used as the same rock. Both of them are mainly grayish black and thin-layered, and are mainly composed of micrites, with muddy laminae or muddy strips (Fig. 46). Bioturbate structures are widely developed in both of the two dolomites. As a result of the dense lithology, no pores develop in either of the two dolomites. The pores are mostly developed in the low-energy reductive sedimentary water environment below the wave base inside the restricted platform and are the main rocks of dolomitic muddy lagoon in the lagoon of the restricted platform.

2.1.2 Sedimentary Facies Combination and Characteristics

The above sedimentary background of the Sichuan basin shows that the Cambrian Longwangmiao Formation is deposited in an environment of a carbonate platform.

Fig. 46 Petrological characteristics of micritic dolomite/muddy micritic dolomite. **a** Dark-gray micritic dolomite, with bioturbation structure, Well Tongtan 1, 4997.7–4998.3 m. **b** Dark-gray micritic dolomite, Well Guangtan 2, 5628.17 m

Restricted-half restricted platforms are mainly developed in the Hechuan-Tongnan research area, mainly identified as intra-platform beach, inter-beach sea and storm beach (Table 2).

Intra-platform Granular Beach

The intra-platform granular beach is mainly developed with the lithology of doloarenite and oolitic dolomite, the color of gray-light to gray, medium-thick layer, and large amount of dissolution pores. It is the product of high-energy hydrodynamic conditions near the average wave base. According to the core characteristics of Well Tongtan 1 and Guangtan 2, it can be seen that the intra-platform granular beach is mainly composed of doloarenite, with a small amount of oolitic dolomite occasionally, and the single-layer thickness of the beach body is large.

Table 2 Sedimentary facies and petrological characteristics of Longwangmiao and Xixiangchi Formation in Cambrian

Open-half restricted platform		
Intra-platform beach	Storm beach	Inter-beach sea
Dominated by doloarenite and oolitic dolomite occasionally. Characterized by light gray, medium-thick layers, and dissolution pores	Doloarenite or oolitic dolomite is symbiotic with muddy laminae, storm bedding and grading bedding, thin layers, low development of pores	Dark-gray muddy micritic dolomite, micritic dolomite and micritic silty dolomite, horizontal bedding, bioturbation and storm bedding
Longwangmiao, Xixiangchi		

Intra-platform Storm Beach

The storm beach is mainly composed of doloarenite and oolitic dolomite and primarily cemented by micrites. The color is usually dark-gray or interbedding of dark gray and light gray. The intra-platform storm beaches are formed with thin layers, and mostly symbiotic with muddy lamina, storm bedding and grading bedding. In the sedimentary environment below the average wave base, the intra-platform sediments transformed by storm or gravity are transported to the deep water area. According to the cores of Well Tongtan 1 and Guangtan 2, and the cores in Xixiangchi Formation of Well Guangtan 2 and He 12, the core characteristics can be identified as the color of dark gray, thin layers, the symbiosis of doloarenite and muddy laminae, the symbiosis of storm bedding and grading bedding, and the underdevelopment of dissolution pores.

Inter-beach Sea

The inter-beach sea is mainly developed by the lithology of micritic dolomite and micritic silty dolomite, the color of dark gray, thin layers, horizontal beddings, biological disturbance and storm beddings. According to the core characteristics of Well Tongtan 1 and Guangtan 2, the color of inter-beach sea is dark. This difference may be related to paleoclimate, sedimentation rate and the scale of sea level change.

2.2 *Sedimentary Characteristics of Backbone Section*

2.2.1 Sedimentary Characteristics of Cores in Well Tongtan 1

Well Tongtan 1 was cored four times in the Cambrian Longwangmiao Formation, with a depth of 5010–4985 m. The coring locations are the middle and upper parts of Longwangmiao Formation.

The sedimentary characteristics of the core from bottom to top are mainly as follows (Fig. 47). For the 6th core, the bottom is gray algal dolomite, the dolomite is dirty with low degree of automorphism, and the upward dissolution pores increase. The upper part is gray-light and gray algal dolomite and doloarenite, with highly developed dissolution pores. Algal clot dolomite with a thickness of about 0.4 m is developed at the bottom. The top is dark gray micritic silty dolomite. In the 15th coring, the lower part is gray-light and gray algal doloarenite with dissolution pores, forming an upward shallowing high-frequency cycle with the thick-layer micritic silty dolomite in the lower part. A cycle composed of dark-gray micritic silty dolomite and algal granular dolomite is developed in the upper part, forming another upward shallowing high-frequency cycle. For the 4th and the 3rd coring, the lower part is dark-gray thick-layered fine-grained dolomite mixed with thin-layered algal dolomite, the

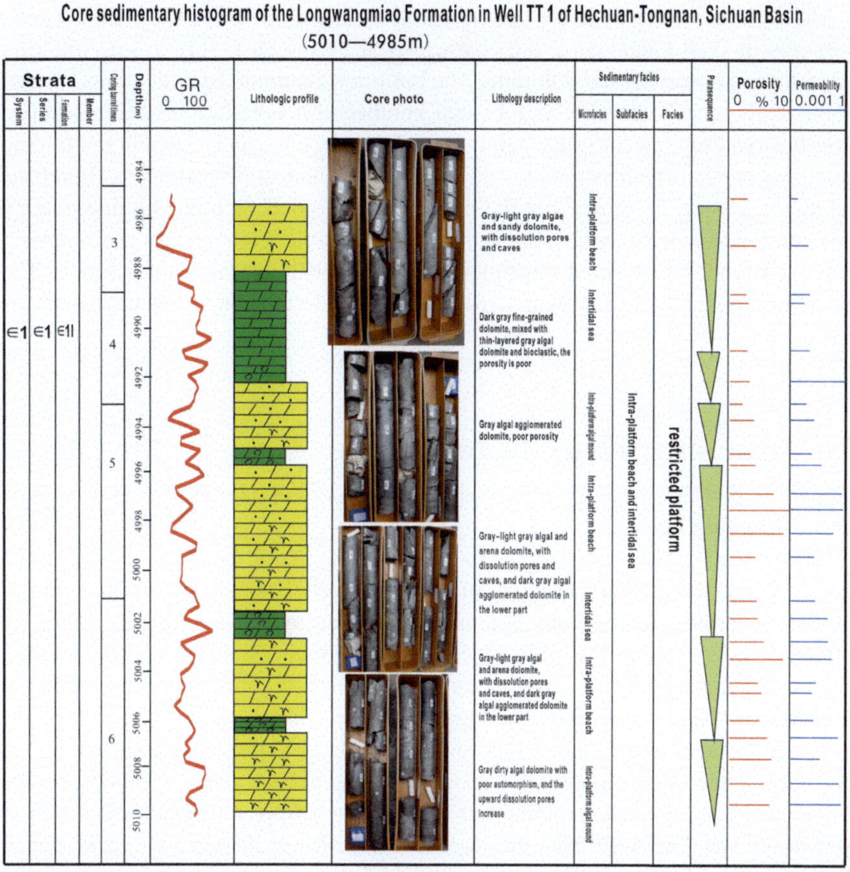

Fig. 47 Comprehensive column map of core sedimentary characteristics in Cambrian Longwangmiao Formation of Well Tongtan 1

upper part is light-gray doloarenite with dissolution pores, and the whole is an upward shallowing high-frequency cycle.

Several high-frequency cycles composed of micritic silty dolomite and algal granular dolomite are developed in Longwangmiao Formation, and dissolution pores are widely developed in granular intra-platform. On the whole, the devclopment of granular beach and dissolution pore in Well Tongtan 1 is high.

2.2.2 Sedimentary Characteristics of Cores in Well Guangtan 2

Well Guangtan 2 was cored three times in Longwangmiao Formation, with a depth of 5360–5314 m and a cumulative core thickness of 46 m. The coring locations are in the middle and upper parts of Longwangmiao Formation.

The core characteristics from bottom to top are as follows (Fig. 48). On the whole, the cores in Well Guangtan 2 are characterized by the dark-gray to gray micritic dolomite and power crystal dolomite. The bottom is intercalated with a thin-layered (about 1 m) sandy dolomite, and a small number of needle-like dissolution pores. Thin-layered gray micritic silty dolomite and doloarenite can be seen in the middle and upper parts, as well as muddy banded beddings and storm beddings. Therefore, sedimentary analysis shows that the coring profile reflects a relatively low-energy intra-platform environment, which is deposited in subtidal low-energy environment, with occasional thin-layered intra-platform beach, and common storm beach. The dissolution pores are generally underdeveloped due to the whole low-energy facies belt.

2.3 Sedimentary Facies Analysis

2.3.1 Logging Facies

Based on the sedimentary analysis of cores in Well Tongtan 1 and Guangtan 2, the core location is carried out, the natural gamma-ray curve with high lithology sensitivity is selected as the main series of logging facies analysis, and the logging facies characteristics of microfacies combination in Longwangmiao Formation are summarized (Figs. 49 and 50). The logging curve is mainly characterized by high-amplitude dentation and funnel type, which can be interpreted as interbedding of thin intra-platform beach and inter-beach sea. A small amount of toothed positive rhythm can also be seen, explained as the transition from intra-platform beach to inter-beach sea. Different from the large number of box-shaped and block-shaped logging facies in Dengying Formation, this high-frequency dentation indicates the influence of superposed pattern difference caused by the change of intra-platform beach and inter-beach sea inside the platform according to the high-frequency sea level change.

2.3.2 Sedimentary Facies Analysis of Single Wells

Based on the analysis of petrological characteristics and logging facies, the sedimentary facies of single wells are analyzed. For example, Well Gaoshi 16 is located in the north section of Gaoshiti-Hechuan in Sichuan Basin, and south of Longnvsi structure, which belongs to the typical deposition environment of intra-platform beach. Longwangmiao Formation in Well Gaoshi 15 is developed with the depth of 5021–5132 m and the thickness of 111 m. In the early stage, it was developed as a carbonate platform, and a large set of thin-layered micritic dolomite with the color of gray-black and dark-gray was recorded. It was a typical subfacies of low-energy inter-beach sea. With the change of sea level, many sets of dark gray thin-layered intra-platform granular beach were developed, forming the subfacies of intra-platform beach. In

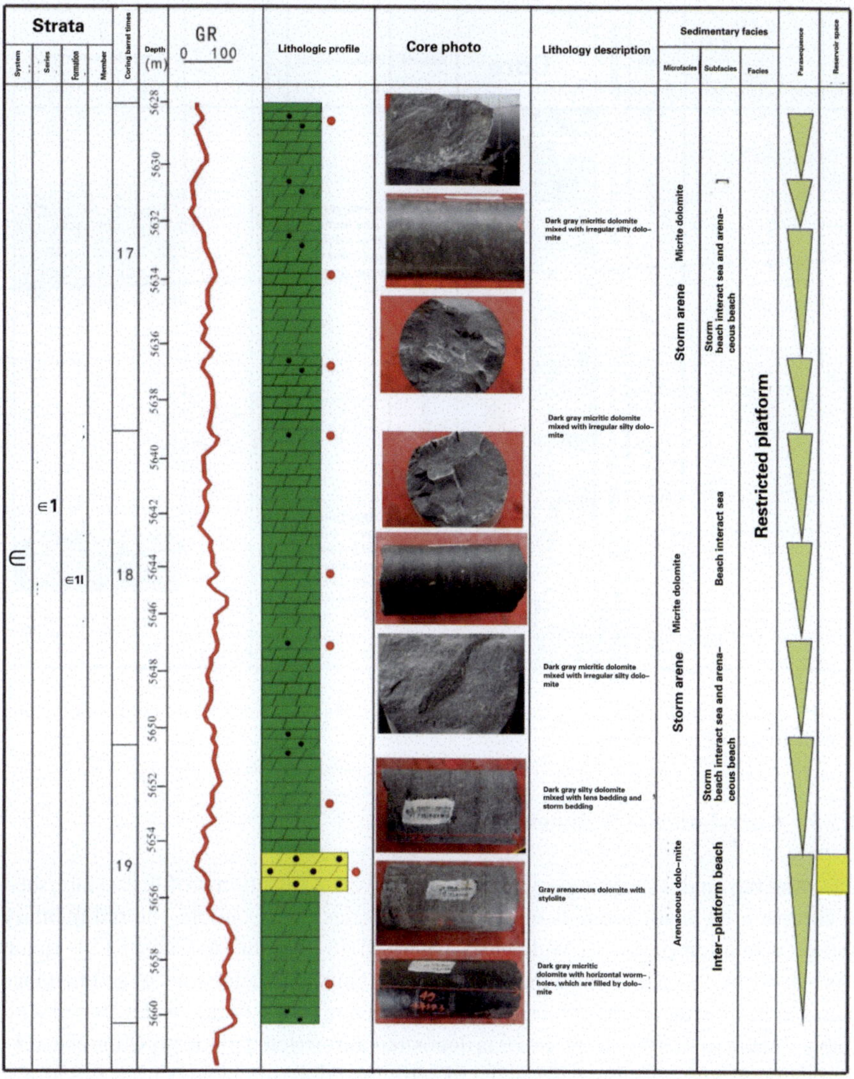

Fig. 48 Comprehensive column map of sedimentary characteristics of core in Cambrian Longwangmiao Formation of Well Guangtan 2

the middle and late stages of the highstand system, the high-frequency sea-level decreased slowly. Multiple sets of thin-layered intra-platform granular beach were developed to form the combination of low-energy belt and beach, with a thickness of 2–3 m. Based on the characteristics of lithologic combination, Well Gaoshi 16 is identified as the facies of restricted platform.

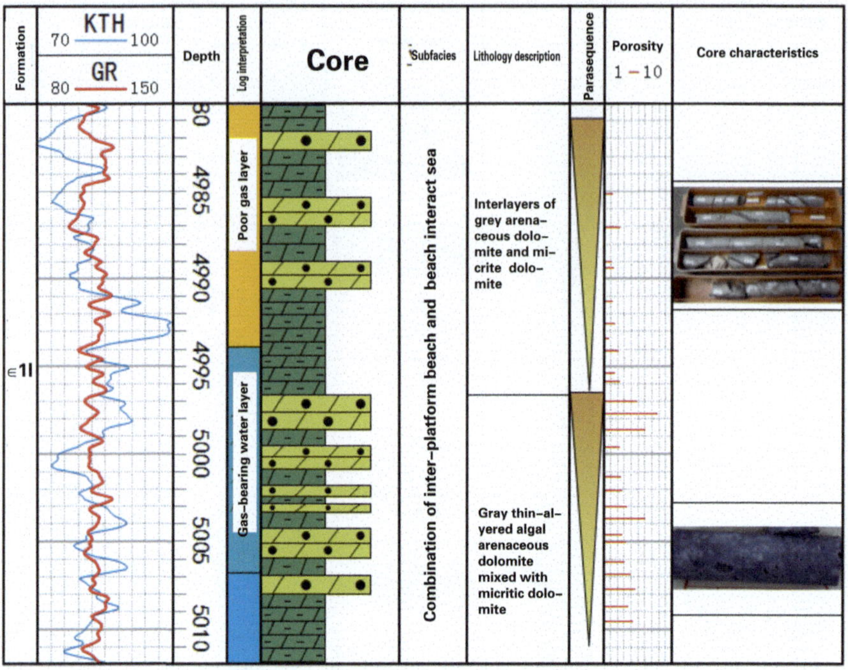

Fig. 49 Logging facies characteristics of microfacies combination in Longwangmiao Formation of Well Tongtan 1

2.3.3 Correlation of Sedimentary Facies

Based on the analysis of single-well facies, the well connection profile is established to compare the characteristics of sedimentary facies. For example, the sedimentary facies correlation profile of Well Moxi 16–Moxi 46–Gaoshi 16–Gaoshi 113–Hetan 1 is established. According to the macro sedimentary characteristics of sedimentary facies in Longwangmiao Formation, it can be seen that the stratum is developed with stable thickness of 79–120 m, and it is deposited as restricted platform with relatively stable environment. In the early stage of sedimentation, the facies of intra-platform beach are developed in Well Moxi 46 and Hetan 1. A set of carbonate platforms is developed in other wells, deposited as a typical subtidal low-energy sedimentation. With the change of sea level, the intra-platform beach is widely developed in the profile, and multiple sets of medium-thin beach bodies can be seen in each well with relatively good connectivity. In general, the development frequency and scale of beach body in the middle and late sedimentary period of the Longwangmiao Formation are larger than those in the early stage. It can be considered that the middle and late sedimentary periods of the Longwangmiao Formation are the main period for beach body formation. The single-layer thickness of high-energy beach in this profile ranges from 1 to 4 m, 2.6 m on average. And the cumulative thickness

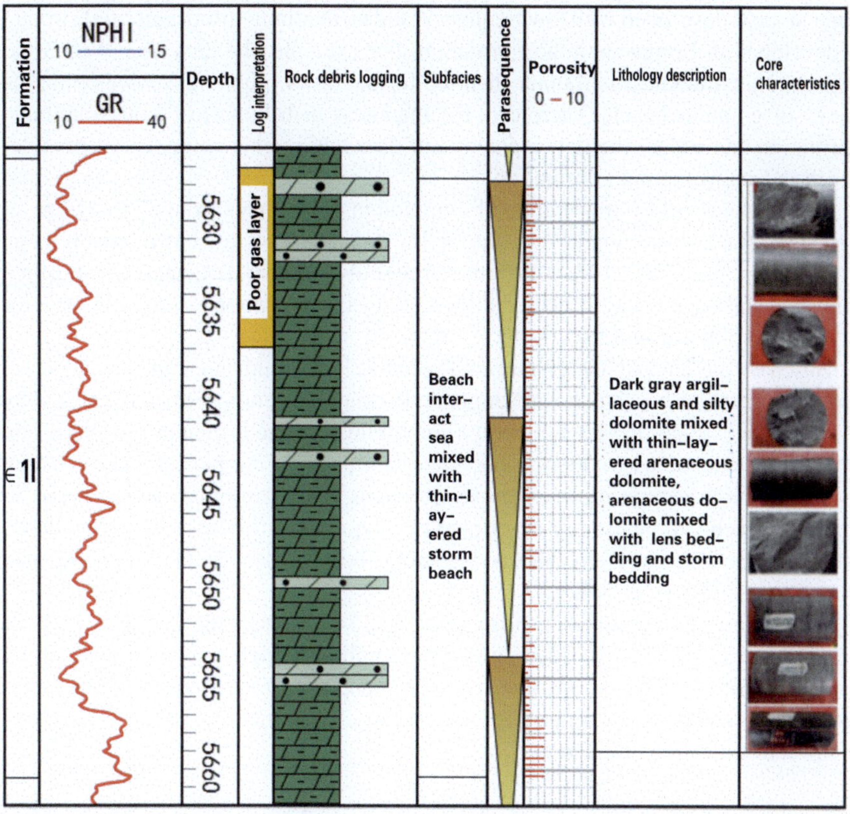

Fig. 50 Logging facies characteristics of microfacies combination in Longwangmiao Formation of Well Guangtan 2

of single wells ranges from 28 to 42.7 m, 36.2 m on average. The intra-platform beach in Well Gaoshi 113 is developed with the largest single-layer and cumulative thickness, while in Well Moxi 46 it is developed with the smallest single-layer and cumulative thickness.

2.4 Seismic Facies Characteristics

According to the seismic reflection characteristics of the Longwangmiao Formation (Fig. 51), the thickness of Longwangmiao Formation increases in the gentle inner slope where granular beach is developed. It can be seen from the seismic reflection profile that a set of trough with medium-strong amplitude is developed in the Longwangmiao Formation with strong continuity, such as the area near Well Gaoshi 16. The seismic frequency is generally higher than 36 Hz. The seismic facies of gentle inner slope is developed with low frequency, and no medium-strong amplitude trough is developed in Longwangmiao Formation. For example, the area near Well Moxi 107. And the thickness is thinner than the formation of other wells developing the facies of granular beach. Therefore, the thickness distribution of Longwangmiao Formation can reflect the development and distribution characteristics of granular beach in Longwangmiao Formation.

The thickness of Longwangmiao Formation mainly ranges from 87 to 119 m, of which the thickness around Well Gaoshi 16–Hetan 1–Gaoshi 21 is relatively large (Fig. 52), ranging from 108 to 120 m. Through the comparative analysis of drilled thickness and seismic interpretation thickness, the interpretation results with the error of less than 5 m account for 100%.

Based on the 2D and 3D seismic data volumes in the study area, the development profile of granular beach in Longwangmiao Formation is compared and analyzed by the attribute extraction, such as average amplitude, average instantaneous phase and average instantaneous frequency after determining the time period corresponding to the reservoir section on the time section. It is considered that the reservoirs in granular beach can be exactly descripted by frequency attribute as the coincidence rate between frequency attribute and granular beach reservoir of Longwangmiao Formation is 93% (Fig. 53).

The seismic facies analysis shows that (Fig. 54): the seismic facies of granular beach are characterized by medium-strong amplitude, high frequency and strong continuity, which is the warm-tone area in the map. The gentle slope seismic facies are characterized by weak amplitude, low frequency, and medium continuity, located in the cold-tone area in the map.

2.5 *Lithofacies Paleogeographic Characteristics*

According to the thickness, seismic attributes and seismic facies of Longwangmiao Formation in Central Sichuan, combined with sedimentary facies characteristics and drilling results, the lithofacies paleogeographic distribution of Longwangmiao Formation is drawn (Fig. 55). From the distribution of lithofacies paleogeographic in Longwangmiao Formation, it can be seen that the sedimentary trend of Longwangmiao Formation is consistent with the axial trend of Longshan-Longnvsi paleouplift anticline, which shows that the deposition of Longwangmiao Formation is controlled by Leshan-Longnvsi paleouplift during the deposition period. The lagoonal facies is developed near the Huayingshan, with gypsum salt developed locally. In the NW direction, controlled by the Leshan-Longnvsi paleouplift, the inner gentle slope is developed, and the thickness of Longwangmiao Formation increases. Especially in the line of Well Gaoshi 111–Gaoshi 21–Tongtan 1, the thickness of Longwangmiao Formation is the thickest, indicating the development of granular beach in the gentle inner slope. In addition, a small amount of low-energy granular beaches are developed near Well Guangtan 2 at the inclined end of Leshan-Longnvsi paleouplift.

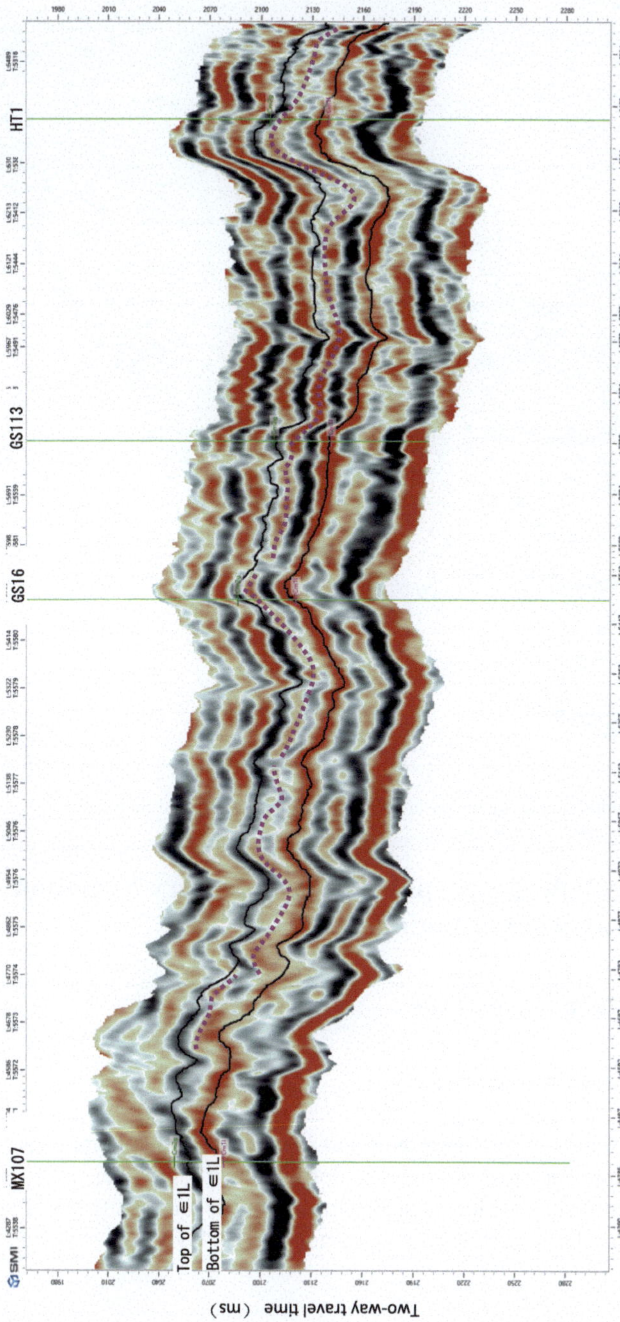

Fig. 51 Seismic reflection characteristics of Longwangmiao Formation in Well Moxi 107–Gaoshi 16–Gaoshi 113–Hetan 1

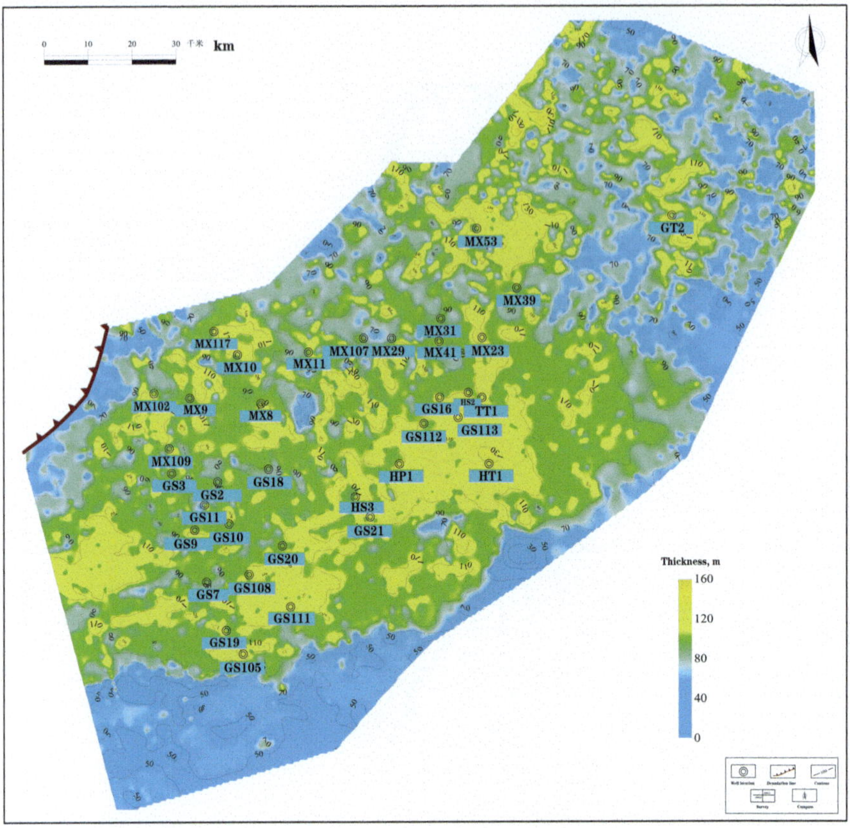

Fig. 52 Stratum thickness map of Longwangmiao Formation in Central Sichuan

3 Lithofacies Paleogeography and Sedimentary Evolution of the Xixiangchi Formation

3.1 Sedimentary Facies Type

3.1.1 Petrological Characteristics

The above basin sedimentary background shows that carbonate platforms are developed in Longwangmiao, and Xixiangchi Formation in Cambrian, restricted-half restricted platforms are developed in Central Sichuan, and the two sedimentary microfacies are similar in petrological characteristics and types.

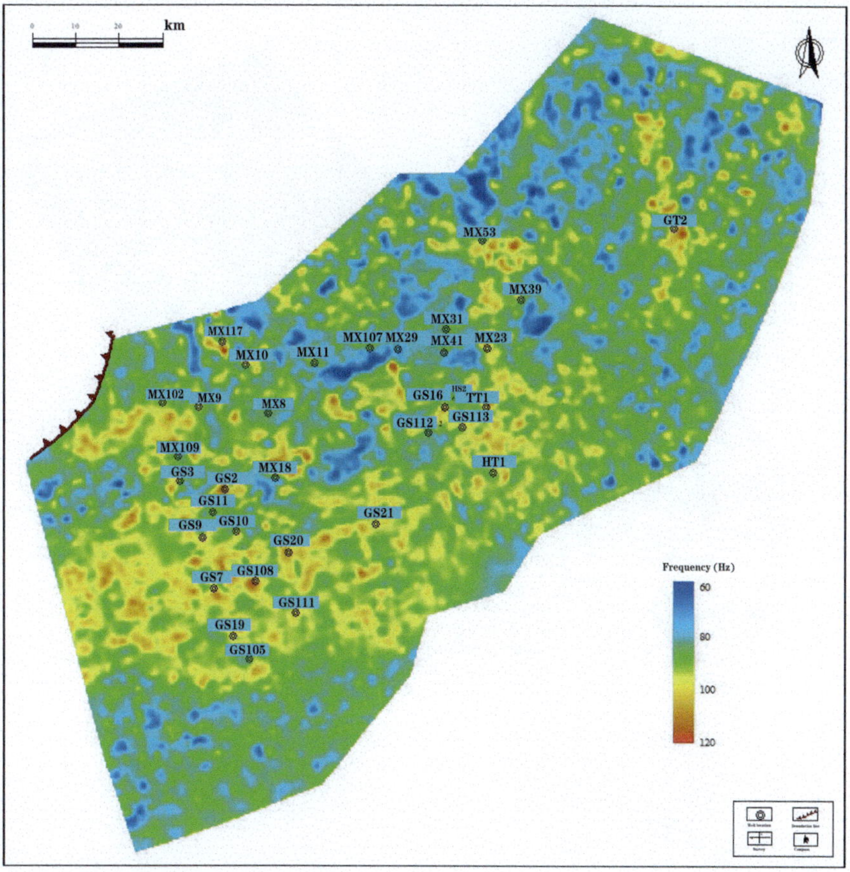

Fig. 53 Frequency attribute map of Longwang Formation in Central Sichuan

Doloarenite

The doloarenite mainly occurs in the color of gray-dark gray and thin-medium layers. The particles of doloarenite are mainly composed of sand debris, occasionally of gravel debris, and sparry dolomites cement the inter particles. Cross beddings are developed in the doloarenite, with common needle-like dissolution pores and holes, and the general-good porosity (Fig. 56). Doloarenite is well developed in Well Hetan 12 and Guangtan 2, but not all can be developed as good dissolution pores. Doloarenite is mainly developed in the high-energy water environment above the wave base in the restricted platform, forming the intra-platform shoal which is the main component of the restricted platform granular shoal.

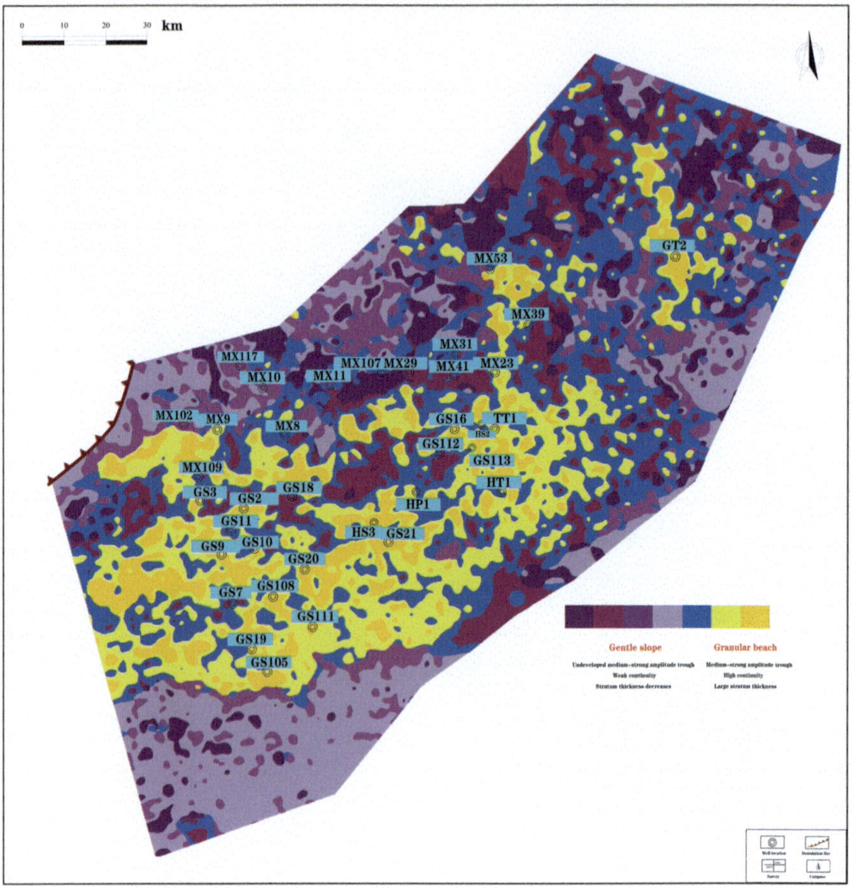

Fig. 54 Seismic facies map of Longwangmiao Formation in Central Sichuan

Micritic Doloarenite

The core of micritic doloarenite shows an obvious sense of granularity, and it generally coexists with grading bedding, storm bedding and muddy lamina, occurs in thin layers with underdeveloped dissolution pores (Fig. 57). It is mainly developed below the average wave base and is mostly transported and deposited in different places under the action of storm or gravity.

Micritic Silty Dolomite

Micritic silty dolomite mainly occurs in gray and dark-gray, and thin layers. It is developed with muddy laminae and muddy strips, biological disturbance structures,

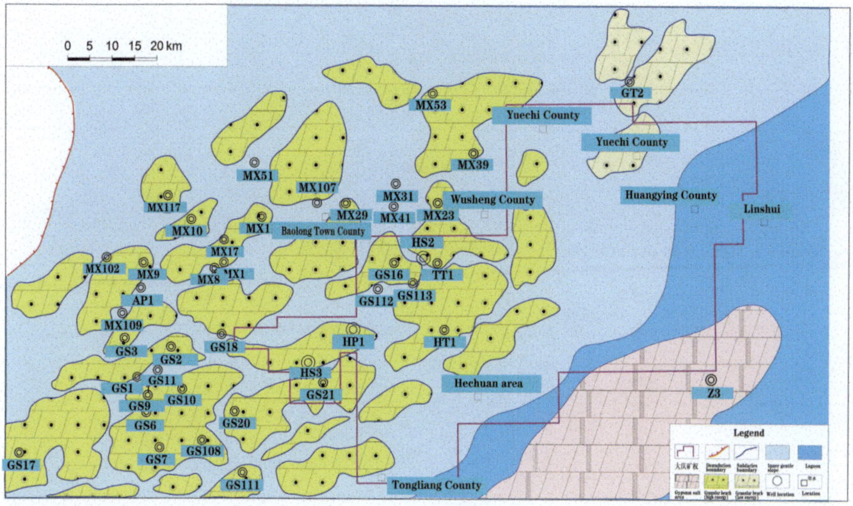

Fig. 55 Lithofacies paleogeographic distribution map of Longwangmiao Formation in Central Sichuan

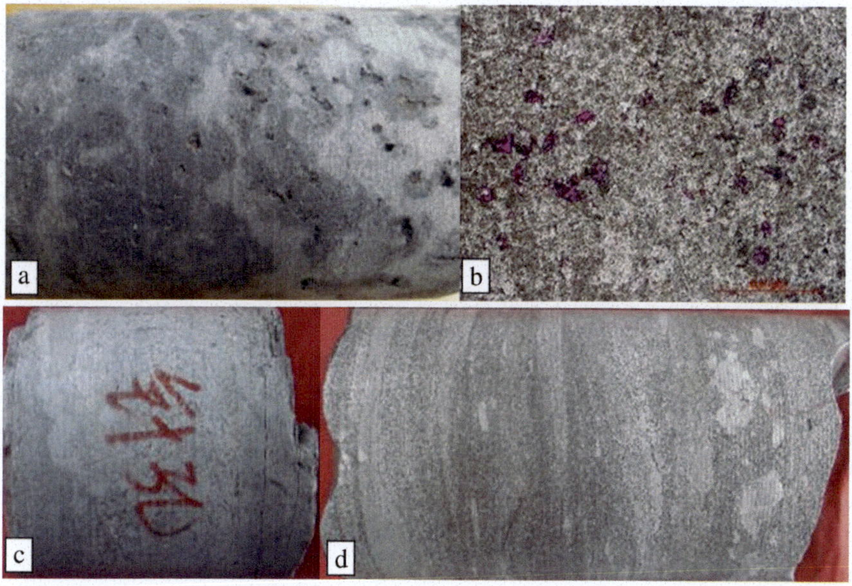

Fig. 56 Petrological characteristics of doloarenite in Xixiangchi Formation. **a** Spot doloarenite, dissolution pores, Well Guangtan 2, 5336.5 m. **b** Residual doloarenite, dissolution pores, Well Guangtan 2, 5336.5 m. **c** Granular dolomite, needle-like dissolution pores, Well He 12, 1–9/24. **d** Granular dolomite with gravel debris, Well He 12, 6–49/59

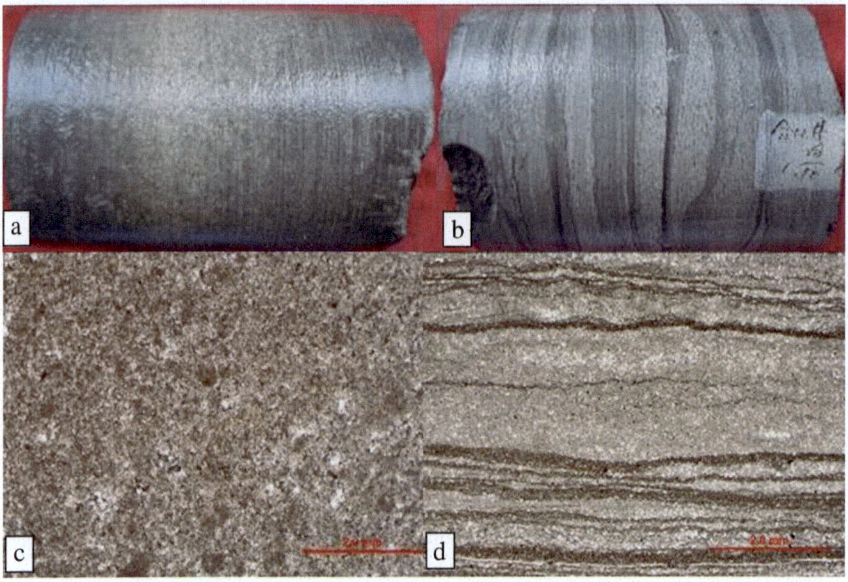

Fig. 57 Petrological characteristics of micritic doloarenite in Xixiangchi Formation. **a** Micritic doloarenite with grading bedding, Well He 12, 8–6/12. **b** Micritic doloarenite with storm cross-bedding, Well He 12, 1–28/50. **c** Micritic silty doloarenite with residual arena, Well He 12, 8–6/12. **d** Micritic dolomite, with micritic laminae, Well He 12, 1–28/50

dense lithology and no pores (Fig. 58). It is mainly developed in the low-energy reductive sedimentary water environment below the wave base in the earth.

3.1.2 Sedimentary Facies Combination Types and Characteristics

The above basin sedimentary background shows that carbonate platforms are developed in Longwangmiao, and Xixiangchi Formation of Cambrian, restricted-half restricted platforms are developed in Central Sichuan, and the two sedimentary microfacies are similar in petrological characteristics and types. They are mainly identified as intra-platform beach, inter-beach sea and storm beach (Table 2).

3.2 Sedimentary Characteristics of Backbone Section

3.2.1 Sedimentary Characteristics of Cores in Well He 12

Well He 12 is coring 5 times in Longwangmiao Formation, with a depth of 4679–4608 m and a cumulative thickness of 71 m. The coring location is the upper part of the Longwangmiao Formation (Fig. 59). The core characteristics of the Longwangmiao

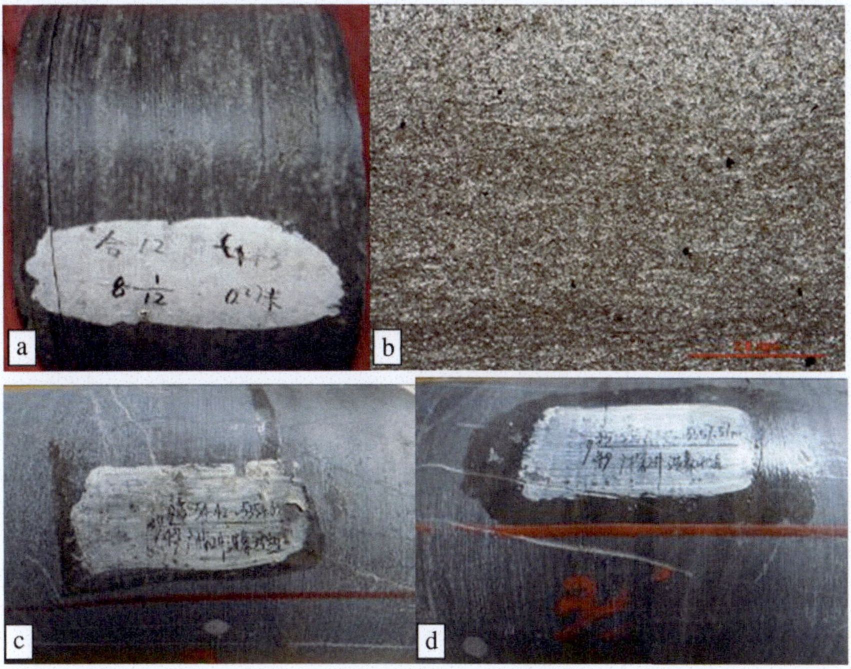

Fig. 58 Petrological characteristics of micritic silty dolomite in Xixiangchi Formation. **a** Dark-gray micritic dolomite, horizontal bedding, bioturbation, Well He 12, 8–1/12. **b** Micritic dolomite, horizontal lamination, Well He 12, 8–1/12. **c** Dark-gray silty dolomite, Well Guangtan 2, 5354.6 m. **d** Dark-gray silty dolomite, Well Guangtan 2, 5357.6 m, 8–1/12

Formation from bottom to top are as follows. The 10th and 9th coring, the whole is light-gray doloarenite, mixed with gypsum and gray doloarenite with maroon spots. The upper part is developed with dark-gray micritic dolomite, with muddy strips, horizontal beddings and lens beddings. As a whole, two upward shallowing high-frequency cycles are developed. Needle-like dissolution pores are developed in doloarenite, which is interpreted as intra-platform beach deposition. For the 8th and 7th coring, the whole is dark-gray micritic dolomite with thin-layered micritic oolitic dolomite and micritic doloarenite, developing muddy strips, lens beddings and horizontal beddings, which can be interpreted as multi-layer storm beaches deposited in deeper-water platform. In the 6th and 5th coring, the whole is light-gray doloarenite, intercalated with a gray thin-layered micritic silty dolomite. Needle-like dissolution pores are developed in some doloarenite, forming multiple upward shallowing cycles composed of thin-layered micritic dolomite and thick doloarenite as a whole. It also indicates that the sea level continued to decline slowly in the late deposit period of the Longwangmiao Formation. The dissolution pores are mainly developed in doloarenite, and dissolution pores with high porosity are only developed in few doloarenites.

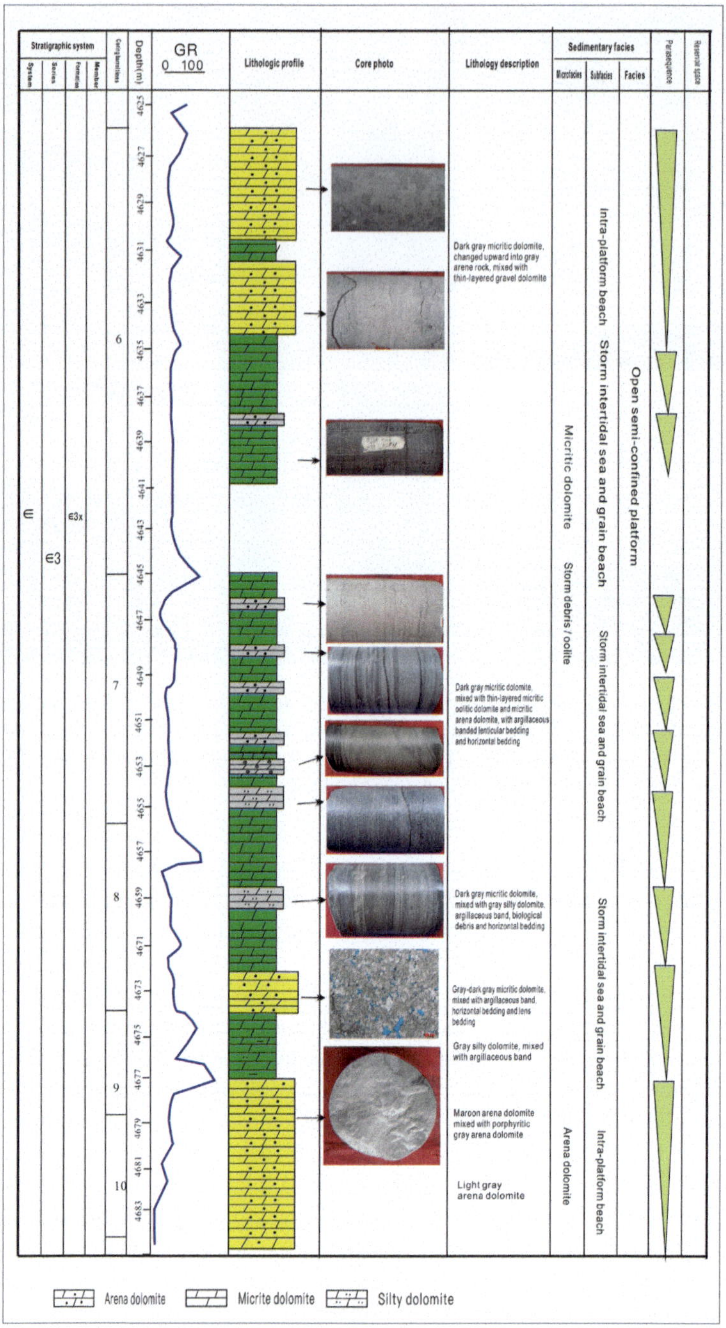

Fig. 59 Comprehensive column map of sedimentary characteristics of cores in Cambrian Xixiangchi Formation of Well He 12

3.2.2 Sedimentary Characteristics of Cores in Well Guangtan 2

There are 5 cores in Xixiangchi Formation of Well Guangtan 2, with a depth of 5360–5314 m and a cumulative thickness of about 46 m (Fig. 60). The core characteristics from bottom to top are mainly as follows. In the 9th and 8th core, the lower part is gray dark and gray micritic dolomite, intercalated with thin-layered storm bedding, doloarenite and muddy stripped dolomite, which is deposited as relatively deep-water platform. The upper part is gradually transformed into doloarenite, with dissolution pores and dissolution holes filled with asphalt and quartz, which can be interpreted as intra-platform beach deposited by sea level decline. 3–4 upward shallowing high-frequency cycles can be explained on the whole. The 7th coring shows the interbedding of doloarenite, doloarenite with storm bedding and micritic dolomite, indicating the frequent change of sea level, which can be interpreted as the frequent interbedding of subtidal low-energy belt, storm beach and intra-platform sandy beach. For the 6th and the 5th coring, the lower part is dark-gray thick micritic dolomite mixed with doloarenite developing irregular muddy bands, lens beddings and storm beddings. Two sets of doloarenite are developed in the upper part. They are presented as multiple upward shallowing high-frequency cycles. It should be pointed out that the thickness of intra-platform doloarenite is small, and the needle-like dissolution pores are only developed in a small amount of doloarenites.

3.3 Sedimentary Facies Analysis

3.3.1 Logging Facies

Based on the sedimentary analysis of cores in Well Tongtan 1 and Guangtan 2, the core location is carried out, and the natural gamma-ray curve with high lithology sensitivity is selected as the main series for logging facies analysis. The logging facies characteristics of microfacies combination in Xixiangchi Formation are summarized (Fig. 61). It is mainly characterized by high-amplitude dentation and funnel type, which can be interpreted as interbeddings of thin intra-platform beach and inter-beach sea. Small dentation positive rhythms are also developed in the cores of Xixiangchi Formation, which can also be interpreted as the transition from intra-platform beach to inter-beach sea. Unlike many box-shaped and block-shaped logging facies of Dengying Formation, this high-frequency dentation indicates the influence of superposed pattern difference caused by the change of intra-platform beach and inter-beach sea inside the platform caused by the high-frequency change of sea level.

3.3.2 Single Well Facies

Consistent with the establishment idea of the single-well sedimentary facies profile of Sinian, the sedimentary characteristics of Cambrian Xixiangchi Formation is

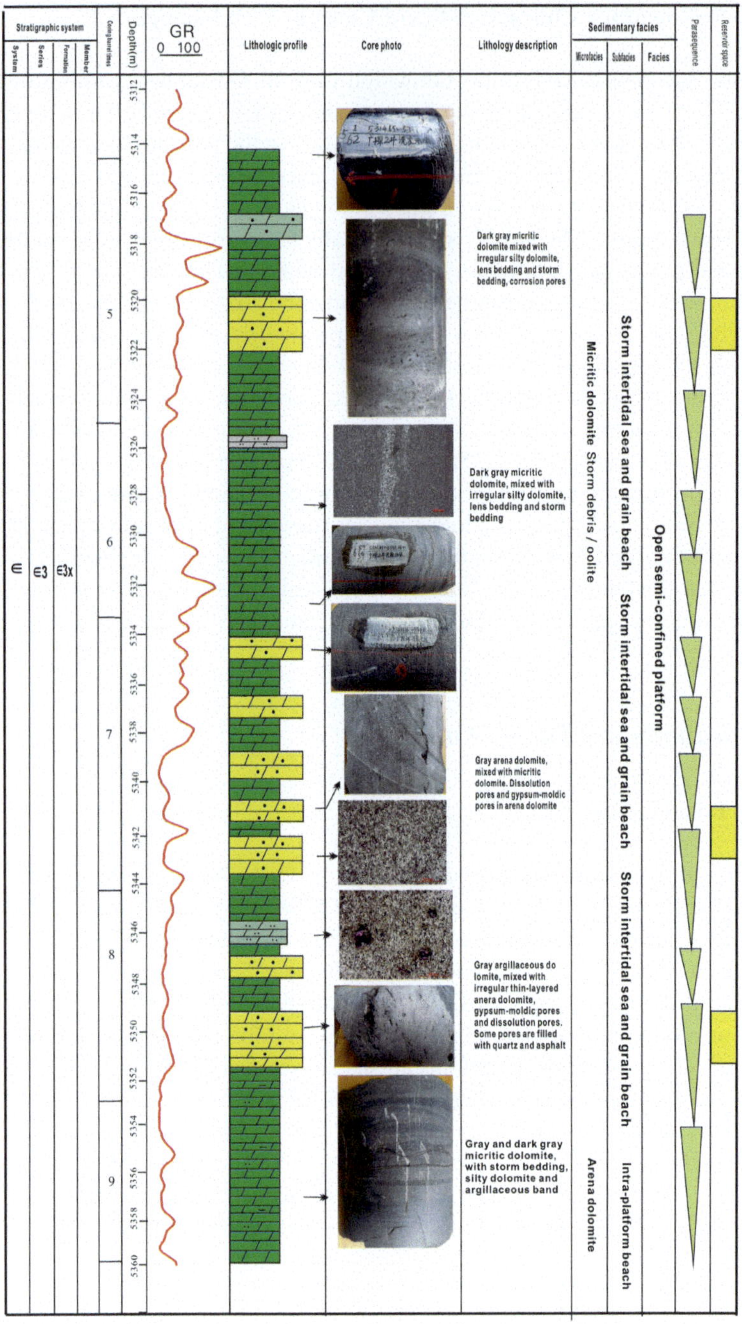

Fig. 60 Comprehensive column map of sedimentary characteristics of cores in Cambrian Xixiangchi Formation of Well Guangtan 2

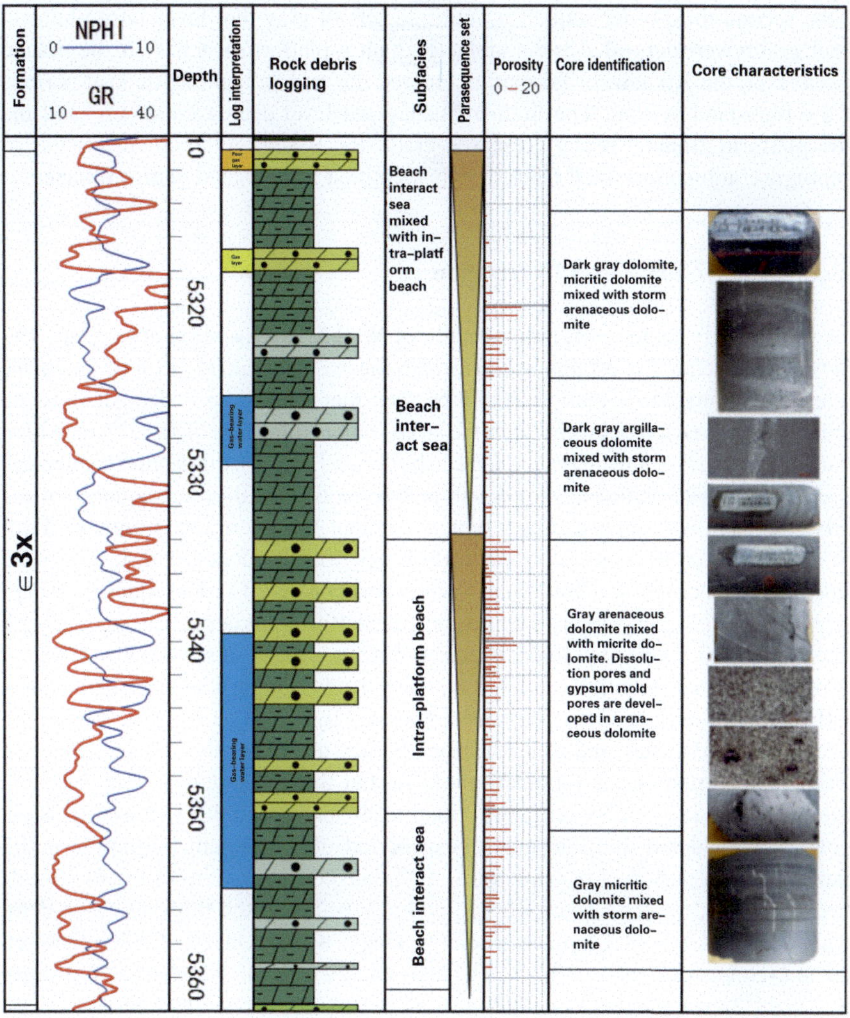

Fig. 61 Logging facies of microfacies combination in Xixiangchi Formation of Well Guangtan 2

analyzed by the sedimentary interpretation of logging lithology. For example, Well Gaoshi 16 is located in the north profile of Gaoshiti-Hechuan block, the south of Longnvsi structure, Sichuan Basin, which is deposited as the typical intra-platform beach. The Xixiangchi Formation is developed with a depth of 4730–4936 m, and a thickness of 206 m. It can be divided into two third-order sequences, the first third-order sequence is developed with the thickness of 113 m, and the second is developed with the thickness of 93 m. Multiple sets of gray thin-layered intra-platform beach grains are deposited in the first third-order sequence, forming the combination of low-energy belt and beach, with the thickness of 0.5–3 m. In the early deposit period of the second third-order sequence, multiple sets of gray thin-layer intra-platform beach grains were deposited as the subfacies of intra-platform beach. With the change in sea level, the Xixiangchi Formation entered the middle and late deposit periods of the highstand system. The high-frequency sea level decreased slowly, and the intra-platform granular beach decreased gradually. Based on the characteristics of lithologic combination, Well Gaoshi 16 is identified as restricted platform facies.

3.3.3 Sedimentary Facies Correlation

The sedimentary facies correlation profile of Well Moxi 16–Moxi 46–Gaoshi 16–Gaoshi 113–Hetan 1 is established. Through macroscopic analysis on sedimentary facies in Xixiangchi Formation, it can be seen that the profile is deposited as the subfacies of restricted platform, with the stratigraphic thickness of 69–213 m, which changes greatly. Xixiangchi Formation can be divided into two third-order sequences. A set of carbonate platforms was developed in the first third-order sequence in the early stage and was deposited as a typical subtidal low-energy environment. With the change of sea level, the intra-platform beach was developed with larger scale, improved connectivity and thin layers. As for the second third-order sequence, except for Well Gaoshi 113, in which a set of carbonate platform was deposited, the rest were deposited with 2–3 sets of granular beaches. With the change of sea level, Xixiangchi Formation entered the middle and late deposit periods of highstand system, the intra-platform beach decreased and was still dominated by thin layers. On the whole, the development frequency and scale of beach bodies in two third-order sequences of Xixiangchi Formation are relatively uniform, but the development frequency and scale in the late stage of Xixiangchi Formation are larger than that in the early stage. It can be considered that the late deposit period of Xixiangchi Formation is the main formation period of beach bodies in Xixiangchi Formation. In the first third-order sequence, the high-energy beach is developed with the thickness of 1–3 m, an average thickness of 1.5 m, a cumulative single-well thickness of 5.26–22.44 m, and an average thickness of 13.2 m. The high-energy beach in the second third-order sequence is developed with the thickness of 1–3 m, an average thickness of 1.5 m, a cumulative single-well thickness of 8.62–21.87 m, and an average thickness of 16.4 m.

3.4 Seismic Facies

Based on seismic reflection characteristics, the intra-platform granular beaches are developed in Xixiangchi Formation, with the external mound-shape reflection, internal chaotic reflection, and weak amplitude (Fig. 62).

Based on the stratum thickness of Xixiangchi Formation (Fig. 63), it can be seen that the Xixiangchi Formation is mainly developed with the thickness of 8–256 m, of which the thickness in the southeast is large, thinned to the northwest, and pinched

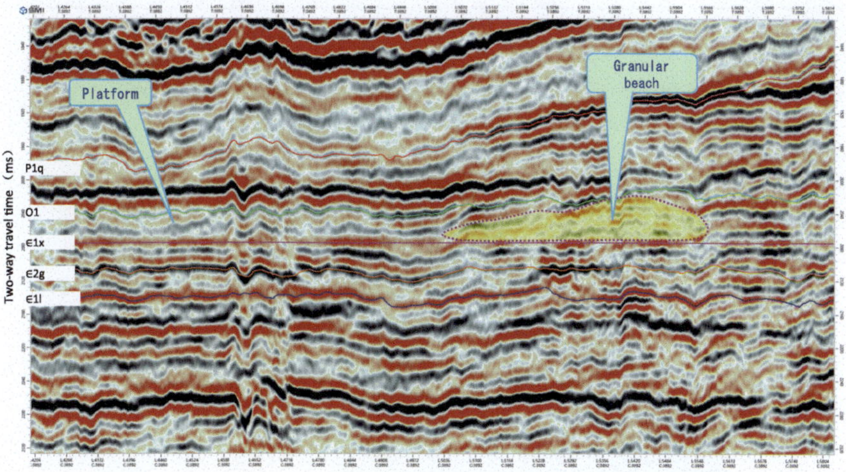

Fig. 62 Reflection characteristics of granular beach of Xixiangchi Formation in Central Sichuan

out near Well Moxi 9. According to the statistical results of formation thickness in Xixiangchi Formation and drilling strata, interpretation results with the error of less than 5 m account for 100%.

Based on Xixiangchi Formation's thickness change in Central Sichuan, there is a slope break with obvious thickness change in Xixiangchi Formation (Fig. 64), and the stratum thickness on both sides of the slope break changed greatly. The strike of this slope break is basically consistent with the axial strike of Leshan-Longnvsi paleouplift. In the westside of the slope break, the thickness of Xixiangchi Formation changes little, while the thickness in the east of the slope break increases sharply. According to the geomorphic characteristics during the deposition of the Xixiangchi Formation, it is considered that above the slope break (west of the slope break), the hydrodynamic force during the deposition of the Xixiangchi Formation should be strong, and the high-energy granular beach should be developed. Under the slope break, when the Xixiangchi Formation was deposited, the water body was deep and the hydrodynamic force was weak. The developed granular beach should be low-energy granular beach.

Based on the 2D and 3D seismic data volumes in the study area, the development section of granular beach in Xixiangchi Formation is compared and analyzed by the attribute extraction, such as average reflection intensity, average instantaneous phase, average amplitude, average instantaneous frequency, RMS amplitude and average absolute amplitude after determining the time period corresponding to the reservoir section on the time section. It is considered that the reservoirs in granular beach can be exactly described by RMS amplitude (Fig. 65). The abnormal change belt of RMS amplitude in the Xixiangchi Formation is a remarkable feature of "external mound-shaped reflection, internal chaotic reflection and weak amplitude", which may symbolise granular beach reservoir.

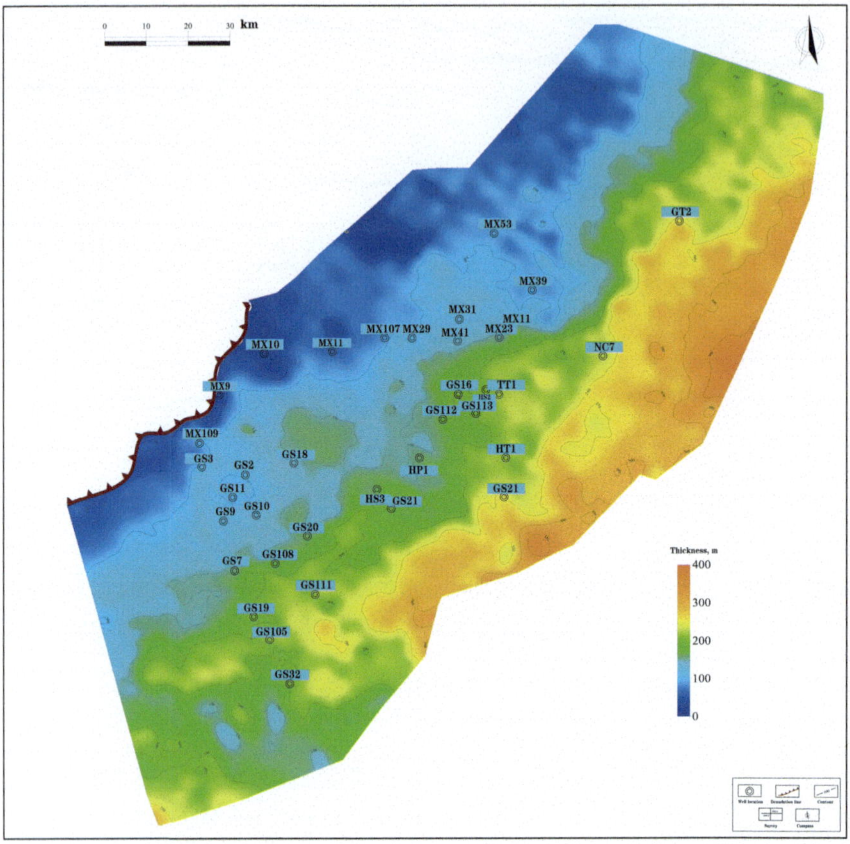

Fig. 63 Stratum thickness map of Xixiangchi Formation in Central Sichuan

The analysis results of seismic facies show that (Fig. 66), the seismic facies of granular beach are characterized by weak amplitude, weak continuity and mound-shaped chaotic reflection, which is distributed in the warm-tone area in the map. And the seismic facies of gentle slope is characterized by medium amplitude and medium-strong continuity, which is parallel to the top and bottom seismic events, and is distributed in the cold-tone area in the map.

3.5 Lithofacies Paleogeographic Characteristics

From the variation thickness of Xixiangchi Formation, combined with sedimentary facies and seismic facies, the lithofacies paleogeography of Xixiangchi Formation is drawn (Fig. 67). The Xixiangchi Formation is mainly deposited with the facies of restricted platform, including the subfacies of intra-platform beach, tidal flat and

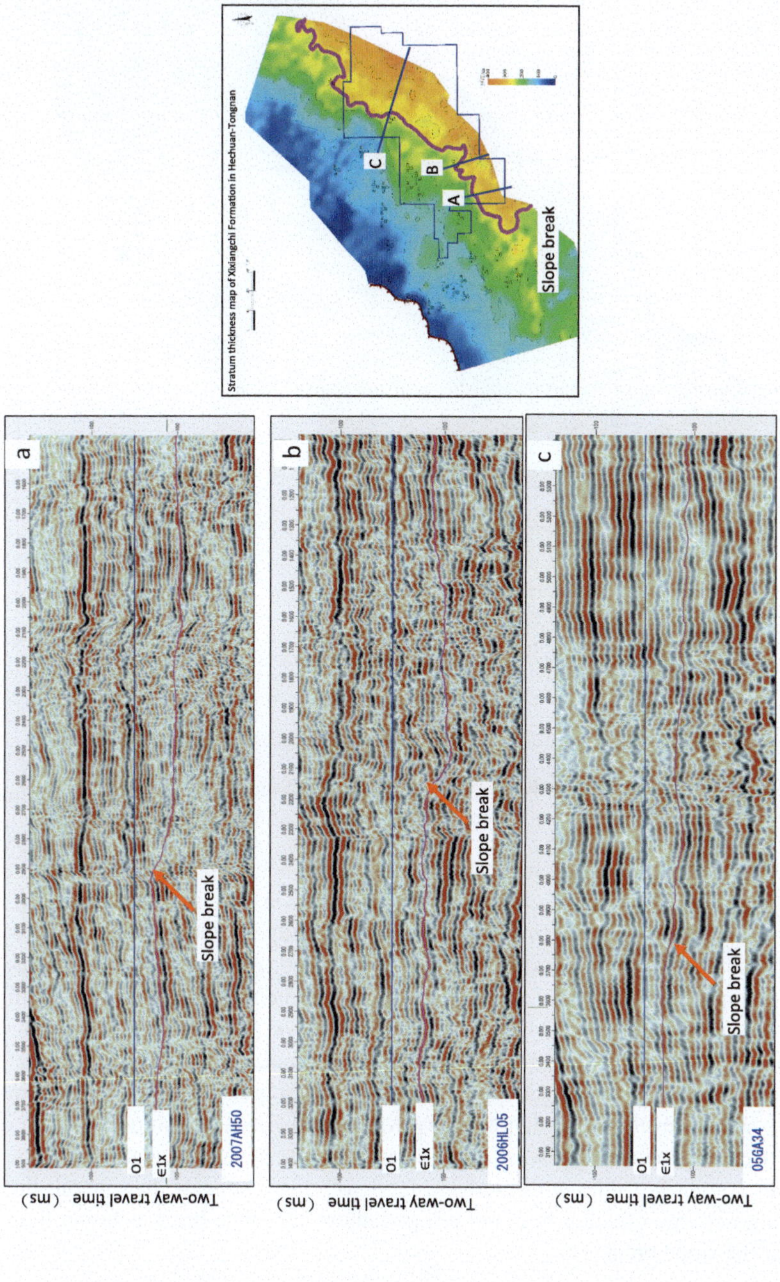

Fig. 64 Seismic profile of slope broken in Xixiangchi Formation of Central Sichuan

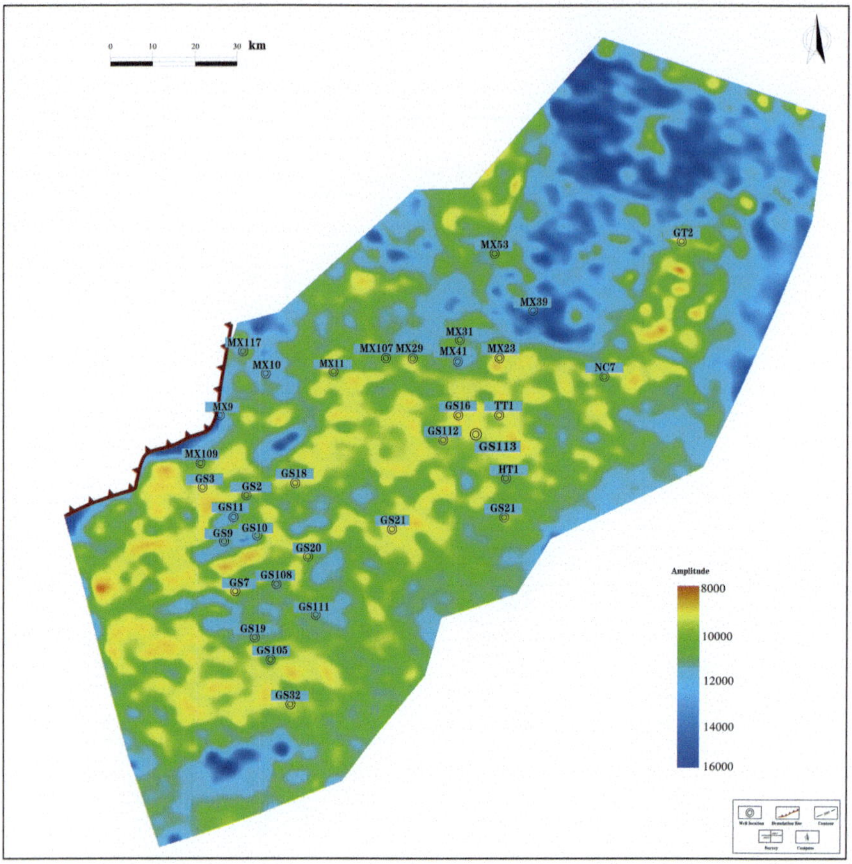

Fig. 65 Amplitude attribute of Xixiangchi Formation in Central Sichuan

restricted bay lake. Among them, intra-platform beach is the main reservoir facies, while the microfacies of sandy beach is mostly developed in the intra-platform beach. Through the analysis of the sedimentary facies in Xixiangchi Formation, the distribution characteristics of dominant facies are determined. According to the lithofacies paleogeographic of Xixiangchi Formation, the sedimentary trend of Xixiangchi Formation is consistent with the trend of Leshan-Longnvsi paleouplift, and the sedimentation of Xixiangchi Formation is affected by the paleouplift. Mixed tidal flat facies is developed in the west. A NE trending high-energy facies of intra-platform granular beach is developed along the line of Xixiangchi Formation slope break, and some NE trending low-energy facies of granular beach are developed under the slope break, which transits to the intra-platform lagoon in the southeast. In general, the scale of high-energy granular beach developed above the slope break is larger than that under the slope break.

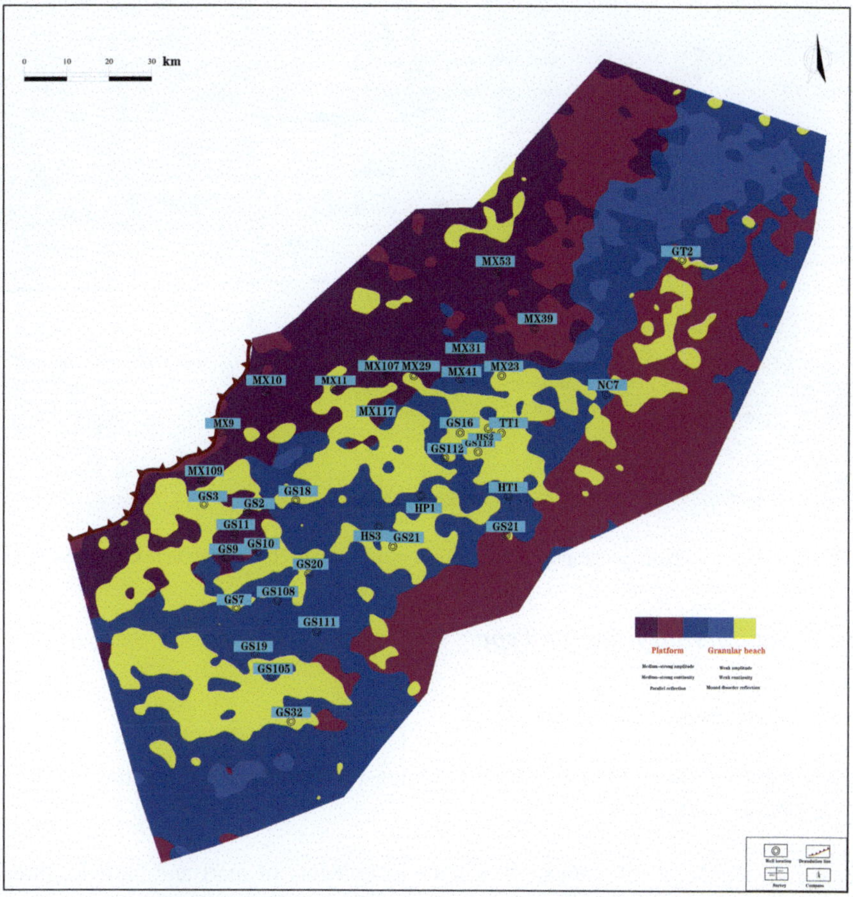

Fig. 66 Seismic facies of Xixiangchi Formation in Central Sichuan

4 Lithofacies, Paleogeography and Sedimentary Evolution of Qixia-Maokou Formation

Qixia Formation and Maokou Formation are both developed as open carbonate platforms which are rich in microorganisms, similar in petrological characteristics and microfacies.

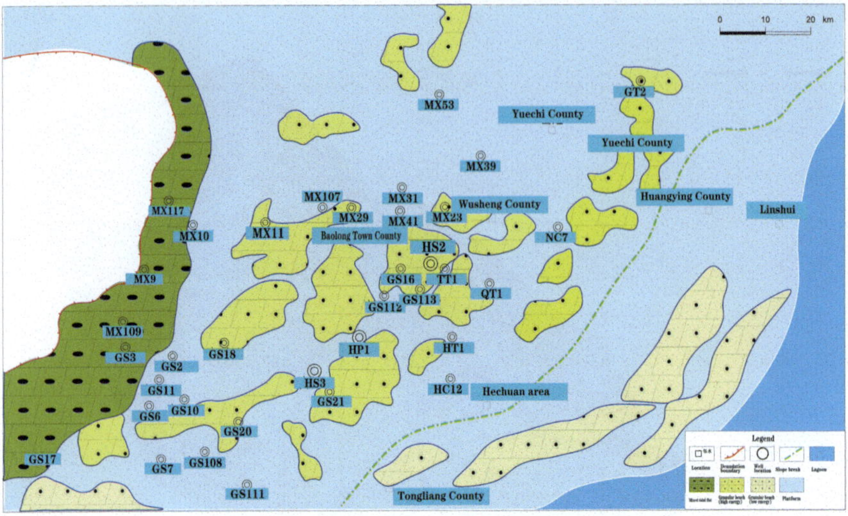

Fig. 67 Lithofacies paleogeographic distribution map of Xixiangchi Formation in Central Sichuan

4.1 *Sedimentary Facies Type*

4.1.1 Petrological Characteristics

Bioclastic Limestone, Micritic Bioclastic Limestone and Muddy Micritic Bioclastic Limestone

Based on Dunham's classification scheme of contents of marl and particle, three petrological types can be identified in Qixia-Maokou Formation: bioclastic limestone, micritic bioclastic limestone and muddy bioclastic micritic limestone (Fig. 68). In the Permian, the carbonate platforms were deposited with abundant organisms, such as green algae (Vermiporella, Pseudovermiporella, Mizzia, Eogoniolina and Sinoporella), red algae (Permocalculus, Gymnocodium, Ungdarella, etc.), foraminifera (Ammodiscus, Textularia, Paraglobivalvulina, Pachyphloia Nankinella, Schwagerina Parafusulina, etc.), brachiopods (Productus, Marginifera, Squamularia, etc.), ostracod and echinoderm. From the analysis of particle content and cement characteristics, the sedimentary energy of sparry bioclastic limestone is the highest, followed by micritic bioclastic limestone, and the energy of muddy micritic bioclastic limestone is the lowest.

Residual Bioclastic Dolomite

The dolomites of Qixia-Maokou Formation are mainly developed in northwest, southwest and central Sichuan. Predecessors have done a lot of research work in

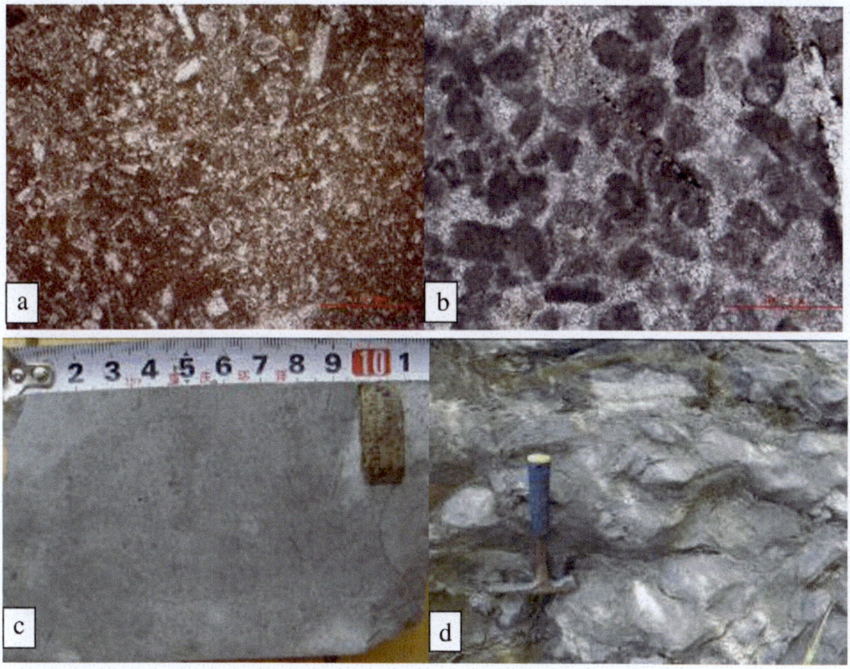

Fig. 68 Petrological characteristics of Qixia-Maokou Formation. **a** Micritic bioclastic limestone, Changjianggou profile, Maokou Formation. **b** Sparry bioclastic limestone, Qixia Formation, Changjianggou profile. **c** Sparry bioclastic limestone, Well Moshen 1, 4279.5 m. **d** Muddy micritic bioclastic limestone (eyeball limestone), Huayingshan, Maokou Formation

these areas, and this study will not be described here. In Hechuan-Tongnan area of Central Sichuan and the adjacent areas, the dolomites of Qixia-Maokou Formation are mainly composed of bioclastic dolomite and residual bioclastic dolomite (Fig. 69). The dolomites are formed in isolation with poor continuity, and are mainly developed in Well Guangcan 2, Nvji and Moxi 42, while are not or rarely developed in other wells.

4.1.2 Types and Characteristics of Sedimentary Facies Combination

Previous studies have shown that Qixia-Maokou Formation is deposited in an environment of open carbonate platform with gentle slope. The facies of inner gentle slope is mainly developed in Central Sichuan, which is equivalent to open platform (Table 3). It is mainly composed of micritic limestone with medium-thin layers, bioclastic micrite limestone, eyeball limestone and mudstone, and a small amount of sparry bioclastic limestone locally. The facies of inner gentle slope can also be further identified into the subfacies of intra-platform beach and intra-platform depression (deeper platform and inter-beach sea). The subfacies of intra-platform beach is

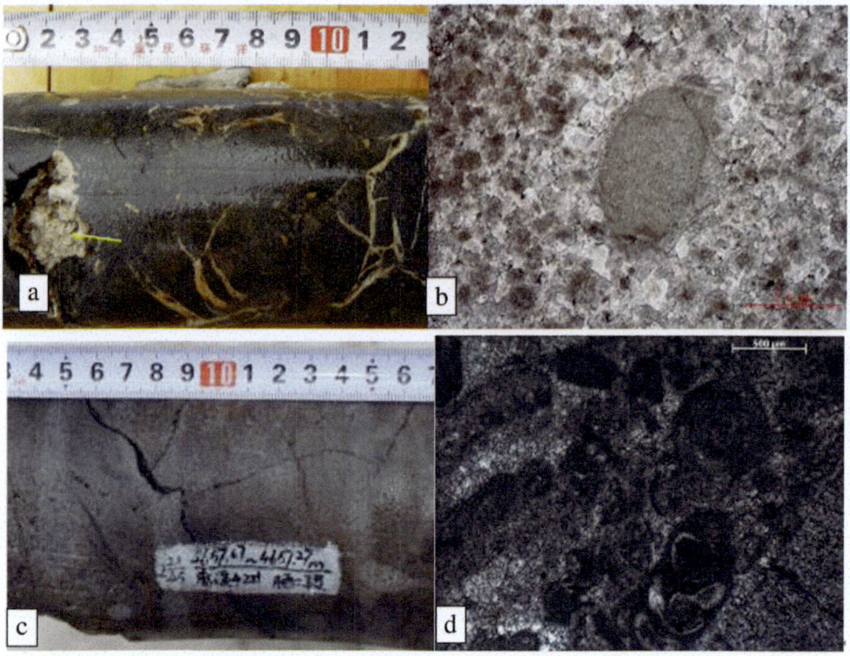

Fig. 69 Petrological characteristics of dolomites in Qixia-Maokou Formation. **a** Dark-gray medium-coarse grained residual bioclastic dolomite, with a large amount of dolomites (indicated by the yellow arrow) filled in the dissolved holes and fractures, and not completely filled, the middle and upper part of Maokou Formation in Well Guangcan 2, 4612 m. **b** Medium-sized dolomite, subhedral-anhedral, some crystal planes are curved, developed with residual bioclastics and residual sand debris, and sparry coarse-grained dolomite, single polarized, in the middle and upper parts of Maokou Formation of Well Guangcan 2, 4612 m. **c** Dolomite with dissolution pores and fractures, Well Moxi 42, 4657.3 m, Qixia Formation. **d** Micritic bioclastic limestone with a small amount of fine-grained dolomites, Well Tongtan 1, 4165 m, Maokou Formation

mainly composed of sparry bioclastic limestone or micritic-sparry bioclastic limestone. The bioclastic is mainly composed of foraminifera, followed by algae, with a small amount of sand debris. The host rocks of residual bioclastic dolomite are mainly bioclastic limestone, so the dolomite can be identified as a relatively high-energy facies belt. The lithology of subfacies of intra-platform depression is mainly "eyeball limestone", which can also be the interbedding of bioclastic micrite limestone and mudstone.

Table 3 Types and petrological characteristics of sedimentary facies in Qixia Maokou Formation

Inner gentle slope	
Intra-platform beach	Inter-beach sea
Sparry bioclastic limestone or sparry-micritic bioclastic limestone, the bioclastic is mainly foraminifera, followed by algae. Residual bioclastic dolomite	"Eyeball limestone", interbedding of bioclastic micrite limestone and argillaceous rock
Qixia Formation, Maokou Formation	

4.2 Sedimentary Characteristics of Backbone Section

4.2.1 Outcrop Sedimentary Characteristics

Changjianggou profile in Shangsi, Jiange County. This profile is located in the Changjianggou of Shangsi Village, in the northwest of Jiange County, Guangyuan City. This profile develops a complete set of strata from Liangshan Formation of Middle Permian to Feixianguan Formation of Lower Triassic. The shale of Bottom Liangshan Formation is obviously contacted with the underlying stratum of Carboniferous. The strata are confirmed, and the top of Upper Feixianguan Formation is not developed.

As can be seen from Fig. 70, taking the Changjianggou profile as the anatomical object, the change of relative water depth can be analyzed by the change of lithofacies, and a third-order sequence can be identified, which includes Liangshan Formation and Qixia Formation. The coastal swamp clastic rocks of Liangshan Formation are the products of the early stage of transgression. With the continuation of transgression, the sedimentary facies in Sichuan Basin was transformed into carbonate clear water platform, and the relative water depth reached the maximum at the bottom of Qixia Formation. A set of dark-gray thin-layered micritic limestone-muddy limestone is deposited, which is characterized by condensation section. In the upper part, the upward shallowing parasequence/parasequence set composed of multiple high-frequency cycles can be seen, which shows as the characteristic of highstand system. It is mainly composed of the combination of bioclastic micritic limestone/muddy limestone-micritic bioclastic limestone. With the gradual shallowing of the relative water depth, a large set of thick-massive gray-white sparry sandy limestone can be seen upward, accompanied by dolomitization. It is deposited as high-energy medium-gentle slope beach with the shallowest sedimentary water body, indicating the deposition end of the sequence. The overall thickness of Maokou Formation is about 220 m. The lower part is deposited with thin-middle-layered bioclastic micritic limestone mixed with mudstone transiting upward to eyeball limestone, which is the condensation section of PSQ2 sequence. Beyond this condensation section, the thickness of single layer gradually increases, as well as the sedimentary energy, until the formation of a set of gray massive sparry bioclastic sandy limestone with a thickness of more than 20 m, indicating the deposition end of the sequence (Fig. 71).

This is the first third-order sequence in Maokou Formation. A large set of bioclastic micritic limestone with medium and thin layers intercalated with mudstone is developed in relatively deep sedimentary water body above the massive sparry bioclastic sandy limestone. The location with thicker mudstone is identified as the condensation section of PSQ3 sequence. Upward, multiple upward shallowing high-frequency cycles with gradually increasing thickness and obvious characteristics of highstand system can also be seen. And the formation of thick layer-massive micritic-sparry bioclastic limestone indicates the deposition end of the sequence and the Maokou Formation. Above this sequence is a thin layer of micrite limestone with the thickness of about 30 m, which can be interpreted as Wujiaping Formation. The weathering crust at the top of Maokou Formation is not obvious.

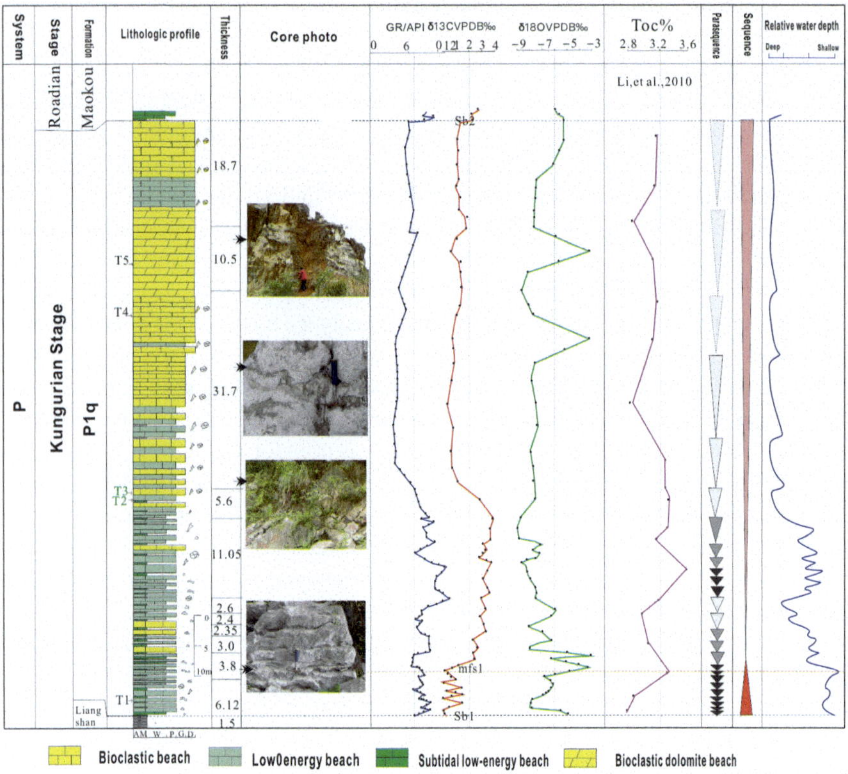

Fig. 70 Column map of sedimentary characteristics of Qixia Formation in outcrop of Jiange

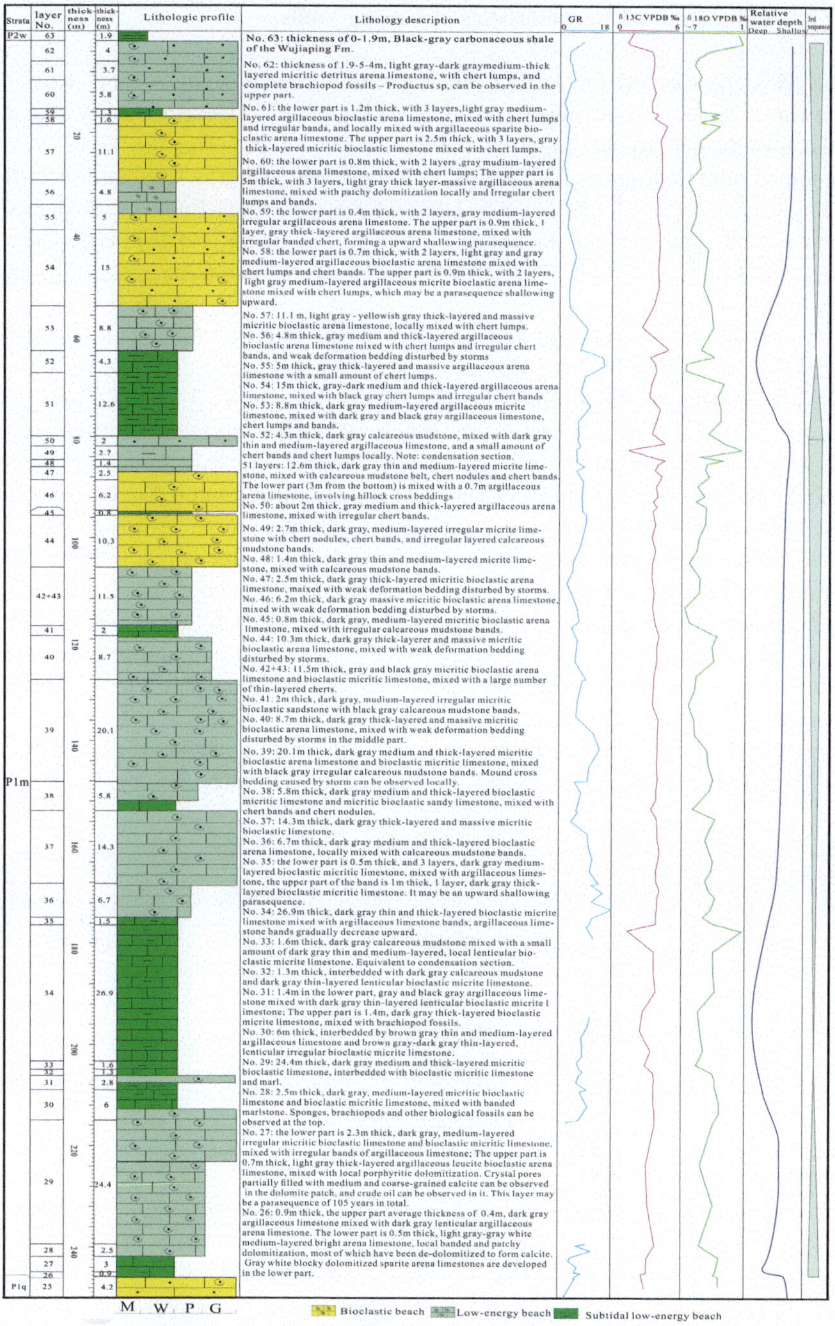

Fig. 71 Column map of sedimentary characteristics of Maokou Formation in outcrop on Jiange

4.2.2 Sedimentary Characteristics of Coring Wells

Well Moxi 42 was cored twice in Qixia Formation, from a depth of 4660–4649 m. The coring location is the upper part of Qixia Formation. The sedimentary characteristics of the core from bottom to top are mainly as follows (Fig. 72). Three cycles combined with dark-gray bioclastic micrite limestone and gray residual bioclastic dolomite are developed, which can be interpreted as three high-frequency sequences upward shallowing, and the upper part of the high-frequency sequence is deposited as relatively high-energy intra-platform beach. There are dissolution pores, holes and fractures in dolomite.

Well Tongtan 1 has two continuous corings in Maokou Formation, with a depth of 4165–4150 m and a cumulative thickness of about 15 m. The coring location is the top Maokou Formation. The overall characteristics are as follows (Fig. 73). Dark-gray to gray micritic bioclastic limestone and bioclastic micritic limestone, with irregular cracks and many dissolution pores, holes, and fractures. A small amount of calcite, asphalt and karst seepage mud and sand are found in the fractures. The core reveals that Maokou Formation is deposited as a deep-water platform in a whole.

Well Guangcan 2 was cored in Maokou Formation with the length of 49 m. The coring locations are the middle and upper parts of Makou Formation. The sedimentary characteristics of the core from bottom to top are mainly as follows (Fig. 74). The lower part is dark-gray bioclastic micrite limestone, muddy micrite limestone and eyeball limestone, representing the deposition of deep-water platform or subtidal low-energy belt. Upward, it is mainly composed of dark-gray micritic bioclastic limestone. It is deposited as inter-beach sea with slightly shallower sedimentary water body. The upper part is micritic bioclastic limestone mixed with a set of gray

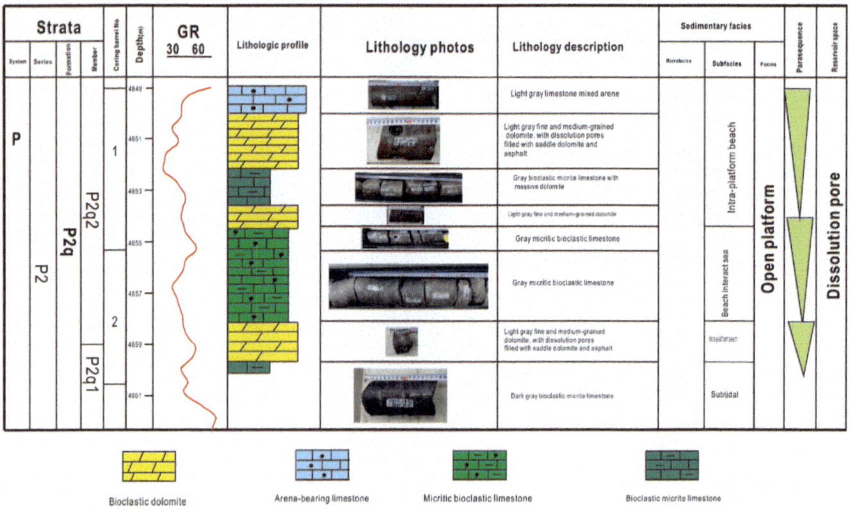

Fig. 72 Column map of sedimentary characteristics in cores of Qixia Formation in Well Moxi 42

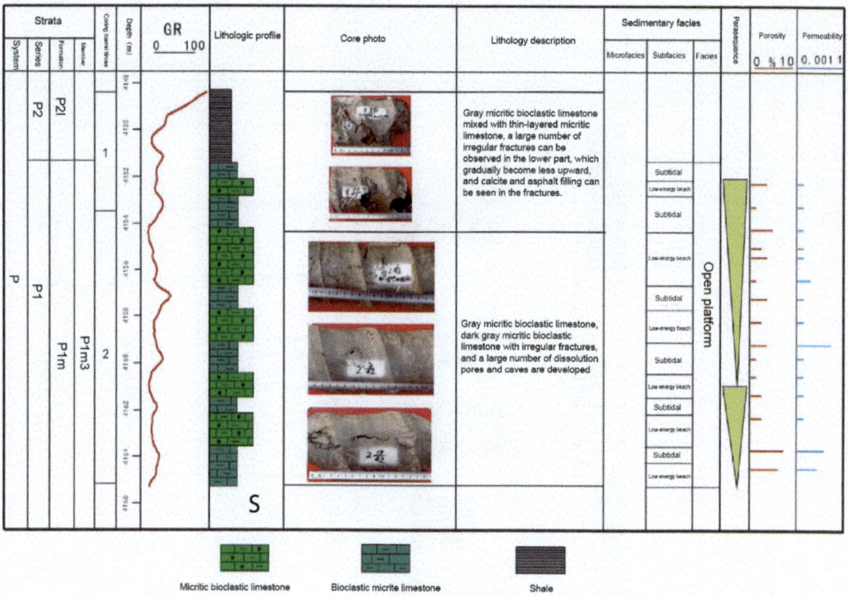

Fig. 73 Column map of sedimentary characteristics in cores of Maokou Formation of Well Tongtan 1

bioclastic dolomite with a thickness of about 4 m. A large number of dissolution pores, holes and fractures are found in the dolomite, and megacrystalline dolomite filling can be found in the dissolution holes.

4.3 Sedimentary Facies Analysis

4.3.1 Logging Facies

On the basis of sedimentary analysis of cores in Well Tongtan 1 and Moxi 42, core location is carried out, and the natural gamma-ray loggings with high lithology sensitivity are selected as the main series in logging facies analysis, in order to summarize the characteristics of logging facies in Qixia-Maokou Formation (Figs. 75 and 76). Similar to the Longwangmiao Formation and Xixiangchi Formation above, the natural gamma-ray curve corresponding to the intra-beach and low-energy belt combination in Qixia and Maokou Formation is also formed with the combination of toothed low amplitude and high amplitude, and several funnel-shaped rhythms. The logging facies association can be interpreted as the response of sea-level changes related to high-frequency sequences.

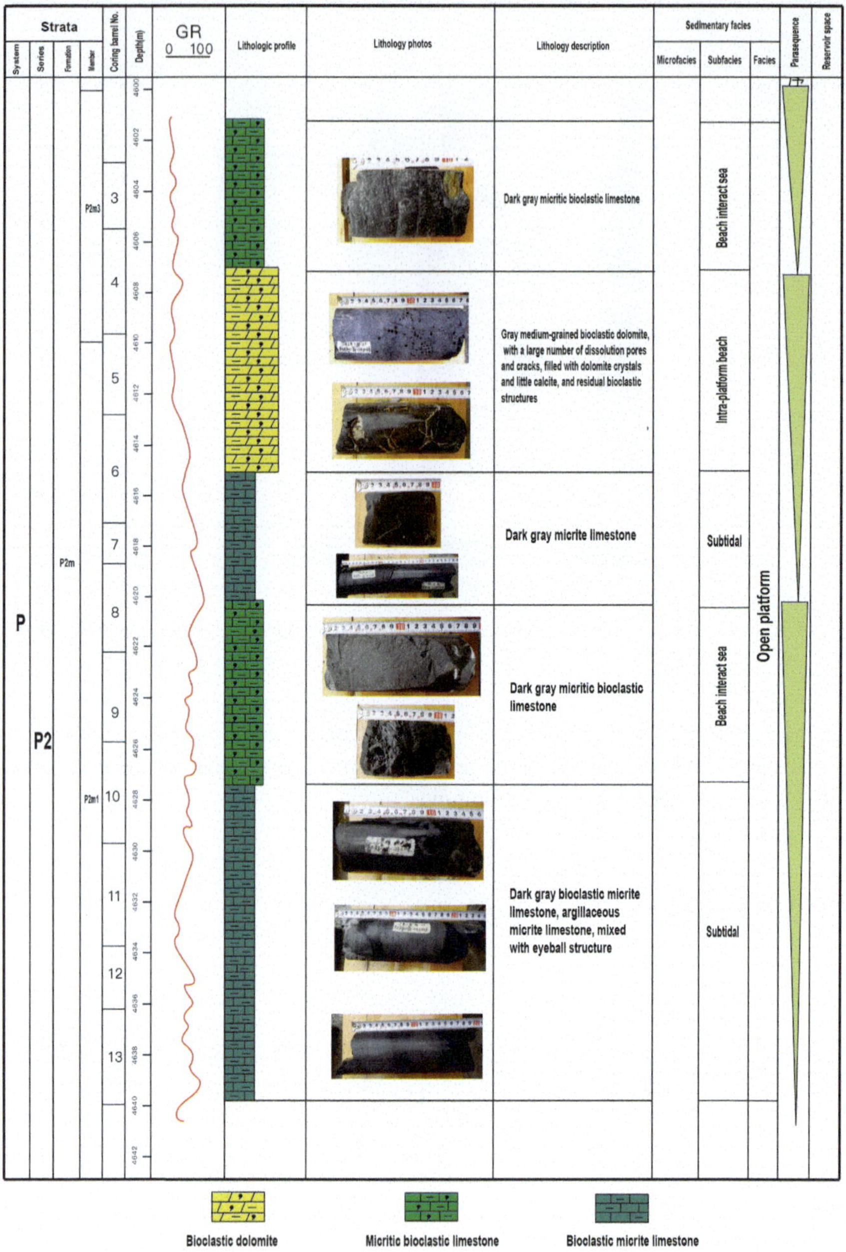

Fig. 74 Column map of sedimentary characteristic of cores in Maokou Formation of Well Guangcan 2

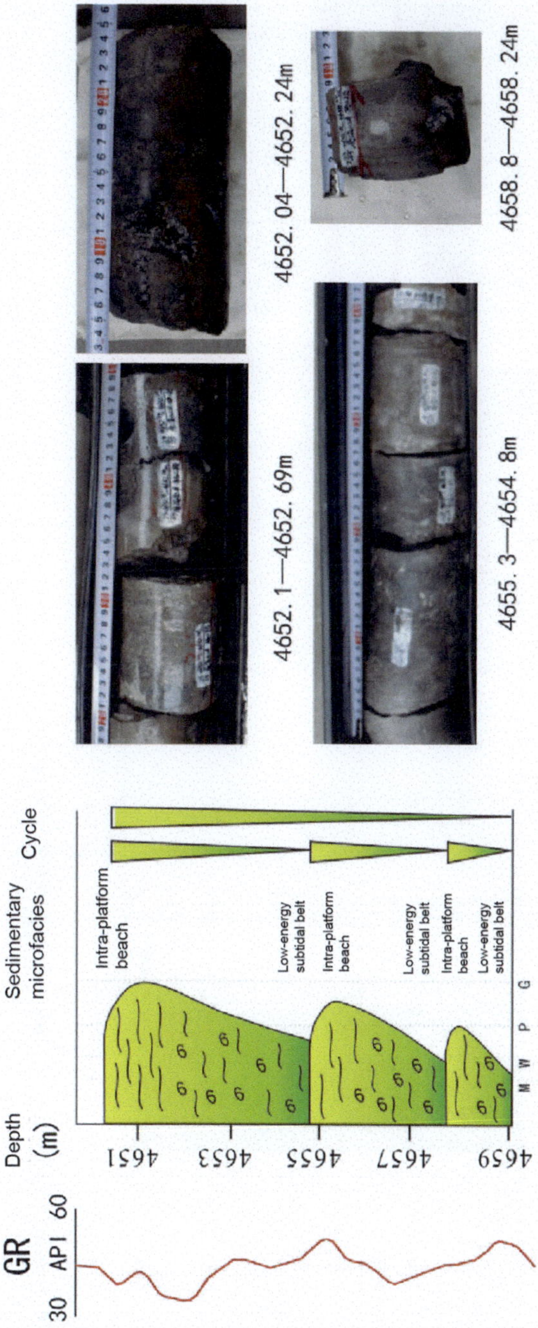

Fig. 75 Logging response map of Qixia Formation in Well Moxi 42

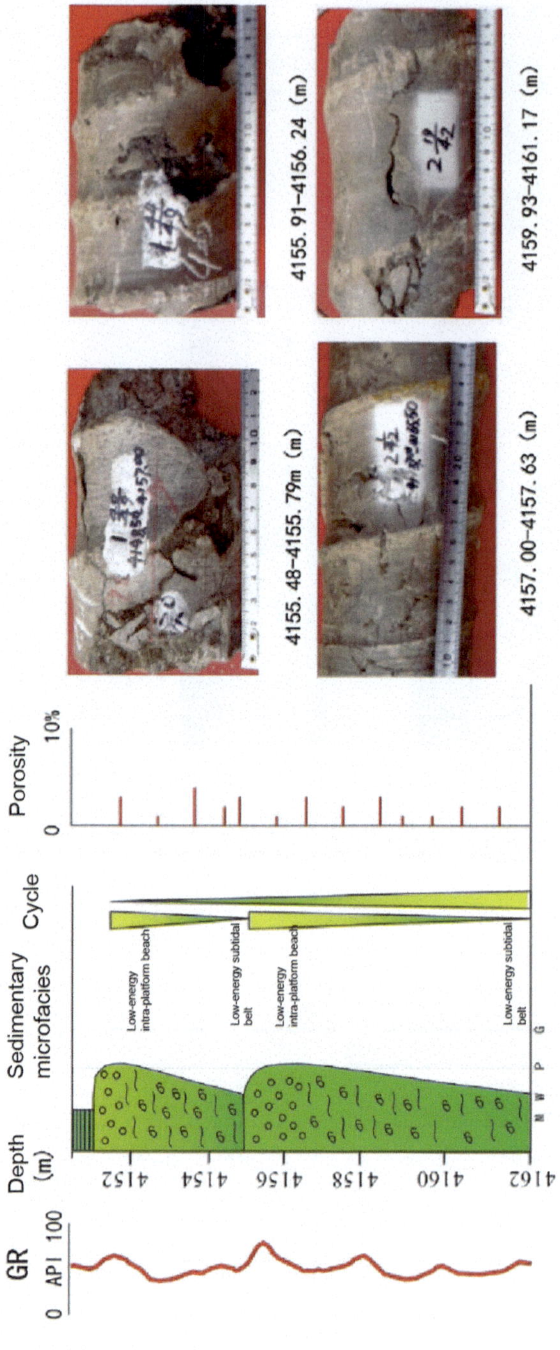

Fig. 76 Logging response map of Maokou Formation in Well Tongtan 1

4.3.2 Sedimentary Facies in Single Wells

Sedimentary Facies in Single Wells of Qixia Formation

For example, Well Gaoshi 16 is located in the northern profile of Gaoshiti-Hechuan block of Sichuan Basin, south of Longnvsi structure, which is deposited as a typical open platform. The well reveals Qi2 Member and Qi1 Member. Qi2 Member of Qixia Formation is developed with the depth of 4501.7–4527.6 m and a thickness of 25.9 m. Qi1 Member is developed with the depth of 4527.6–4624.6 m and a thickness of 97 m. Qi1 Member is deposited with subtidal low-energy belt, and is composed of gray muddy micritic limestone, gray micritic bioclastic limestone, dark-gray algal micritic limestone and dark-gray bioclastic micritic limestone. The early deposit period of Qi2 Member entered the middle and late stages of the highstand system, the high-frequency sea level decreased slowly, and the intra-platform beach was developed rapidly, forming the combination of low-energy belt and intra-platform beach. The combination of intra-platform beach gradually thickened upward, with the thickness of 3–6 m. Affected by the change of sea level, a set of gray-brown gray micrite limestone developed in the late stage of Qi2 Member, which is deposited as subtidal low-energy belt. Based on the lithologic combination characteristics, three sedimentary subfacies can be identified, such as intra-platform, low-energy belt, and subtidal belt.

Another example is Well Guangtan 2, which is located in the high part of the seismic crest structure of Guang'an structure in the slope zone of Leshan-Longnvsi paleouplift, Sichuan Basin, and is deposited as a typical open platform. The well reveals Qi2 Member and Qi1 Member. Qi2 Member of Qixia Formation is developed with the depth of 4901–4916 m and a thickness of 15 m. Qi1 Member is developed with the depth of 4916–5020 m and a thickness of 104 m. Qi1 Member is deposited as a subtidal low-energy belt, and is mainly composed of dark-gray muddy micritic limestone, gray bioclastic micritic limestone and gray micritic bioclastic limestone. The early deposit period of Qi2 Member entered the middle and late stage of the highstand system, the high-frequency sea level decreased slowly, and the intra-beach developed rapidly, forming the combination of low-energy belt and intra-beach. The combination of intra-beach gradually thickened upward with the thickness of 1–3 m. Affected by the change of sea level, a set of gray micritic limestone was developed in the late deposit period of Qi2 Member, which is deposited as a subtidal low-energy belt. Based on the lithologic combination characteristics, the well is identified as three sedimentary subfacies: intra-beach, low-energy beach and subtidal belt.

Sedimentary Facies in Single Wells of Maokou Formation

For example, Well Gaoshi 21 is located at the high point of the seismic crest structure of Wangjiachang hide structure in Leshan-Longnvsi paleouplift, Sichuan Basin, which is typically deposited as a typical deep-water open platform sedimentation.

The well reveals Mao3, Mao2 and Mao1 Member. Mao3 member of Maokou Formation is developed with the depth of 3937–3979 m and a thickness of 42 m. Mao2 Member is developed with the depth of 3979–4063 m and a thickness of 84 m. Mao1 Member is developed with the depth of 4063–4141 m and a thickness of 78 m. Maokou Formation is a deep-water open platform deposited in a subtidal low-energy environment with underdeveloped high-energy beaches. A large number of dark-gray eyeball limestone are developed, "eyeball" is bioclastic micrite limestone, "eyelid" is argillaceous limestone, and there is no shoal patch in this well. Based on the lithologic combination characteristics, two sedimentary subfacies, low-energy beach and subtidal belt, can be identified in Well Gaoshi 21.

Another example is Well Nanchong 7, which is located at the local high point in the east of Longnvsi structure and is deposited as a typical deep-water open platform. The well reveals Mao3r, Mao2 and Mao1 Member. Mao3 Member of Maokou Formation is developed with the depth of 4343–4375 m and a thickness of 32 m. Mao2 Member is developed with the depth of 4375–4479 m and a thickness of 104 m. Mao1 Member is developed with the depth of 4479–4539 m and a thickness of 60 m. Maokou Formation is developed as a deep-water open platform deposited in a subtidal low-energy environment with underdeveloped high-energy beach. A large number of dark-gray eyeball-shaped and eyelid-shaped limestones are developed, "eyeball" is bioclastic micrite limestone, "eyelid" is argillaceous limestone, and shoal patch can be seen locally. Muddy dolomite is developed at the depth of 4434.2–4440.8 m with a thickness of 6.6 m. The logging interpretation of the corresponding well section is gas reservoir, and the logging shows gas invasion. From the lithologic combination characteristics, two sedimentary subfacies, low-energy beach and subtidal belt, can be identified in Well Nanchong 7.

Based on the analysis of sedimentary facies in single wells, the basic characteristics of high-energy bioclastic beach are counted as follows. The beach body in single wells is developed with the average thickness of 10.1 m, a maximum thickness of 16 m and a minimum thickness of 2.05 m. The whole beach body is developed with the average thickness of 3.9 m, a maximum thickness of 6.5 m and a minimum thickness of 1.36 m.

4.3.3 Sedimentary Facies Correlation

Sedimentary Facies Correlation of Qixia Formation

The correlation profile of sedimentary facies across Well Gaoshi 21–Gaoshi 112–Gaoshi 113–Tongtan 1–Nanchong 7–Guangtan 2 is established. According to the macro sedimentary characteristics of sedimentary facies in Qixia Formation, it can be seen that the strata are developed with the thickness of 110–132 m with small range, and a relatively stable sedimentary environment, which is deposited as the facies of open platform. The intra-platform beach is developed in the third-order sequence highstand system, which is mostly found in Qi2 Member and also in the upper part of Qi1 Member. During the deposition period of Qi2 Member, the intra-platform beach is widely developed in the whole profile. Except for Well Nanchong 7, there are 2–3 sets of beach bodies in other wells with good connectivity. The early deposition period of Qi1 Member is dominated by low-energy subtidal environment, and intra-platform beach is not developed. In the late deposition period of Qi1 Member, 1–2 sets of beach bodies can be seen in Well Gaoshi 113 and Tongtan 1 with a certain connectivity. However, on the whole, the development frequency and scale of the beach bodies in Upper Qi1 Member are smaller than those in Qi2 Member. It can be considered that the deposition period of Qi2 Member is the main period for the formation of the intra-platform beach of Qixia Formation. The thickness of high-energy beach in single layers of this profile ranges from 1.5 to 6.05 m, with an average of 3.11. The cumulative single-well thickness of high-energy beach ranges of 4.05–15.45 m, with an average of 9 m. The intra-platform beach in Well Gaoshi 13 is developed with the largest single-layer and cumulative thickness, and the intra-platform beach in Well Guangtan 2 is developed with the smallest single-layer and cumulative thickness.

Sedimentary Facies Correlation of Maokou Formation

The correlation profile of sedimentary facies across Well Gaoshi 21–Gaoshi 112–Gaoshi 113–Tongtan 1–Nanchong 7–Guangtan 2 is established, and the macro sedimentary characteristics of sedimentary facies in Maokou Formation are dissected. It can be seen that the strata in this profile is developed with the thickness of 185.3–219.1 m with small range, and a relatively stable sedimentary environment, which is deposited as the facies of open platform. The Maokou Formation is deposited as a subtidal low-energy environment with an underdeveloped intra-platform beach.

4.4 Seismic Facies

4.4.1 Qixia Formation

From the seismic reflection characteristics of the intra-platform granular beach in the Qixia Formation of Well Gaoshi 112–Gaoshi 16–Tongtan 1, the development area of granular beach is characterized by the seismic reflection of mound-shaped weak amplitude chaotic reflection (Fig. 77).

As can be seen from the stratum thickness map (Fig. 78), the stratum of Qixia Formation is mainly developed with the thickness of 89–130 m, of which the stratum around Well Moxi 107–Moxi 53–Moxi 41–Gaoshi 16–Hetan 1–Lai 1 is developed with relatively large thickness of 110–128 m. According to the statistical results of formation thickness and drilling thickness of Qixia Formation, the interpretation results with the error of less than 5 m account for 100%.

Based on the 2D and 3D seismic data volumes in the research area, various attributes such as RMS amplitude are extracted after determining the time period

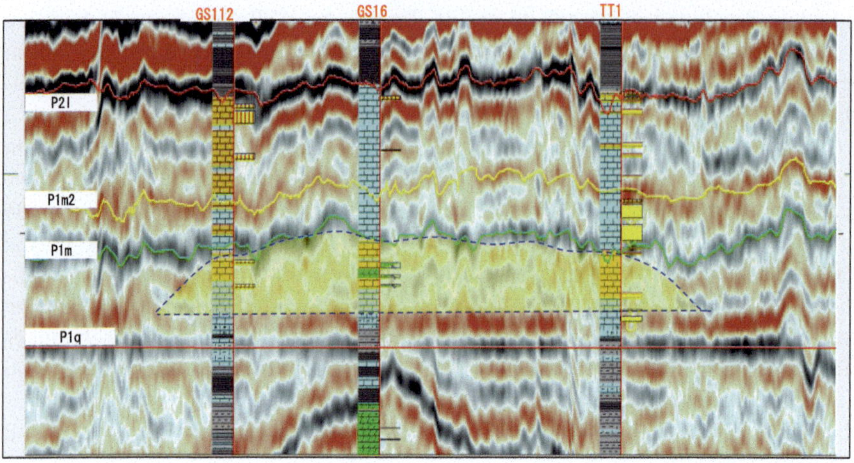

Fig. 77 Seismic reflection characteristics of granular beach in Qixia Formation of Central Sichuan

corresponding to the reservoir section on the time section. Through the comparative analysis for the development section of granular beach in Xixiangchi Formation, it is considered that the granular beach reservoirs can be well described by the attributes such as RMS amplitude (Fig. 79).

The analysis results of seismic facies show that (Fig. 80), the seismic facies of granular beaches are characterized by weak amplitude, weak continuity and mound-shaped chaotic reflection, which is located in the warm-tone area. The seismic facies of gentle slope is characterized by medium amplitude and medium-strong continuity, which is parallel to the top and bottom seismic events and is located in the cold-tone area.

4.4.2 Maokou Formation

After the deposition of the Maokou Formation, affected by the Dongwu movement, Maokou Formation was exposed and denudated. As a result, Mao4 Member is missing in most exploration areas of Hechuan-Tongnan. The current thickness of Maokou Formation can only reflect the residual ancient landform after the deposition, which has no reference for the original paleogeographic pattern. The gas producing layers found in Maokou Formation are mainly distributed in Mao2 and Mao3 Member. Drawing the paleogeographic map of Mao2 and Mao3 Member is of great significance to the exploration of Maokou Formation. Therefore, based on the regional distribution of Mao1 Member, it can be seen that there is no denudation in Mao1 Member, and the deposition of Mao2 and Mao3 Member has a certain inheritance to the sedimentary landform of Mao1 Member. In this research, during the process of mapping the lithofacies paleogeography of Mao2 and Mao3 Member, the

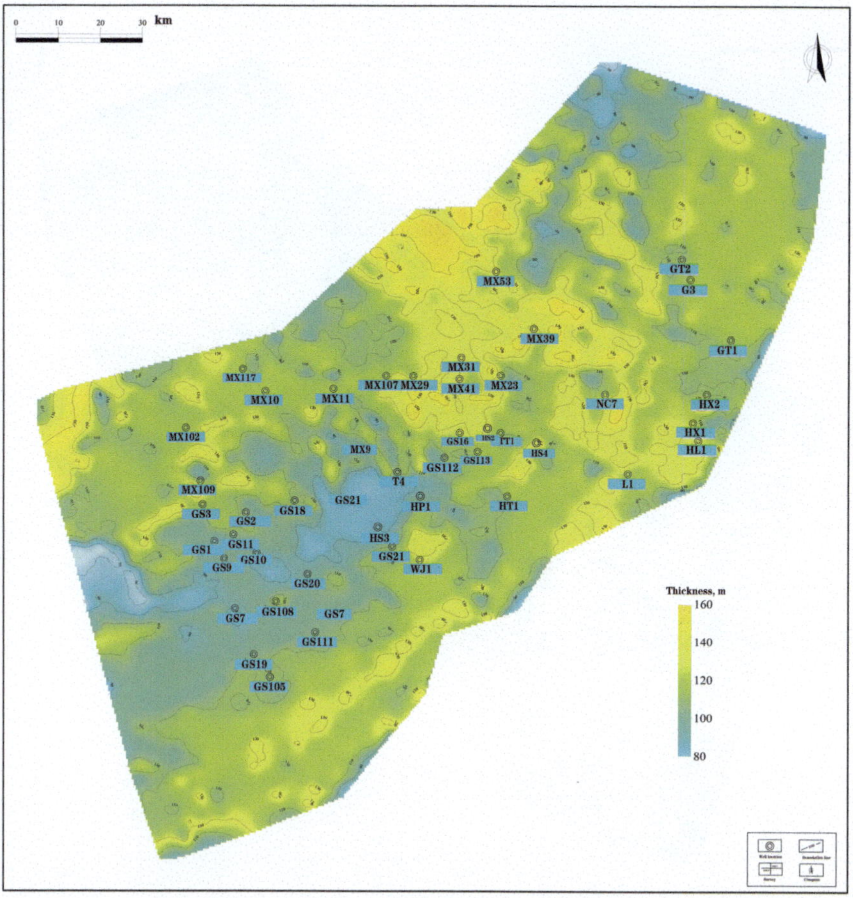

Fig. 78 Thickness distribution map of Qixia Formation in Central Sichuan

sedimentary pattern of Mao2 and Mao3 Member is mainly restored with reference to the thickness of Mao1 Member.

As Mao1 Member is the product of transgression, the sedimentary thickness at the bottom of Mao1 Member is generally larger, while at the high part is generally smaller. As shown in Fig. 81, the thickness of Mao1 Member encountered by drilling now varies among different wells. Based on the thickness of Mao1 Member revealed by drilling, the variation trend of thickness in Mao1 Member is drawn (Fig. 82). According to the stratum thickness of Mao1 Member, a NW trending paleogeomorphic high belt is developed along the line of Well Moxi 51–Moxi 41–Tongtan 1–Lai 1, a NW trending paleogeomorphic low belt is developed in Well Moxi 13–Gaoshi 2–Gaoshi 20.

The layering scheme of top and bottom Maokou Formation is determined, and after fine well-seismic calibration, the seismic interpretation scheme of top and bottom

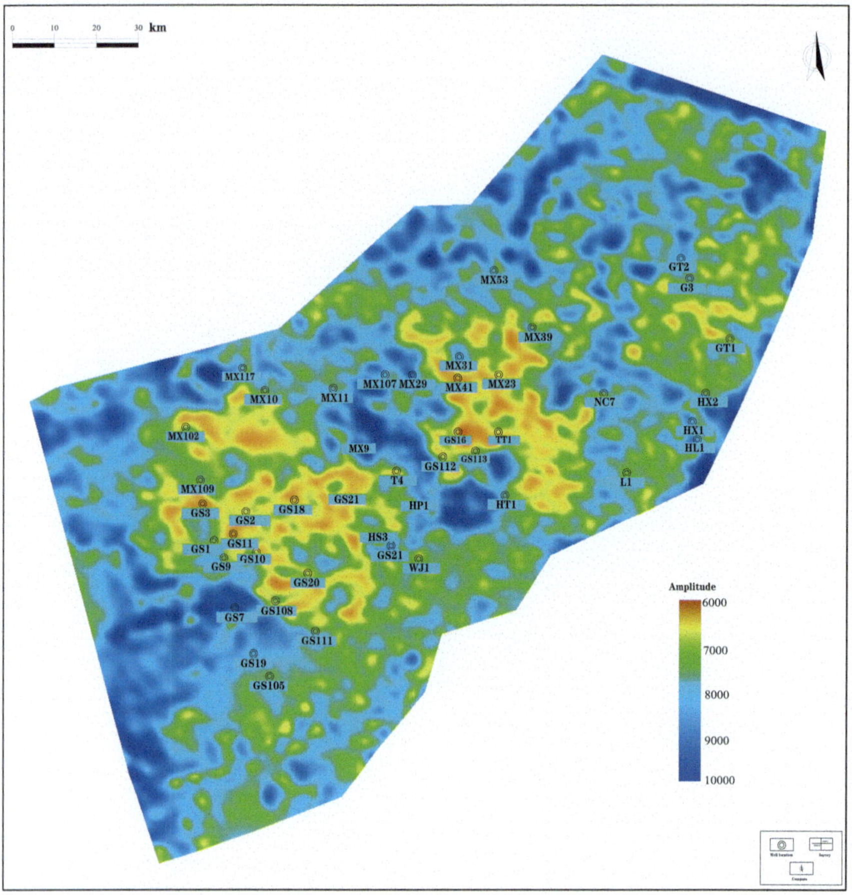

Fig. 79 Amplitude attribute map of Qixia Formation in Central Sichuan

Maokou Formation is determined. Based on the fine structural interpretation and time depth conversion of top and bottom Maokou Formation, the thickness map of Maokou Formation (Fig. 83) is obtained. The Maokou Formation is developed with the thickness of 188–227 m. Compared with the thickness of 36 wells drilled in and around the mining right area, the error less than 5 m accounts for 100%, indicating the high accuracy of formation thickness in the map. Compared with the thickness distribution range of Qixia Formation, Mao1 Member has a certain inheritance to the paleogeomorphology of Qixia Formation. The consistent trend between the two shows that, it is reasonable reflect the paleogeographic pattern of Mao2 and Mao3 Member by the thickness map of Mao1 Member.

The strata in Maokou Formation is generally distributed in a nearly NS direction. Five NW thickness increasing zones are developed in Maokou Formation, including

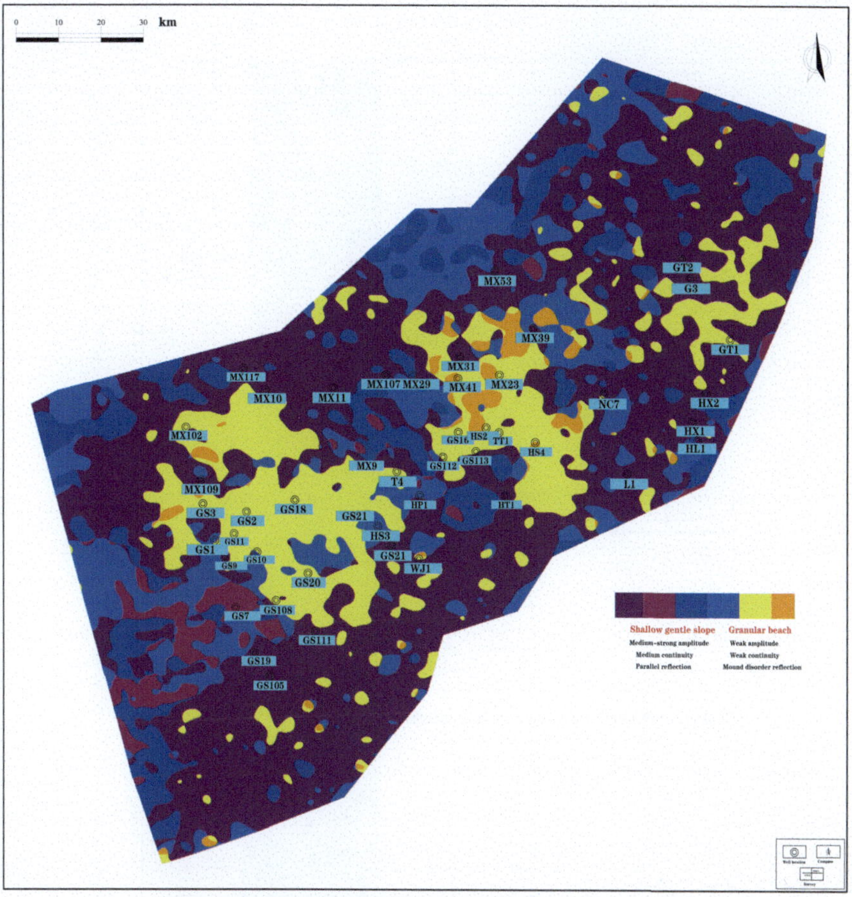

Fig. 80 Seismic facies of Qixia Formation in Central Sichuan

the well area of Gaoshi 3–Gaoshi 111, Moxi 13–Gaoshi 21, Moxi 11–Hechuan 12, Moxi 31–Lai 1 and Guangtan 2.

On the seismic profile, it can be seen that the reservoir of granular beach shows the characteristics of mound-shaped low-frequency weak amplitude reflection and poor continuity, the intertidal sea is characterized by parallel reflection with medium-strong amplitude and strong continuity (Fig. 84).

Based on the 2D and 3D seismic data volumes in the study area, the development section of granular beach in Maokou Formation is compared and analyzed by the attribute extraction, such as RMS amplitude after determining the time period corresponding to the reservoir section on the time section. It is considered that the reservoirs in granular beach can be exactly descripted by RMS attribute (Fig. 85). The abnormal change belt of RMS amplitude in Maokou Formation may be the

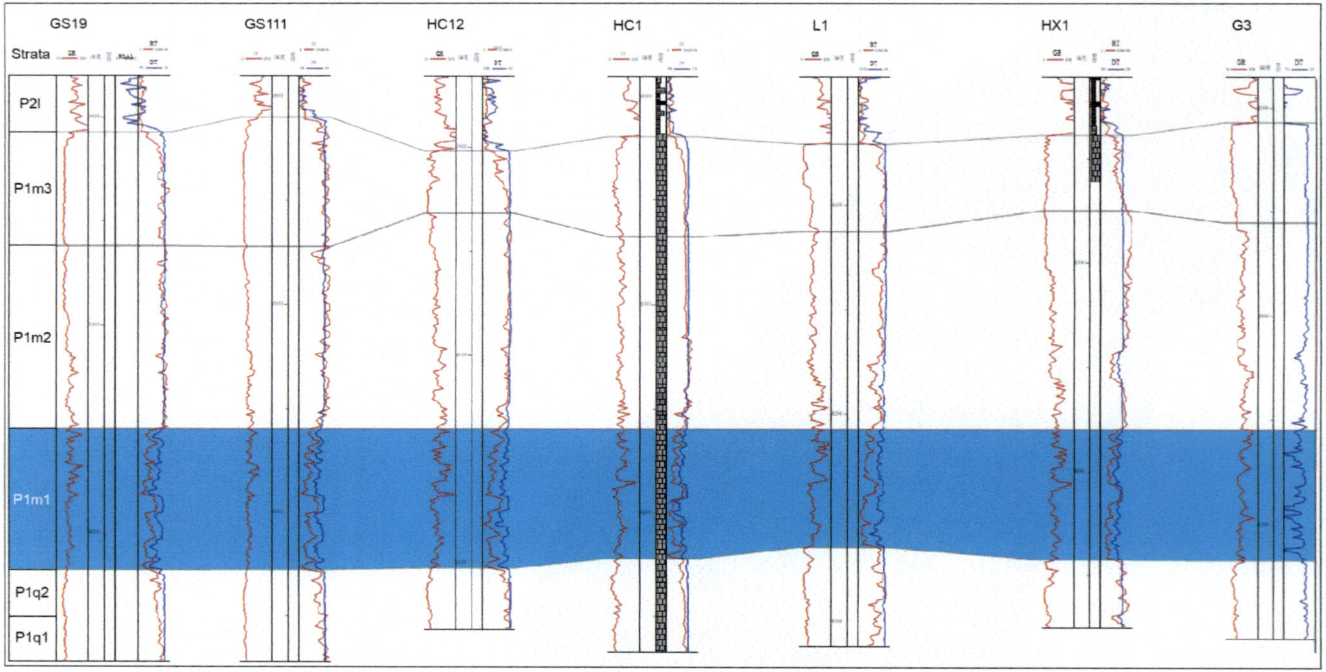

Fig. 81 Stratigraphic correlation profile of Maokou Formation across Well Guogaoshi 19–Gaoshi 111–Hechuan 12–Hetan 1–Lai 1–Huaxi 1–Guang 3

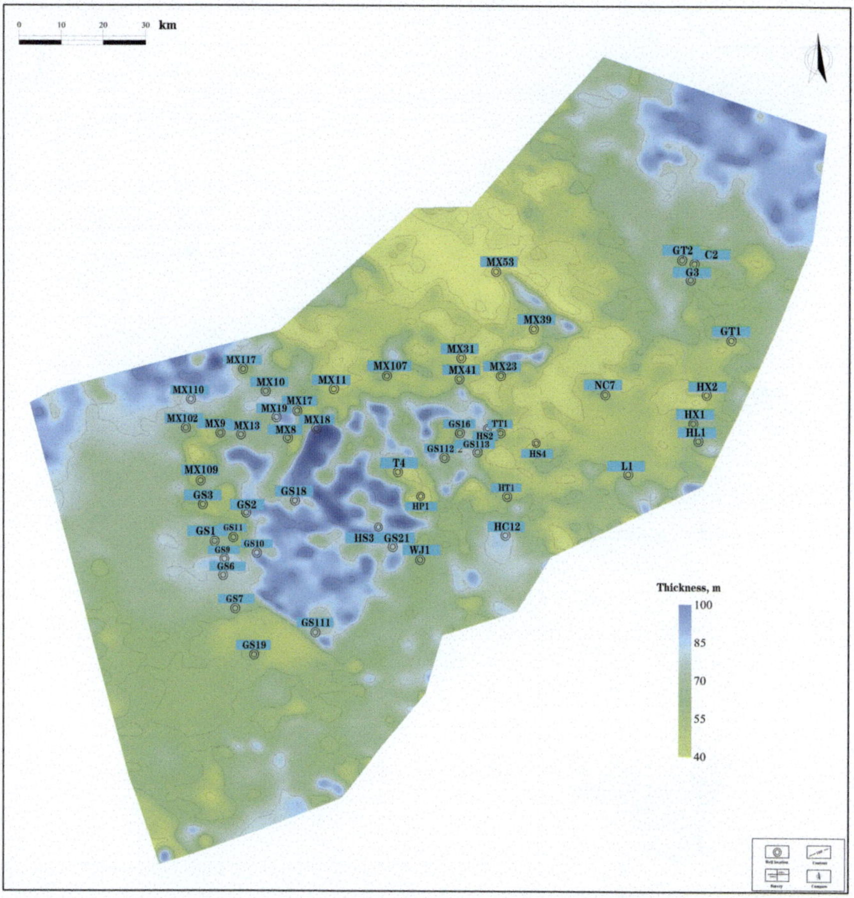

Fig. 82 Thickness distribution map of Mao1 Member revealed by drilling in Central Sichuan

remarkable characteristics of "weak amplitude, internal clutter and mound-shaped reflection structure", and may be the reflection of reservoirs in granular beach.

The analysis results of seismic facies show that the seismic facies of granular beach is characterized by mound-shaped low-frequency weak amplitude reflection and poor continuity. The seismic facies of gentle slope are characterized by medium-strong amplitude, medium-strong continuity and parallel reflection. The seismic facies of the depression are characterized by medium-strong amplitude, medium-strong continuity, parallel reflection and increased stratum thickness (Fig. 86).

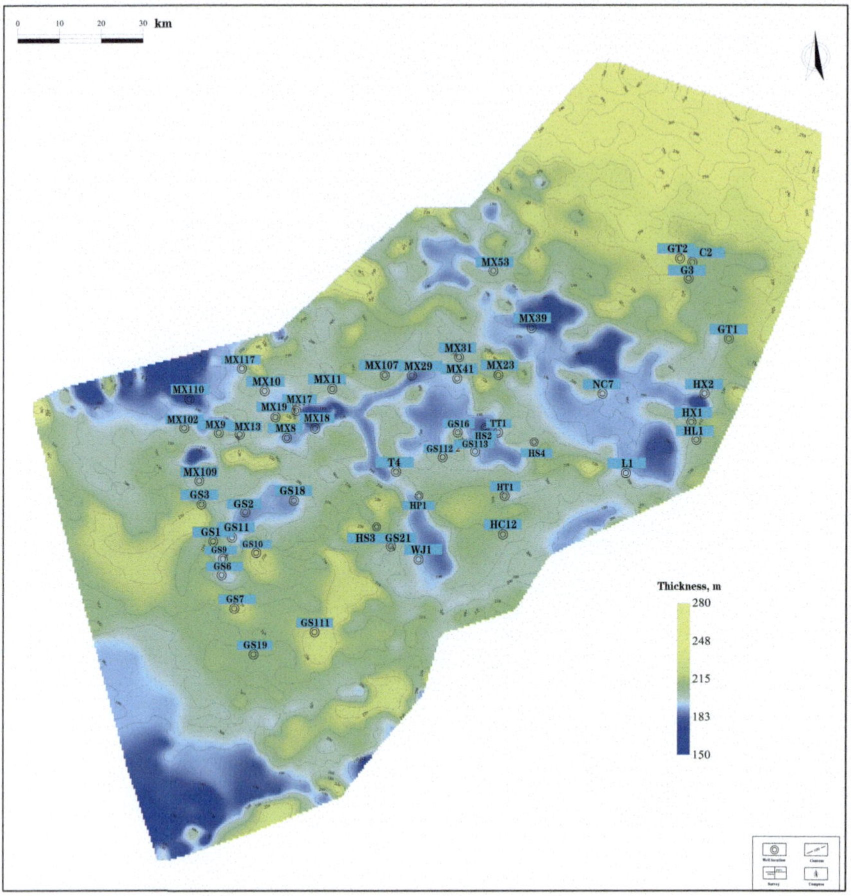

Fig. 83 Thickness map of Maokou Formation in Central Sichuan

4.5 Lithofacies Paleogeographic Characteristics

4.5.1 Qixia Formation

The lithofacies paleogeography map of the Qixia Formation is developed based on the thickness change of the Qixia Formation (Fig. 87). According to the lithofacies paleogeographic map of the Qixia Formation, the Qixia Formation's sedimentary trend is NW, and the granular beach with a shallow gentle slope is primarily distributed along the NW. It becomes a deep-water gentle slope in the northeast of Huayingshan.

Fig. 84 Seismic profile across Well Nanchong 1–Lai 7 (flattening P1m)

4.5.2 Maokou Formation

The lithofacies paleogeography map of the Maokou Formation is constructed using sedimentary and seismic facies in accordance with the formation's thickness variation (Fig. 88). It can be seen from the map that the sedimentary trend of Maokou Formation is NW, and a NW trending intra-platform depression is developed in the shallow gentle slope. A low-energy granular beach is developed in the west of the intra-platform depression, and a high-energy granular beach is developed on the east side of the intra-platform depression. The single beach bodies of high energy granular beach are distributed on a large scale and mostly in isolation, most are distributed along the NW direction. The deposition environment transmits into a deep-water gentle slope in the northeast of Huayingshan.

5 Lithofacies Paleogeography and Sedimentary Evolution of the Changxing Formation

5.1 Sedimentary Facies Type

5.1.1 Petrological Characteristics

Based on the Dunham classification scheme, the rocks of Changxing Formation can be divided into two types: one is bioclastic limestone, micritic bioclastic limestone and muddy bioclastic micritic limestone related to particles and micrites, and the second is in-situ reef limestone/dolomite with skeleton structure (Fig. 89).

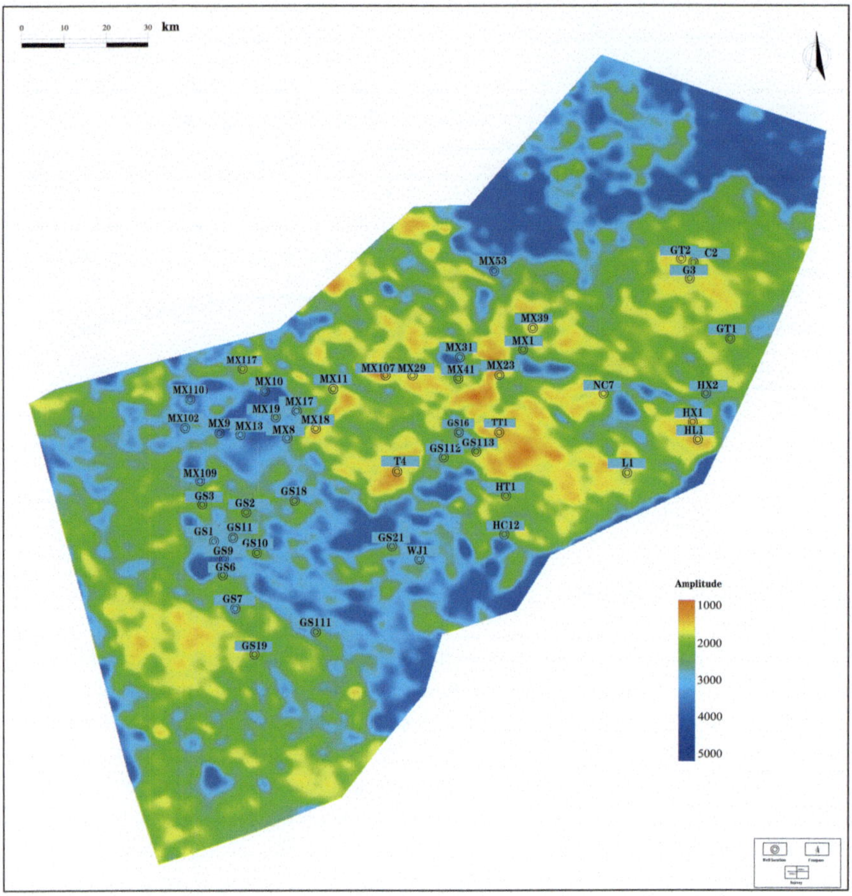

Fig. 85 Amplitude attribute of Maokou Formation in Central Sichuan

Bioclastic Limestone, Micritic Bioclastic Limestone, Bioclastic Micritic Limestone and Muddy Limestone

Bioclastic limestones are composed of sparry cementation, and mostly contain foraminifera, algae and bivalves, which are the products under high-energy hydrodynamic conditions. In micritic bioclastic limestone, bioclastic is similar to sparry bioclastic limestone, the margin of bioclastic is not clear, and micritic cementation is dominant. Bioclastic micritic limestone is formed with less bioclastic content and mainly micrite. In the outcrop profile to the difference of sedimentary energy, the sparry bioclastic limestone is developed as medium-thick layer with the color of light-gray to gray, the micritic bioclastic limestone and bioclastic micritic limestone are developed as thin-medium layers with the color of gray to dark-gray, the muddy

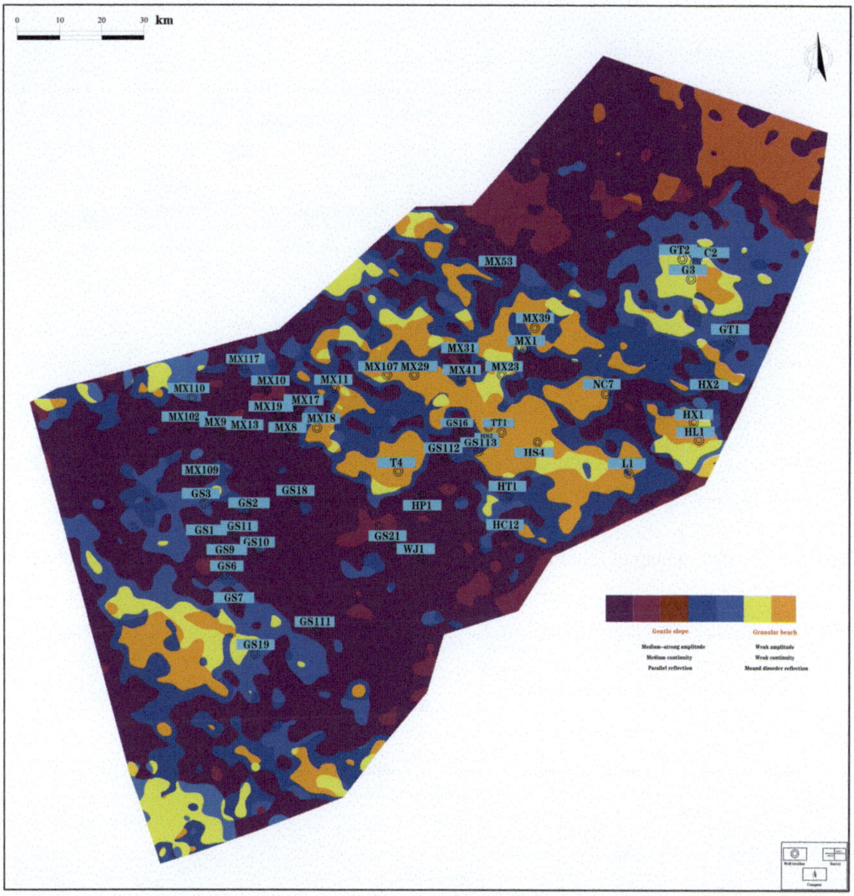

Fig. 86 Seismic facies of Maokou Formation in Central Sichuan

limestone is mostly associated with irregular strips, which is formed as thin layers with the color of dark gray.

Reef Limestone/Dolomite

Reef limestone contains skeleton structure, which is composed of reef-building organisms such as sponge, bryozoan and echinoderm. Most of them are developed with the color of gray to light-gray, and the structure of thick-layered to massive. Reefs can be further divided into microfacies such as reef base, reef core, reef cap/reef front. Hechuan-Tongnan area is located at the southeast margin of the intra-platform depression. The reefs developed in the research area are intra-platform patch reefs.

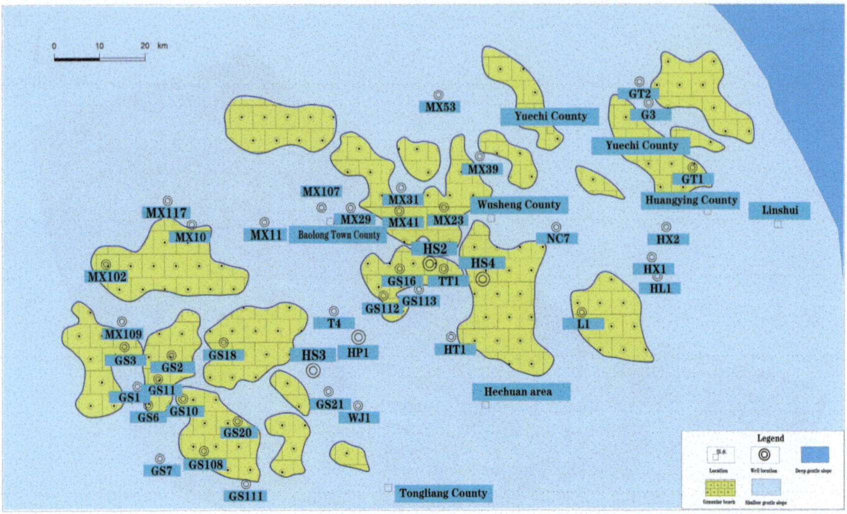

Fig. 87 Distribution map of lithofacies paleogeographic in Qixia Formation of Central Sichuan

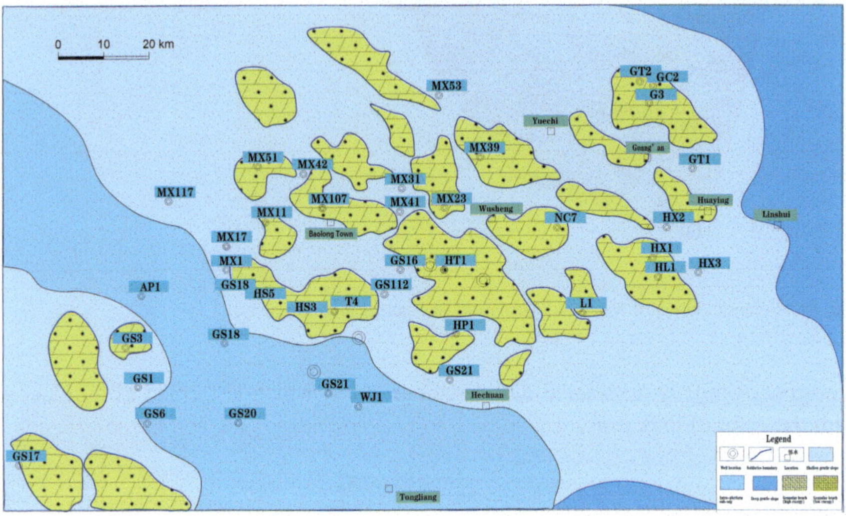

Fig. 88 Distribution map of lithofacies paleogeographic in Maokou Formation of Central Sichuan

Some reef outcrops or drilling reef limestones are dolomitized and changed to be the reef dolomite or dolomite of residual structure.

Fig. 89 Petrological characteristics of Changxing Formation. **a** Sparry bioclastic limestone, foraminifera, algae and bivalves, intra-platform beach, Liziya profile of Huayingshan. **b** Micritic bioclastic limestone, intertidal sea (low-energy subtidal), Liziya profile of Huayingshan. **c** The lower part is gray thick-layered sparry bioclastic limestone, which transits upward to dark-gray thin-middle-layer micritic bioclastic limestone and bioclastic micritic limestone, containing irregular muddy strips and chert nodules, and the upper part is middle-layer micritic bioclastic limestone. The whole generally shows the sedimentary characteristics of intra-platform beach-depression-subtidal low-energy belt, Liziya, Huayingshan. **d** Dark-gray muddy medium-thin-layer bioclastic micrite limestone with muddy strips, intra-platform depression, Chenerwan, Huayingshan

5.1.2 Combination Types and Characteristics of Sedimentary Facies

The carbonate platform of Changxing Formation is rich in organisms. The subfacies combination of intra-platform depression, intra-platform reef, intra-platform bioclastic beach and inter-beach sea is developed in Hechuan-Tongnan work area. The intra-platform depression is mainly composed of dark-gray thin-middle-layered bioclastic micritic limestone containing irregular muddy strips and chert nodules. The intertidal sea/subtidal low-energy belt is mainly composed of gray thin-medium layered micritic bioclastic limestone and bioclastic micritic limestone. The intra-platform reef beach is mainly composed of reef limestone/dolomite and sparry bioclastic limestone (Table 4).

Table 4 Types and petrological characteristics of subfacies in Changxing Formation

Carbonate platform			
Intra-platform beach	Intra-platform reef	Inter-beach sea/ subtidal low-energy belt	Intra-platform depression
Gray-light gray sparry bioclastic limestone or micritic-sparry bioclastic limestone. The bioclastic is mainly composed of foraminifera, followed by algae	Gray medium-thick layered reef limestone/reef dolomite	Gray thin-medium layered micritic bioclastic limestone, bioclastic micritic limestone	Dark-gray thin-medium layered bioclastic micrite limestone, containing muddy strips and chert nodules

5.2 Sedimentary Characteristics of Backbone Section

5.2.1 Sedimentary Characteristics of Outcrops

Outcrop Profile of Changxing Formation, Panlongdong

The Changxing Formation profile in Panlongdong is completely exposed, which is very convenient for observation and measurement, especially the clear sponge reef structure in the river after scouring and weathering. The measured profile is roughly along the direction of stratum dip, and the survey lines are generally distributed in the NE–SW direction and extends along the highway. The strata developed from bottom to top in the profile include Wujiaping Formation, Changxing Formation and Feixianguan Formation. The Changxing Formation is completely exposed with easily identifiable top and bottom boundaries. The medium-thick layered microcrystal dolomite in top Changxing Formation is in contact with the thin-layered muddy dolomite at the bottom of Feixianguan Formation, and the thin-medium layered muddy limestone in bottom Changxing Formation is in contact with the limestone-siliceous interbedded strip at the top of Wujiaping Formation. The profile of Changxing Formation can be divided into 22 layers, with the measured thickness of 249.3 m. The reef is developed in 4–13 layers, in which grain beach or bioclastic beach are often developed interbedded, forming the basement of reef deposition. The 14–22 layers are mainly deposited by shoal in platform margin (Fig. 90).

Layer 1: dark-gray thin-medium layered muddy limestone intercalated with black thin-layered shale, with thin or lenticular siliceous strips occasionally intercalated in the upper part, and the bottom is bounded by black thin-layered shale of Changxing Formation and thick-layered micritic-microcrystal limestone at the top of Wujiaping Formation. Layer 1 is formed with the thickness of 3.2 m, the facies of open platform, the subfacies of subtidal belt, and the microfacies of static mud.

Layer 2 is developed with gray thin-medium layered microcrystal limestone, containing a small amount of bioclastic and sand debris, occasionally intercalated

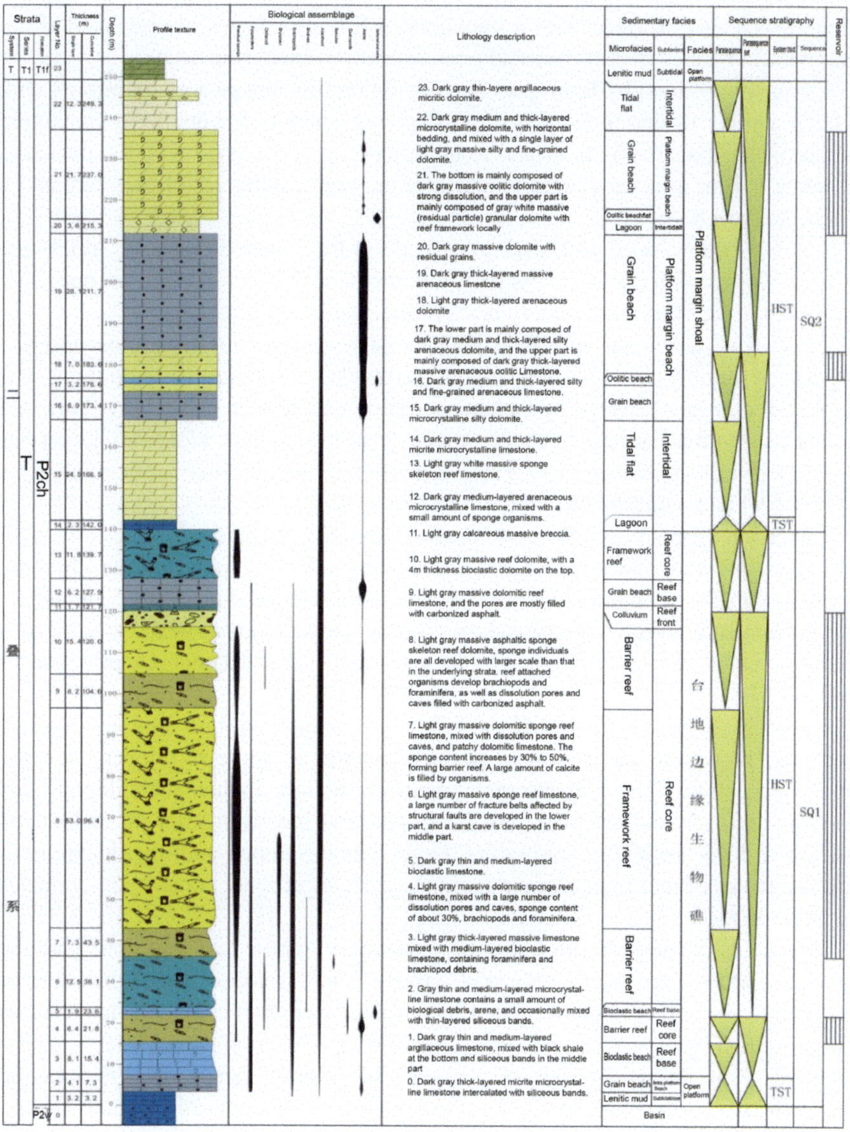

Fig. 90 Comprehensive column map of sedimentary facies and sequence stratigraphy in reef beach profile of Changxing Formation in Panlongdong, Jichang, Xuanhan

with thin-layered siliceous strips. It is formed with a thickness of 4.1 m, a facies of open platform, a subfacies of intra-platform beach and a microfacies of grain beach.

The 3rd Layer is developed with light-gray thick-massive bioclastic limestone intercalated with medium layers, containing bioclastic such foraminifera and brachiopods. It is formed with a thickness of 8.1 m, a facies of reef in platform margin, a subfacies of reef base and a microfacies of bioclastic beach.

Layer 4 is formed by light-gray massive dolomitic sponge reef limestone, with a large number of dissoluted pores and a sponge content of about 30%, containing bioclastic such as brachiopods and foraminifera. It is developed with the thickness of about 6.4 m, a facies of reef in platform margin, a subfacies of reef core and a microfacies of barrier reef.

Layer 5 is developed with dark-gray medium-thin layered bioclastic limestone. It is developed with the thickness of 1.9 m, the facies of reef in platform margin, a subfacies of reef base and a microfacies of bioclastic beach.

Layer 6 is developed with light-gray massive sponge limestone. Reef-inserted organisms mainly include brachiopods, crinoids, fusulinids, foraminifera, etc. Broken-rock zones affected by tectonic activities can be seen in the lower part, and karst caves with a diameter of about 1 m are developed in the middle part. Layer 6 is developed with a thickness of 12.5 m, a facies of reef in platform margin, a subfacies of reef core and a microfacies of barrier reef.

Layer 7 is formed by light-gray massive dolomitic sponge reef limestone, with the development of dissolution pores, and mottled dolomitic limestone is commonly formed with the sponge content of 30–50%, and is filled with a large amount of calcite cement between organisms. Layer 7 is developed with a thickness of 7.3 m, a facies of reef in platform margin, a subfacies of reef core and a microfacies of barrier reef.

Layer 8 is developed with light-gray block asphaltic reef dolomite with sponge skeleton, in which the individual scale of sponge is more developed than the underlying strata. The reef-inserted organisms mainly include brachiopods and foraminifera. Dissolution pores are developed and mostly filled with carbonized asphalt to form ancient oil reservoirs. Layer 8 is developed with the thickness of 53 m, a facies of reef in platform margin, a subfacies of reef core and a microfacies of skeleton reef.

Layer 9 is developed with light-gray massive dolomitic reef limestone, and the pores are often filled with carbonized asphalt. It is developed with a thickness of 8.2 m, a facies of reef in platform margin, a subfacies of reef core and a microfacies of barrier reef.

Layer 10 is formed by light-gray block reef dolomite, and bioclastic dolomite with a thickness of 4 m at the top. It is developed with the thickness of about 15.4 m, a facies of reef in platform margin, a subfacies of reef core and a microfacies of barrier reef.

Layer 11 is developed with light-gray massive lime breccia filled with a large amount of calcite cement. It is developed with a thickness of 1.7 m, a facies of reef in platform margin, a subfacies of reef front and a microfacies of colluvium.

Layer 12 is developed with dark-gray middle-layered sandy microcrystal limestone, containing a small amount of sponge organisms. It is developed with the thickness of 6.2 m, a facies of reef in platform margin, a subfacies of reef base and a microfacies of grain beach.

Layer 13 is formed by gray-white massive reef limestone with sponge skeleton. It is developed with a thickness of 11.8 m, a facies of reef in platform margin, a subfacies of reef core and a microfacies of skeleton reef.

Layer 14 is developed with dark-gray medium-thick layered micritic-microcrystal limestone. It is developed with a thickness of 2.3 m, a facies of shoal in platform margin, a subfacies of inter beach and a microfacies of the lagoon.

Layer 15 is developed with dark-gray medium-thick layered microcrystal dolomite. It is developed with a thickness of 24.5 m, a facies of shoal in platform margin, a subfacies of inter beach and a microfacies of tidal flat.

Layer 16 is developed with dark-gray medium-thick layered power-fine crystal sandy limestone. It is developed with the thickness of 6.9 m, a facies of shoal in platform margin, a subfacies of beach in platform margin and a microfacies of sandy beach.

In Layer 17, the lower part is mainly composed of dark-gray medium-thick layered power crystal sandy limestone, and the upper part is mainly composed of dark-gray thick-massive grain oolitic limestone. Layer 17 is developed with a thickness of 3.2 m, a facies of shoal in platform margin, a subfacies of beach in platform margin and a microfacies of oolitic beach.

Layer 18 is developed with light-gray thick-layered sandy dolomite. It is developed with a thickness of 7.0 m, a facies of shoal in platform margin, a subfacies of beach in platform margin and a microfacies of sandy beach.

Layer 19 is developed with dark-gray thick-massive sandy limestone. It is developed with a thickness of 28.1 m, a facies of shoal in platform margin, a subfacies of beach in platform margin and a microfacies of sandy beach.

Layer 20 is developed with light-gray massive grain dolomite with the powder-fine crystal structure. It is developed with the thickness of 3.6 m, a facies of shoal in platform margin, a subfacies of beach in platform margin and a microfacies of sandy beach.

In Layer 21, the bottom is composed of dark-gray massive oolitic dolomite with strong dissolution, and most oolitic particles are dissolved and filled with carbonaceous asphalt. The upper part is composed of grayish-white massive grain (with residual grain structure) dolomite, with reef skeleton locally. Layer 21 is developed with the thickness of 21.7 m, a facies of shoal in platform margin, a subfacies of beach in platform margin, a microfacies of oolitic beach and sandy beach.

Layer 22 is formed by dark-gray medium-thick layered microcrystal dolomite, and is developed with parallel bedding, intercalated with light-gray block powder-fine crystal dolomite having 2.5 m thickness. The top of Layer 22 is divided by the microcrystal dolomite in middle part and thin-layered muddy dolomite of Feixianguan Formation. It is developed with a thickness of 12.3 m, a facies of shoal in platform margin, a subfacies of inter beach and a microfacies of tidal flat.

Outcrop Profile of Changxing Formation in Wenxingchang

Wenxingchang profile is geographically located in Tianfu town, Beibei District, Chongqing. Paleogeographically, it is located in the southeast of Pengxi-Wushengtai intra-platform sag and adjacent to intra-platform high belt in Suning, which is a typical profile of intra-platform reef (Fig. 91).

The typical reef profile of Changxing Formation, Wenxingchang is formed with small scale of reef, small number of reef-building organisms, single layers with thin thickness and no obvious dolomitization. In addition, Wenxingchang profile is also characterized by the development of chert nodules in the middle and upper parts, some beddings can also be observed. Combined with the previous researches in this area, it is considered that Wenxingchang profile is mainly deposited by open platform, as well as the subfacies of intra-platform reef and beach, and the microfacies of bioclastic beach, barrier beach and inter beach. Among them, reef is developed at the bottom and upper parts of the profile with small scale, and is mainly composed of barrier reef. Secondly, the rock in this profile is mainly developed with single type, and is mainly composed of limestone, including micritic limestone, reef limestone, bioclastic limestone and bioclastic limestone containing chert nodules.

5.2.2 Sedimentary Characteristics of Coring Wells

There are few coring wells in Changxing Formation of Central Sichuan. Only Well Lai 1 is cored in Changxing Formation with the depth of 3824–3778 m, and the coring location in the upper part of Changxing Formation (Fig. 92).

The lithology of the core progressively changes from bottom to the top, from dark-gray micritic limestone and bioclastic micritic limestone to gray micritic bioclastic limestone and bioclastic limestone, indicating an upward shallowing cycle coupled with subtidal low-energy belt and intra-platform beach. Above this cycle, micritic bioclastic limestone with a thickness of about 1 m was deposited, and then evolved into sparry bioclastic limestone and reef limestone. This all can be interpreted as the first reef-building cycle, which is composed of reef base (bioclastic limestone) and reef core. Above this cycle, two reef cycles with similar reefs grow and develop, which are also composed of reef base of bioclastic limestone and reef core of biological reef limestone. The core as a whole has no obvious dissolution pores and no obvious dolomitization.

5.3 *Sedimentary Facies Analysis*

5.3.1 Logging Facies

Based on the sedimentary analysis of Well Lai 1, core location is carried out, and the natural gamma-ray logs with high lithology sensitivity are mainly selected in

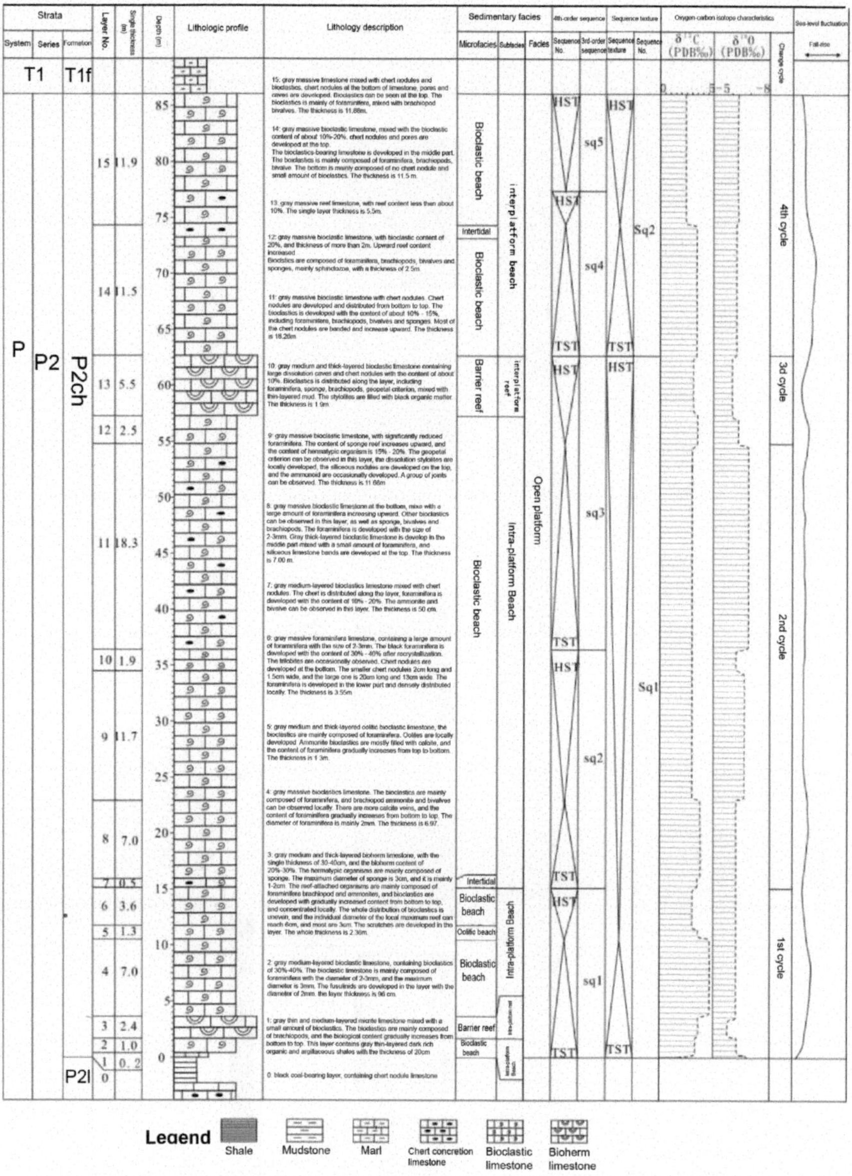

Fig. 91 Comprehensive column map of sedimentary facies and sequence stratigraphy of reef beach profile in Changxing Formation of Wenxingchang

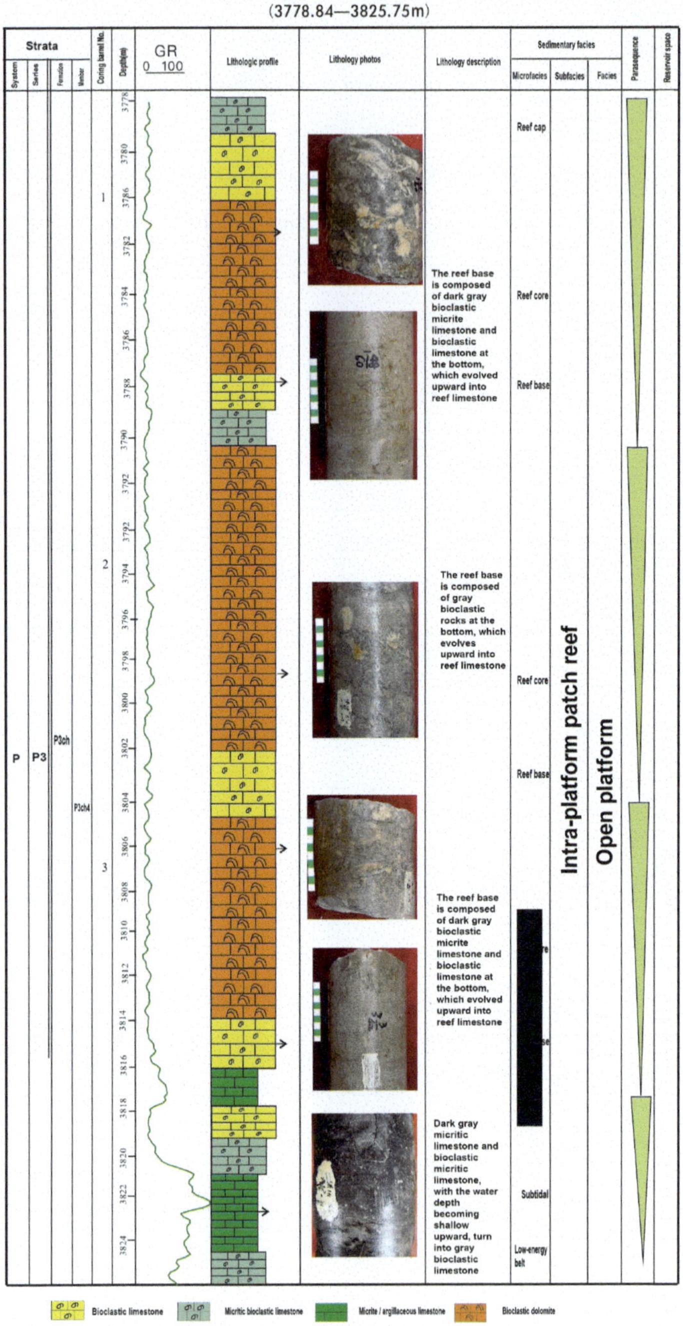

Fig. 92 Comprehensive column map of sedimentary characteristics in cores of Changxing Formation in Well Lai 1

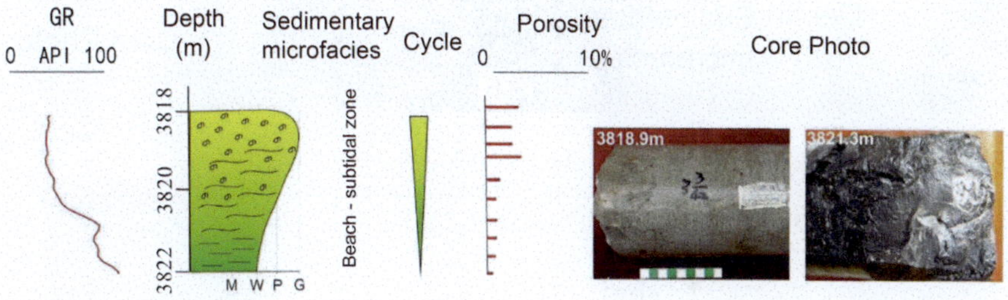

Fig. 93 Logging response of the combination of subtidal low-energy belt and intra-platform beach in Changxing Formation

logging facies analysis. The logging facies characteristics of Changxing Formation are summarized as follows (Figs. 93 and 94). (1) The combination of subtidal low-energy belt and intra-platform beach. Natural gamma-ray is characterized by toothed high-low amplitude, with funnel-shaped characteristics. The logging facies combination can be interpreted as the response of sea-level changes related to high-frequency sequences. (2) Reef. The whole natural gamma-ray is characterized by low amplitude box-type and micro dentation, representing a high-energy sedimentary environment.

5.3.2 Sedimentary Facies in Single Wells

Well Guangtan 1 is located at the south high point of Guang'an structure in Sichuan Basin, and is deposited as the depression margin in the open platform. The strata of Changxing Formation are revealed by this well, with the depth of 4073–4284 m, and a thickness of 211. Thin-layered bioclastic micritic limestone is deposited in Changxing Formation under the sedimentary environment of intra-platform depression. Bioclastic limestone and reef limestone are also deposited as the combination of intra-platform reef and beach. The intra-platform beach is developed with the thickness of 1–5 m, the intra-platform reef is developed with the thickness of 2–9 m, and there is no obvious dolomitization. The reservoir space of intra-platform reef beach is underdeveloped and the high-energy belt is mainly developed in the highstand system of third-order sequence. Based on the characteristics of lithologic combination, four sedimentary subfacies were identified in Well Guangtan 1, including intra-platform beach, low-energy beach, subtidal and inter-beach depression.

Well Lai 1 is located in Laitan structure deposited as the depression margin in the open platform. The strata of Changxing Formation were revealed by this well, with the depth ranging from 3766 to 3956 m, and a thickness of 190 m. Thin-layered bioclastic micritic limestone is deposited in Changxing Formation under the sedimentary environment of intra-platform depression. Bioclastic limestone and reef limestone are also deposited as the combination of intra-platform reef and beach. The intra-platform beach is developed with the thickness of 1–5 m, the intra-platform reef is developed with the thickness of 7–11 m, and there is no obvious dolomitization.

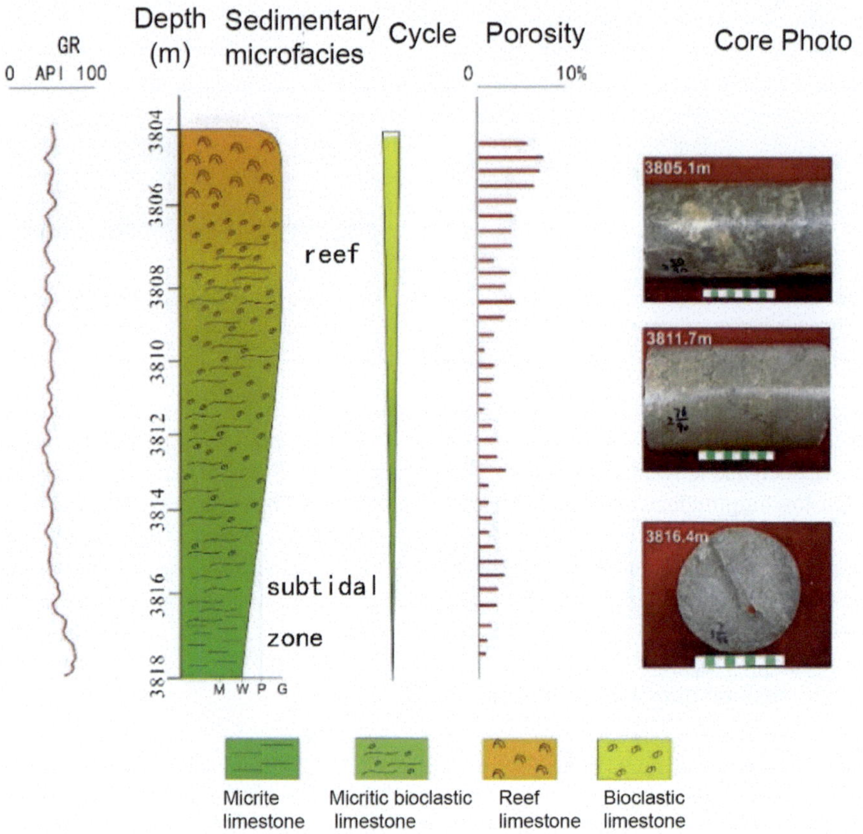

Fig. 94 Logging response of combination of subtidal low-energy belt and reef in Changxing Formation

The reservoir space of intra-platform reef beach is underdeveloped and the high-energy belt is mainly developed in the highstand system of third-order sequence. Four sedimentary subfacies are recognized in Well Guangtan 1 based on their lithologic combination characteristics which are: intra-platform beach, low-energy beach, subtidal, and inter-beach depression.

Based on the analysis of sedimentary facies in single wells, the basic characteristics of high-energy bioclastic beach are counted as follows. The beach bodies in single wells are developed with the average thickness of 10.1 m, a maximum thickness of 16 m and a minimum thickness of 2.05 m. The beach bodies are developed with an average thickness of 3.9 m, a maximum thickness of 6.5 m and a minimum thickness of 1.36 m.

5.3.3 Sedimentary Facies Correlation

Based on the correlation profile of sedimentary facies across Well Gaoshi 21–Hetan 1–Lai 1–Hualai 1–Huaxi 1–Guangtan 1, the macro characteristics of sedimentary facies in Changxing Formation are distinguished. It can be seen that the stratigraphic profile is developed with the thickness of 86.88–211.3 m, and two complete third-order sequences. The overall Changxing Formation is developed as an open platform, and the upper sedimentary facies of Well Hualai 1 and Huaxi 1 are intra-platform depressions. The intra-platform beach is mainly developed in the highstand system of the third-order sequence. The beach body at the top of the first third-order sequence is developed with good connectivity, but the beach bodies are mostly distributed in isolation in the whole profile. There are 3–4 sets of reefs developed in Well Lai 1 and Guangtan 1, with a thickness of 2.1–10.93 m, an average thickness of 7.19 m, and an average cumulative thickness of 24.21 m. The reefs are all developed in the upper part of Changxing Formation. The high-energy beaches are widely developed in this profile, with a single-layer thickness of 1.05–4.6 m, an average thickness of 2.67 m, a single-well cumulative thickness of 13.45–32.25 m, and an average thickness of 21.43 m. The intra-platform beach in Well Guangtan 1 is developed with the largest single-well thickness, while in Well Gaoshi 1 it is developed with the smallest single-well thickness. The intra-platform beach in Well Hetan 1 is developed with the largest cumulative thickness, while in Well Hexi 1 it is developed with the smallest thickness.

According to the correlated profile of sedimentary facies across Well Moxi 39–Nanchong 7–Huaxi 1–Hualai 1, the macro characteristics of sedimentary facies in Changxing Formation are distinguished. It can be seen that the stratigraphic profile is developed with the thickness of 70–154.8 m, and two complete third-order sequences. The first third-order sequence of this profile is deposited as open platform, and the second third-order sequence is developed as an intra-platform depression. The intra-platform beach is mainly developed in the highstand system of the first third-order sequence with good connectivity in the top part. And the isolated beach body is mainly developed in the upper part of the transgressive system and the lower part of the highstand system. The high-energy beach of the wells in the profile is developed with a single-well thickness of 1.2–3.86 m, an average thickness of 2.46 m, a single-well cumulative thickness of 4.7–15.31 m, and an average thickness of 10.53 m. The intra-platform in Well Hualai 1 is developed with the largest single-well and cumulative thickness, in Well Moxi 39 it is developed with the smallest single-well thickness, and in Nanchong 7 it is developed with the smallest cumulative thickness.

5.4 Seismic Facies

Firstly, the characteristics of the top and bottom reflection interface of Changxing formation are determined through the fine well-seismic calibration. The reflection interface of Top Changxing Formation in Upper Permian is developed with strong–weak energy, which changes greatly. Due to the influence of reef developed in Changxing Formation, the energy in reef development area changes greatly with disordered reflection events. And the reflection events in non-reef area are developed with good continuity. Generally speaking, the Top Changxing Formation is difficult to compare, but it can still be tracked in the whole area. The bottoms of Changxing Formation and Dalong Formation in Upper Permian are also characterized as strong reflection events with good continuity, which can be compared and tracked in the whole region. Due to the thin interbed at the boundary between Changxing Formation and Longtan Formation, this phase is generally stable, and changed in some areas. Generally speaking, it can be considered as the tuning reflection formed by the tuning of limestone intercalated with marine shale in the two members of Longtan Formation. Deposited as deep-sea facies, it is formed with good stability. The phase change mainly occurs as the transition from slope facies to deep-sea basin, where this phase disappears. Based on the good stability of shale sedimentary facies, it is speculated that the disappearance of the phase is caused by the thinning of limestone interlayer in Longtan Formation, which may be the deepest sedimentary facies belt in Longtan Formation. The reflection events of Longtan Formation at the bottom of Upper Permian is a regional standard reflection wave. By the determination of the reflection characteristics of the top and bottom interfaces in Changxing Formation, the seismic reflection characteristics of different subfacies in Changxing formation are determined.

Reef: mound shape, high amplitude, disordered reflection structure, weak amplitude and poor continuity. Drilling reveals that it is mainly deposited as reef facies, of which Well Lai 1 is characterized by typical reef reservoir (Fig. 95).

Bioclastic beach: mound shape, low amplitude, medium-weak amplitude and medium continuity. The drilling reveals that it is mainly deposited as the bioclastic beach facies. The greater the thickness of bioclastic beach, the better the gas content and the stronger the seismic reflection amplitude. The drilling reveals that the bioclastic beach in Well Wangjia 1 is developed with a thickness of 20.95 m, and a gas production of 5.87×10^4 m^3 (Fig. 96).

Trough: it is developed in the local deepening area of water body in open platform, below the wave base, which may be formed by tectonic basement subsidence. It is characterized by the interbedding of dark-gray limestone containing chert nodules and thin-layered micritic limestone, and the biological species are monotonous. Trough subfacies in the seismic are characterized by strong amplitude, strong continuity and parallel reflection.

According to the drilling data observation, reefs and beaches in the Changxing Formation exhibit two main characteristics: Guang 3 and Lai 1 have reefs with high-amplitude mound forms and weak reflection, whereas Wangjia 1 and Moxi 7 have bioclastic beaches with low-amplitude mound shapes and middle-weak amplitude continuous reflection (Fig. 97).

Seismic facies analysis shows that the seismic facies of reefs are characterized by high-amplitude mound shape, disordered reflection structure, weak amplitude and poor continuity (Fig. 98). The bioclastic beaches are characterized by low-amplitude

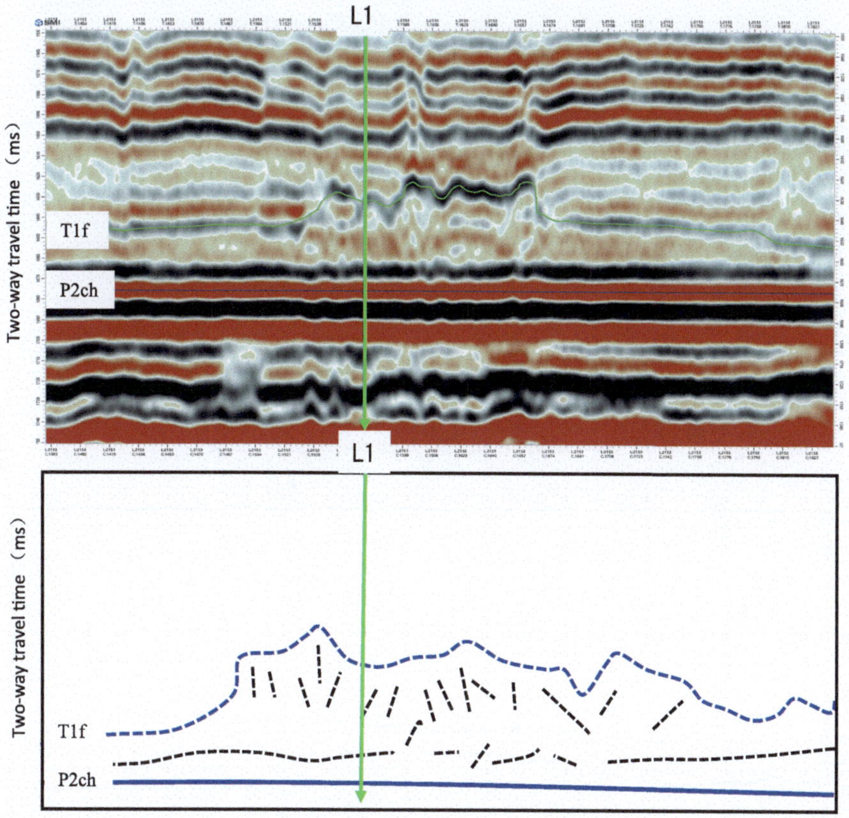

Fig. 95 Logging and seismic characteristics of reef in Changxing Formation of Well Lai 1

mound shape, medium-weak amplitude and medium continuity (Fig. 99). The trough is characterized by strong amplitude, strong continuity and parallel reflection. As shown in Fig. 100, warm-tone areas represent the seismic facies of trough and cold-tone areas represent the seismic facies of reef beach.

In the process of reef beach identification, in addition to the reflected wave identification methods in the conventional amplitude profile, such as the same sign phase, enhanced energy, similar waveform and continuity, the following technical methods are mainly adopted in the correlation process of Top Changxing Formation based on the special conditions of reef oil and gas reservoir.

① The correlation and tracking combined with conventional seismic amplitude profiles, instantaneous phase profiles, and instantaneous amplitude profiles. The strong amplitude is difficult to continuously track in the conventional amplitude profile due to the unstable and greatly changed energy at the abnormal reef location. Given this difficulty, during the correlation process, the instantaneous phase profile is combined with the conventional amplitude profile to improve the

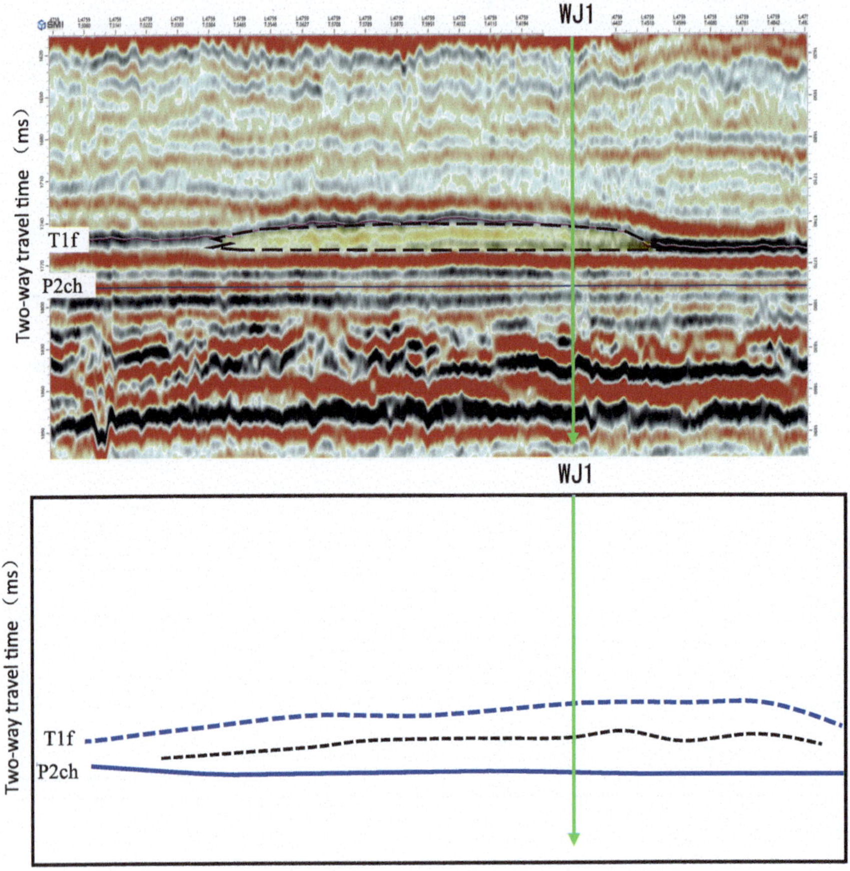

Fig. 96 Logging and seismic characteristics of bioclastic beach in Changxing Formation of Well Wangjia 1

reliability of horizon comparison and tracking with the advantage that the instantaneous phase attribute is not affected by the change of reflected wave energy and only reflects the change of lithology. Figure 101 shows the conventional amplitude profile of seismic reflection, Fig. 102 shows the instantaneous phase, and Fig. 103 shows the instantaneous amplitude. It can be seen that the reef anomaly body of Top Changxing Formation is developed with weak reflection amplitude in the conventional amplitude profile, and is difficult to comparison and tracking, while it is developed with good continuity in the instantaneous phase profile. At the same time, on the instantaneous amplitude profile, it can be more clearly to describe the development characteristics of the reef boundary with the change of energy.

② The special seismic reflection structure of reef should be considered in the process of comparison and tracking. Usually, the abnormal bodies of reef are formed as

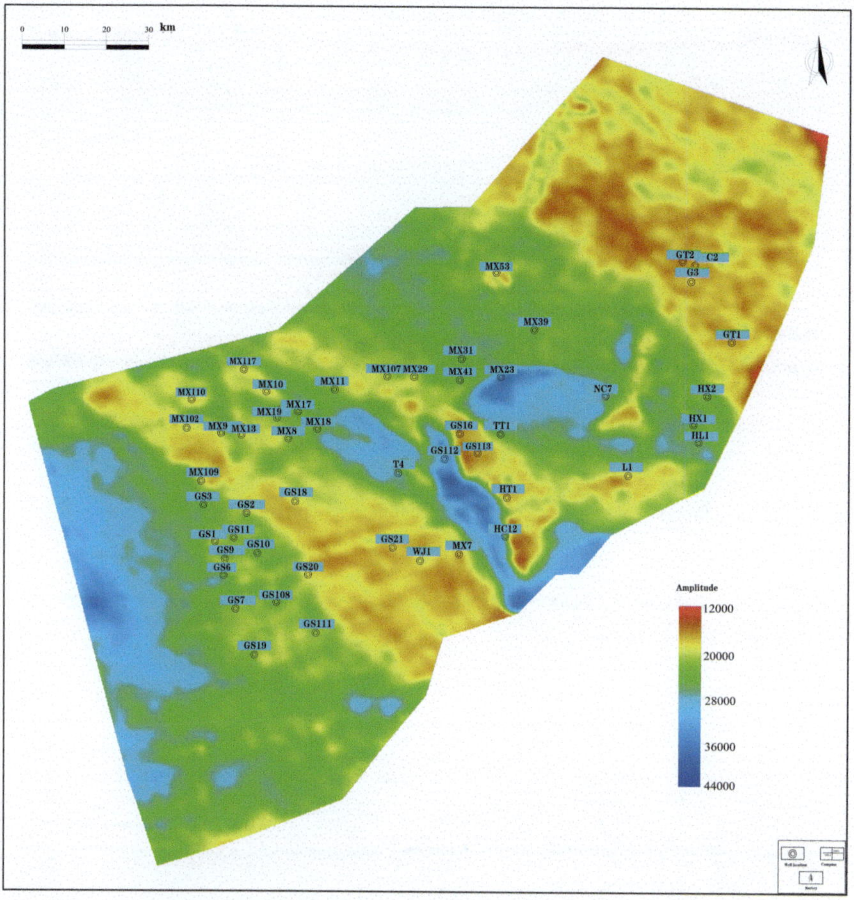

Fig. 97 RMS amplitude map of Changxing Formation in Central Sichuan

the shape of mound, lens or almond. The special reflection shape of the reef should be fully considered when the seismic events weaken or are interrupted during comparison and tracking. It is possible to cross the phase on the top of the Changxing Formation. This shape characteristics of modern reefs can confirm this. It should be noted that according to the regional geological data, the structure in the Upper Permian is stable, and the faults are not developed, as a result, the faults do not cause the interruption of the seismic events. The possibility of reef anomalies should be fully considered.

The fine comparison and tracking of the Top Changxing Formation in the Upper Permian are realized through the above methods. Generally speaking, affected by the development of reefs, the thickness of reef is significantly thicker than that of non-reef. Further combined with the well-seismic calibration, the thickness of Changxing Formation in Central Sichuan is restored. Three NW–SW high belts with

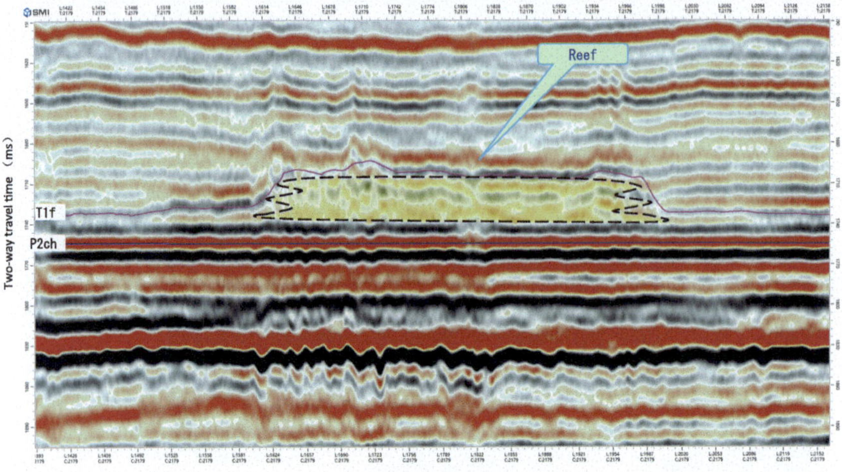

Fig. 98 Seismic facies characteristics of reef in Changxing Formation

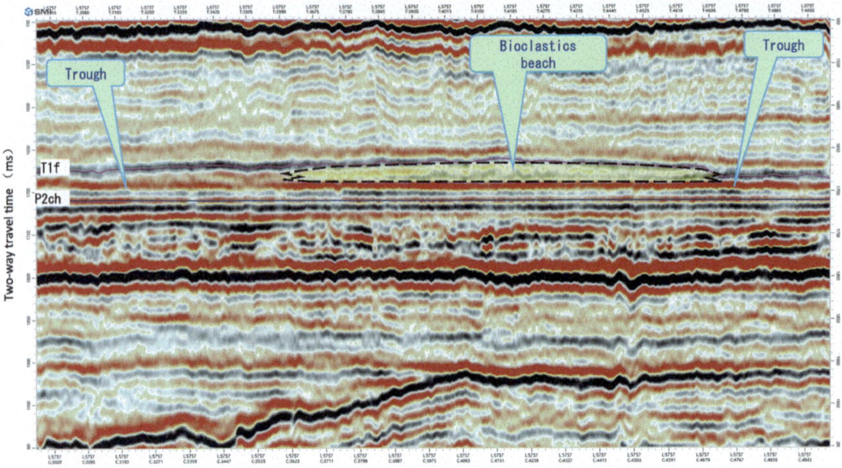

Fig. 99 Seismic facies characteristics of bioclastic beach in Changxing Formation

increasing thickness are developed in Changxing Formation, namely Guang'an high belt, Suining No. 1 high belt and Suning No. 2 high belt. The boundary between the intra-platform high belt and the trough is obvious, and the intra-platform high belt is a favorable area for reef beach development (Fig. 104).

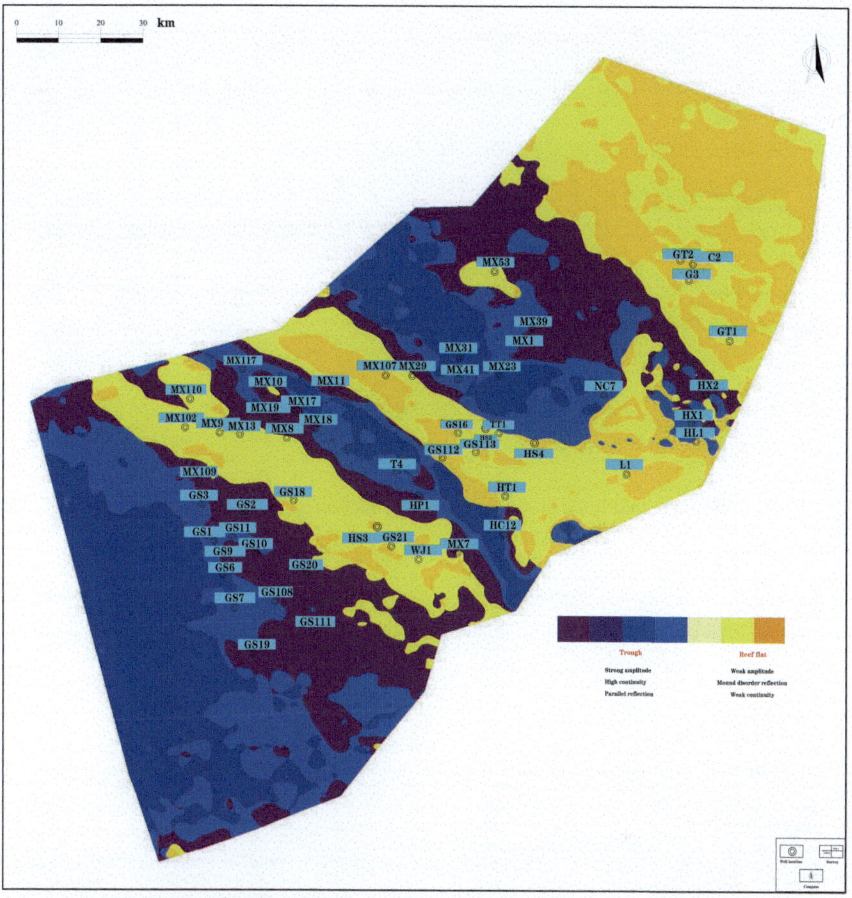

Fig. 100 Seismic facies of Changxing Formation in Central Sichuan

5.5 *Lithofacies Paleogeographic Characteristics*

Based on the thickness variation of Changxing Formation, combined with sedimentary facies and seismic facies, the lithofacies paleogeography map of Changxing Formation is drawn (Fig. 105). According to the lithofacies paleogeographic map of Changxing Formation, the sedimentary trend of Changxing Formation is NW, and Pengxi-Wusheng trough with NW trending is developed as a whole. Two platform margin belts with NW trending are developed on both sides of Pengxi-Wusheng trough, among which, Guang'an high belt is developed in the northeast and Suining high belt is developed in the southwest. A platform basin with NW trending is developed in Suining high belt. The reef beaches of Changxing Formation are mainly distributed on the side of the platform margin belt near the trough, while

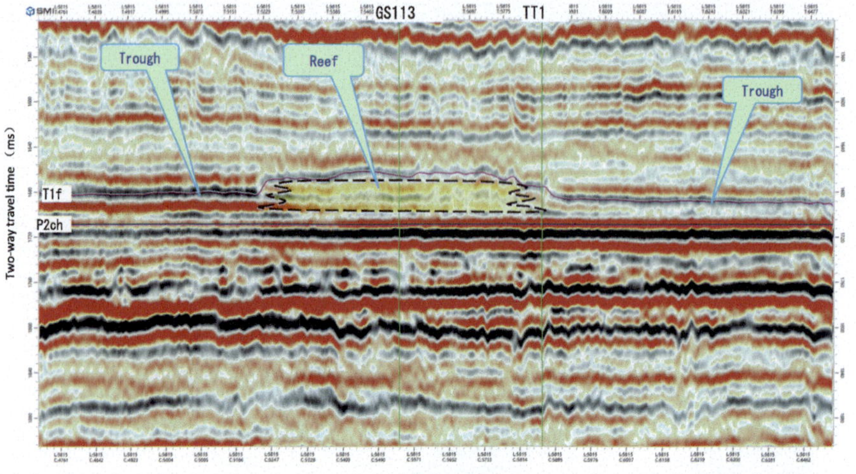

Fig. 101 Conventional amplitude profile of seismic reflection of reefs

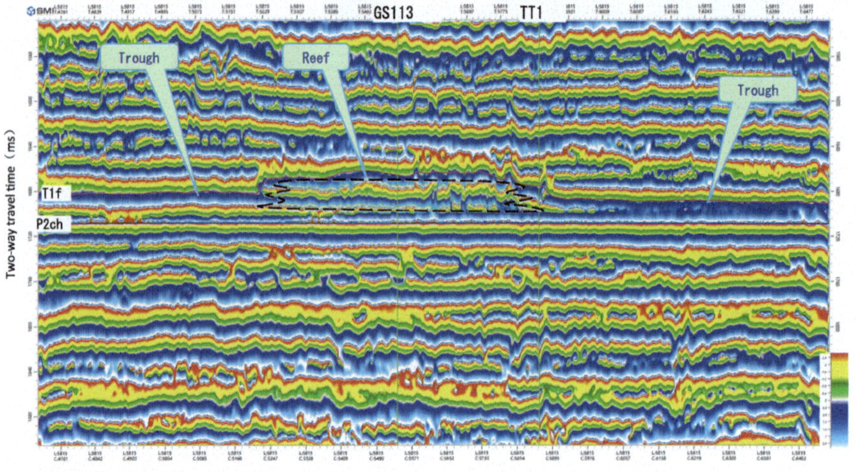

Fig. 102 Instantaneous phase profile of reefs

the bioclastic beaches are mainly distributed on the side of the Suining high belt near the intra-platform belt.

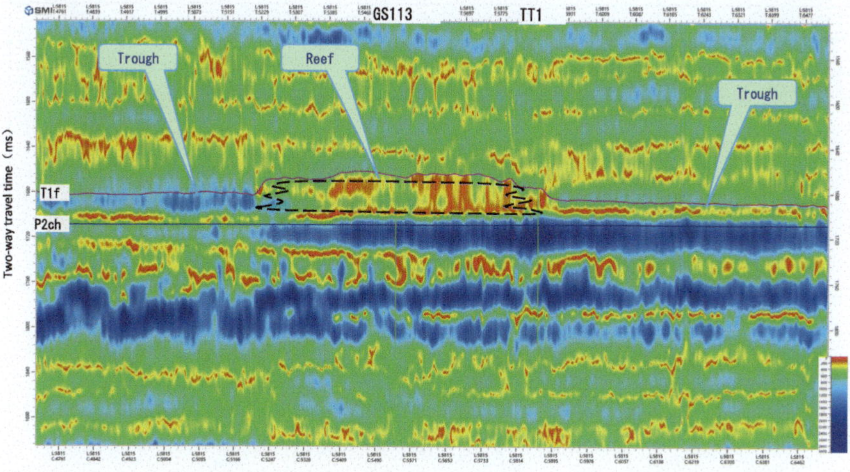

Fig. 103 Instantaneous amplitude profiles of reefs

6 Lithofacies Paleogeography and Reservoir Characteristics of the Leikoupo Formation

6.1 *Sedimentary Facies Type*

6.1.1 Petrological Characteristics

Formation is developed with complex lithology, which can be summarized as "complicated" and "fine". Among them, "complicated" refers to the complex rock composition, including muddy, lime, dolomitic and gypsum rocks, and "fine" refers to the main particle of the rock, which is generally fine. When cores and outcrops from the research region and nearby areas were analyzed, the lithology of the Leikoupo Formation could be characterized as dolomitic rock, muddy rock, and gypsum rock.

Dolomitic Rock

Leikoupo Formation is deposited as an open/restricted platform-evaporation platform and is mainly composed of dolomite. The common dolomites are granular dolomite and grain dolomite. Among them, granular dolomite is mainly composed of sandy dolomite, and grain dolomite is mainly composed of micrite and powder crystal, with residual structure (Fig. 106). Dissolution pores and caves are developed in some granular and grain dolomites.

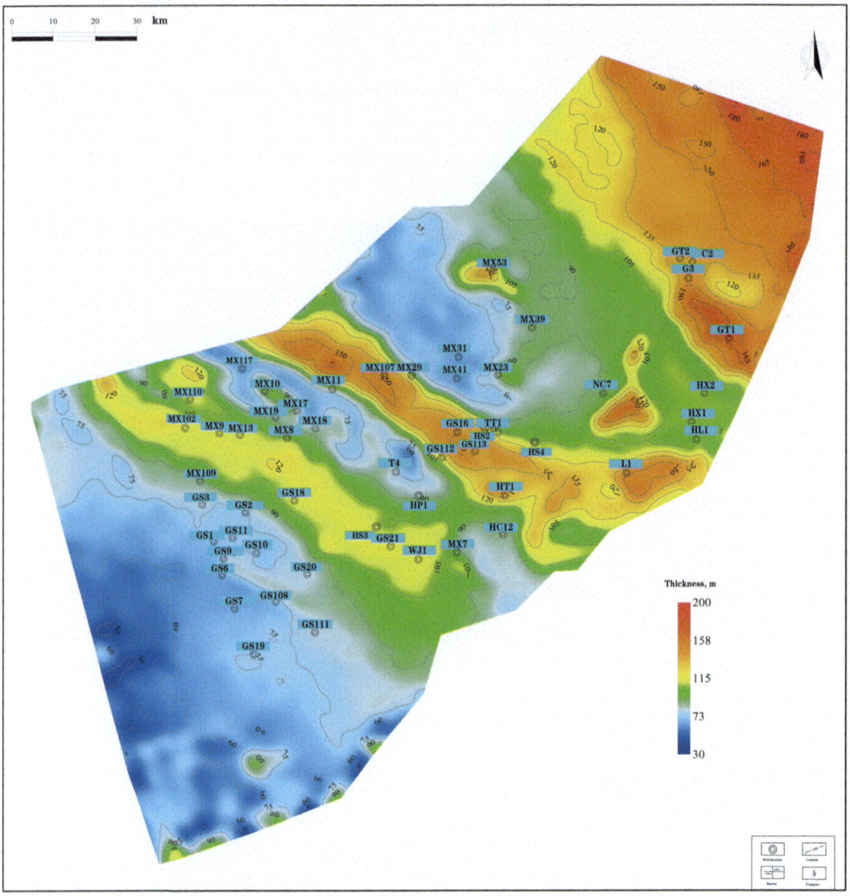

Fig. 104 Thickness map of Changxing Formation in Central Sichuan

Muddy and Gypsum-Bearing Rocks

This kind of lithology mainly appears in the form of content. The main lithology of the rock is generally carbonate rocks, such as muddy dolomite, muddy limestone, gypsum dolomite and muddy gypsum dolomite. These rocks are typically developed in the tidal flat of the evaporation platform or restricted lagoon with nearly no reservoir space (Fig. 107).

6.1.2 Combination and Characteristics of Sedimentary Facies

Previous studies have shown that the Leikoupo Formation in Sichuan Basin is mainly deposited as carbonate restricted platform-lagoon/evaporation platform. Vertically,

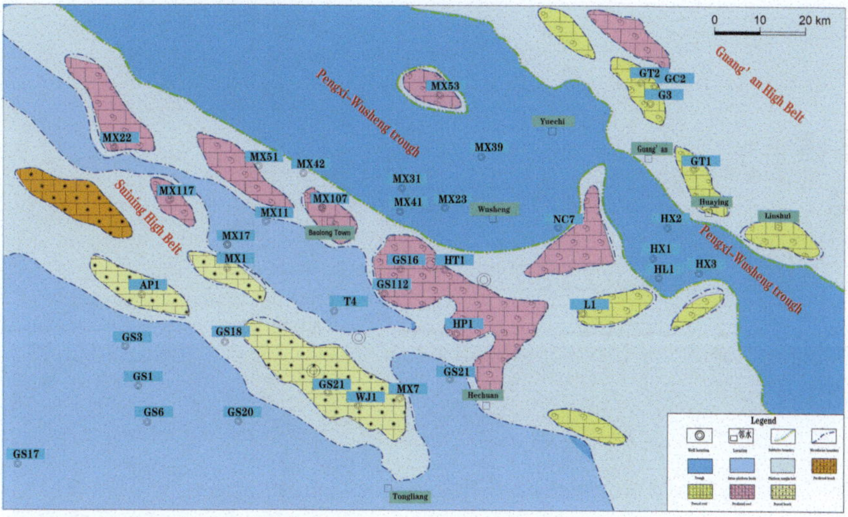

Fig. 105 Lithofacies paleogeography map of Changxing Formation in Hechuan-Tongnan exploration area

the facies belt migration caused by sea-level change is developed in the four members of Leikoupo Formation, as a result, the Leikoupo Formation is deposited with the change of platforms types, such as open platform, restricted platform and evaporation platform. In Hechuan-Tongnan work area, Leikoupo Formation is deposited as restricted platform as a whole. The work area can identify microfacies types and sedimentary background as well as the core sedimentary characteristics (Table 5). Intra-platform is mainly developed with the lithology of sandy dolomite, the inter-beach sea is generally symbiotic with the intra-platform beach, and is mainly composed of fine-grained micritic silty dolomite and mud-bearing silty dolomite. Lagoon is mainly developed with the lithology of gypsum-bearing micritic dolomite and gypsum-salt rock.

6.2 *Sedimentary Characteristics of Backbone Section*

6.2.1 Sedimentary Characteristics of Outcrops

Wangcang Lujiaba profile is located in the north of Sichuan Basin, with complete exposure, clear top and bottom boundaries, and a sedimentary thickness of 360 m. This profile is mainly composed of restricted carbonate platform, locally intercalated with open platform facies, and is generally characterized by the interbedding of subfacies of tidal flat in restricted platform and intra-platform shoal. Dolomite is generally developed on the profile, with high shale content in the lower part. The

Fig. 106 Petrological characteristics of dolomite in Leikoupo Formation. **a** Sandy dolomite, parallel bedding, Well Moxi 27, 3–29/63. **b** Gray medium-layered fine-grained dolomite, dissolution caves, profile of Hanzeng, Jiangyou. **c** Sandy dolomite with cracks, Well Tongtan 1, 2586 m. **d** Silty dolomite, residual structure, filled by large amount of asphalt, Well Tongtan 1, 2585.6 m

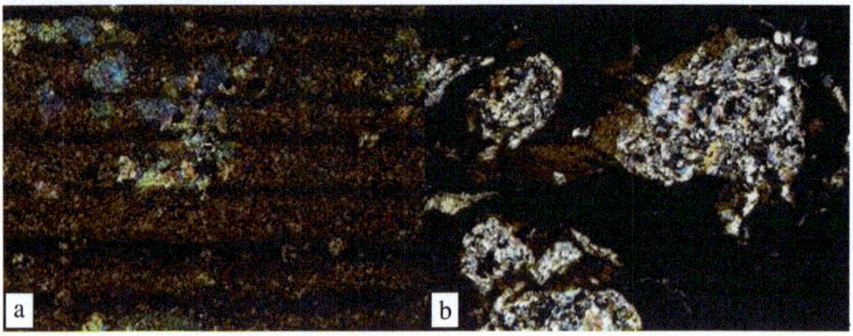

Fig. 107 Petrological characteristics of muddy and gypsum rocks of Leikoupo Formation. **a** Gypsum-bearing muddy micritic dolomite, muddy laminae, Well Tongtan 1, 2376.3 m. **b** Gypsum-bearing micritic dolomite, Well Tongtan 1, 2373.9 m

Table 5 Sedimentary facies types and petrological characteristics of Leikoupo Formation

Restricted platform		
Evaporation lagoon	Inter-beach sea	Intra-platform beach
Gypsum-bearing micritic dolomite and gypsum salt rock	Thin-layered micritic silty dolomite, mud-bearing micritic silty dolomite	Mainly sandy dolomite
Lei2 and Lei3 Member	Lei1, Lei2 and Lei3 Member	Lei1 and Lei3 Member

lithology in Lujiaba profile transits upward to grain dolomite, bioclastic dolomite, oolitic dolomite mixed with limestone, and the upper part of the profile is composed of grain dolomite. According to the law of lithofacies change, the profile can be divided into four third-order sequences, numbered Sq1, Sq2, Sq3 and Sq4, which are all composed of TST and HST. Each sequence corresponds to four lithologic segments, i.e. Sq1–Sq4 corresponds to Lei1, Lei2, Lei3 and Lei4 Member respectively. TST is generally composed of subtidal sediments of restricted platform or open platform, and the intra-platform shoals in the platform are generally developed in this sedimentary period. HST is mainly composed of tidal flat in restricted platform (Fig. 108).

6.2.2 Characteristics of Sedimentary and Logging Facies in Coring Wells

Well Mo 27 was cored 5 times in Lei3 Member, with a depth of 2425–2360 m (Fig. 109). The bottom is composed of dark-gray mudstone and grayish-brown bioclastic sandstone, the middle-lower part is composed of gray micrite limestone with dark-gray mudstone and gypsum-bearing mudstone, mixed with gray sandy limestone and sandy dolomite, and the upper part is grayish-brown sandy dolomite. It shows as the change from open platform to restricted platform, and is mainly deposited as the subfacies of intra-platform beach and inter-beach sea. Well Mo 29 was cored 6 times in Lei1 Member, with a depth of 2835–2785 m (Fig. 110). The lower part is composed of gray gypsum dolomite, the middle-lower part is composed of gray gypsum, dark-gray gypsum dolomite and mudstone, and the middle-upper part is composed of gray sandy dolomite and gray-brown thin-layered micrite limestone, containing gypsum dolomite. The upper part is composed of gray gypsum. The sedimentary environment is developed as the restricted platform, including intra-platform beach, inter-beach sea and lagoon. The natural gamma-ray logging is characterized as large amplitude variation and strong dentition. Among them, gypsum-salt rock is identified as low-amplitude toothed box type, mudstone is identified as thin-layered high-amplitude dentition, dolomite and limestone are identified as medium-amplitude toothed funnel shape or finger shape.

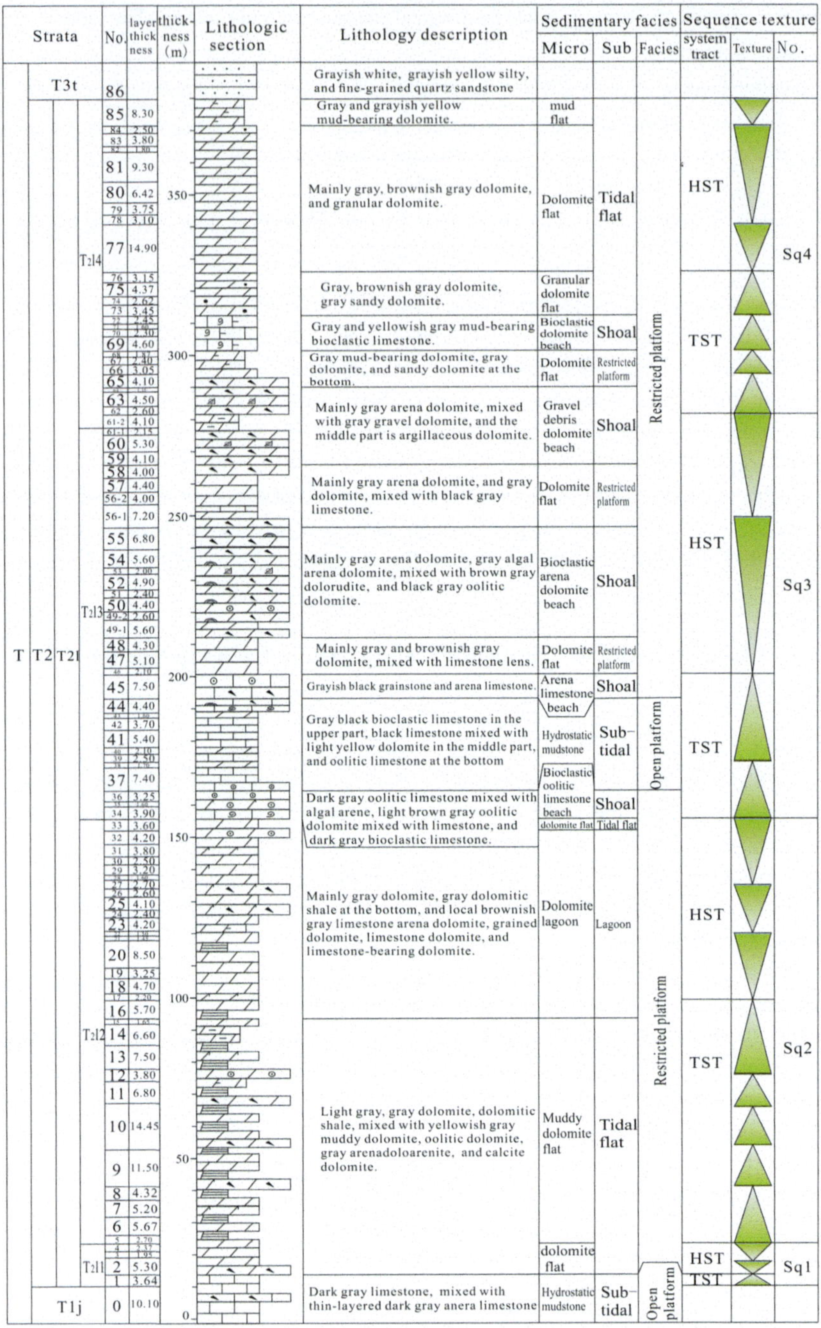

Fig. 108 Sedimentary characteristics of Leikoupo Formation in Middle Triassic of Lujiaba profile, Wangcang

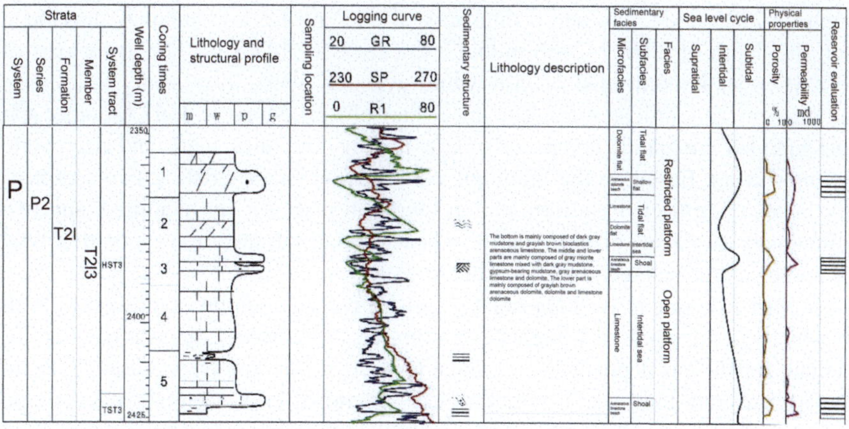

Fig. 109 Sedimentary characteristics of cores in Leikoupo Formation of Well Mo 27

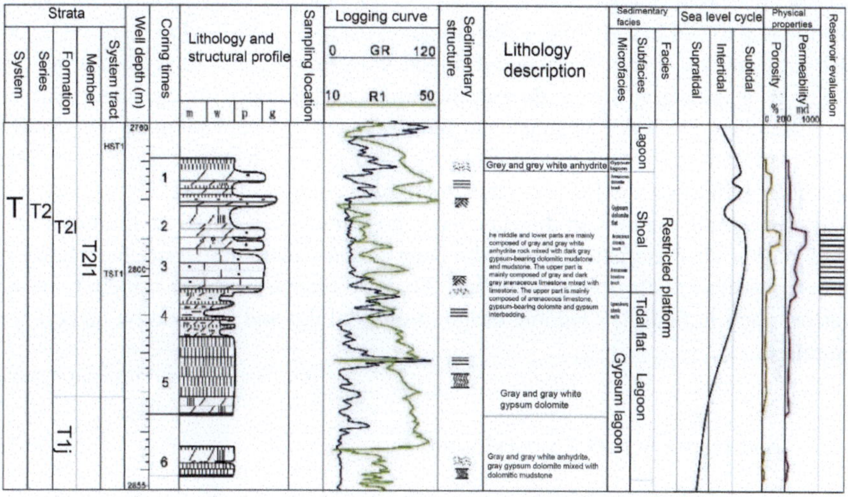

Fig. 110 Sedimentary characteristics of cores in Leikoupo Formation of Well Mo 29

6.3 Sedimentary Facies Analysis

6.3.1 Sedimentary Facies in Single Wells

For example, Well Moxi 7 is located in Moxi structure of gentle structural belt in Central Sichuan Basin, which is deposited as a typical restricted platform-evaporation platform. There is no Lei4 Member in the well, only Lei3, Lei2 and Lei1 Member are revealed. Lei3 Member of Leikoupo Formation is developed with the depth of 2228.6–2275 m and a thickness of 46.4 m. Lei2 Member is developed with the depth of 2275–2338.7 m and a thickness of 63.7 m. Lei1 Member is developed with the depth of 2338.7–2486 m and a thickness of 147.3 m. The lower part of Lei1 Member is composed of gray-white gypsum layers, gray sandy limestones and gray micritic doloarenites. The upper part of Lei1 Member, Lei2 Member and Lei3 Member are mainly composed of gray muddy dolomites, marls and gray-white gypsum layers. Based on the lithologic combination characteristics, two sedimentary facies of restricted platform and evaporation platform are identified, and five sedimentary subfacies of intra-platform beach, tidal flat, lagoon, evaporation tidal flat and evaporation lagoon are identified.

Another example is Well Gaoshi 112, which is located in the southwest of Well Gaoshi 16 in the south of Longnvsi structure on the Leshan-Longnvsi paleouplift, Sichuan Basin. It is deposited as a typical restricted platform-evaporation platform. The well reveals Lei4, Lei3, Lei2 and Lei1 Member. Lei4 Member of Leikoupo Formation is developed with the depth of 2442–2447 m and a thickness of 5 m. Lei3 Member is developed with the depth of 2447–2660 m and a thickness of 213 m. Lei2 Member is developed with the depth of 2660–2718 m and a thickness of 58 m. Lei1 Member is developed with the depth of 2718–2907 m and a thickness of 189 m. The lower part of Lei1 Member is mainly composed of grayish-white gypsum layer, dark-gray muddy dolomite and marl mixed with thin-layered sandy limestone. A large set of dark-gray sandy micritic dolomite is developed in the upper part of Lei1 Member. The Lei2 Member and the lower part of Lei3 Member are mainly developed with gray muddy dolomite, marl and grayish-white gypsum layer. The Upper Lei3 Member and Lei4 Member are mainly composed of a set of gray micritic sandy limestone mixed with thin-layered sandy limestone. Based on the lithologic combination characteristics, two sedimentary facies including restricted platform and evaporation platform are identified. Five sedimentary subfacies include intra-platform beach, tidal flat, lagoon, evaporation tidal flat and evaporation lagoon are identified.

6.3.2 Sedimentary Facies Correlation

Based on the correlation profile of sedimentary facies across Well Gaoshi 21–Hetan 7–Lai 1–Hualai 1–Huaxi 1–Guangtan 1, the macro characteristics of sedimentary facies in Leikoupo Formation are dissected. It can be seen that the stratigraphic profile is developed with the thickness of 271.5–482.2 m, and two complete third-order sequences. And it is deposited as restricted platform. The intra-platform beach is mainly developed in the bottom of the first three-order sequence and the top of the second third-order sequence, and is widely developed in the Lower Lei1 Member, Upper Lei3 Member and Lei4 Member. No beach body is developed in Lei2 Member and the lower part of Lei3 Member. The high-energy beach of the wells in the profile is developed with the single-well thickness of 1.05–7.4 m, an average thickness of 3.6 m, single-well cumulative thickness of 11.8–45.26 m, and an average thickness of 24.4 m. The intra-platform beach in Well Huaxi 1 is developed with the largest single-layer thickness, in Well Gaoshi 21 it is developed with the smallest single-layer thickness. And the intra-platform beach in Well Guangtan 1 is developed with the largest cumulative thickness, in Well Hetan 1 it is developed with the smallest cumulative thickness.

Based on the analysis and comparison of sedimentary facies in single wells, the basic characteristics of high-energy platform are counted as follows. The single-well beach body is developed with the average thickness of 27.4 m, the maximum thickness of 48.2 m and the minimum thickness of 11.8 m. The beach body is developed with an average thickness of 2.62 m, a maximum thickness of 6.4 m and a minimum thickness of 0.4 m.

6.4 Seismic Facies

The Lei1 Member of Leikoupo Formation in Well Tongtan 1 and Gaoshi 112 shows that the reservoirs in beach facies of Leikoupo Formation are favorable reservoirs. Among them, the gas logging of Layers 16 and 17 in the middle part of Lei1 Member of Well Tongtan 1 shows good porosity of about 3.0–6.0%, and an average of about 4.5%. It also shows a certain muddy content, about 10–30%. Layer 16 is developed with the depth of 2591.8–2597.9 m, and the average resistivity of greater than 100 Ω. Combined with the curve characteristics and chart, Layer 6 can be interpreted as gas and water in the same layer. Layer 17 is developed with relatively low resistivity, which can be interpreted as gas-bearing layer. Layers 18 and 19 are developed with low resistivity and negative difference characteristics. Combined with the curve characteristics and chart, they are interpreted as water layers.

The reservoir of granular beach on the seismic profile shows an obvious increase of thickness, and chaotic reflection with low frequency and weak amplitude (Fig. 111).

The bottom of Leikoupo Formation and $Lei1^1$ Submember are calibrated by logging and seismic data in the research area. It can be seen from the seismic profile of Well Moxi 107–Gaoshi 16–Gaoshi 113 (Fig. 112) that the seismic events of the bottom boundary of Leikoupo Formation and the top boundary of $Lei1^1$ Submember can be tracked continuously, which are convenient for tracking and interpretation in the whole region.

The thickness of $Lei1^1$ Submember of Leikoupo Formation is 50–90 m, among which the thickness of Well Tong 4–Hechuan 12–Lai 1–Guangtan 2–Guangcan 2 is relatively large (Fig. 113). According to the error statistics between the formation thickness of 39 wells in $Lei1^1$ Submember of Leikoupo Formation and the drilling formation thickness, the errors less than 5 m account for 100%.

Based on the 2D and 3D seismic data volumes in the study area, the development section of granular beach in $Lei1^1$ Submember is compared and analyzed by the attribute extraction, such as average reflection intensity, average instantaneous phase, average amplitude, average instantaneous frequency, RMS amplitude and average absolute amplitude after determining the time period corresponding to the reservoir

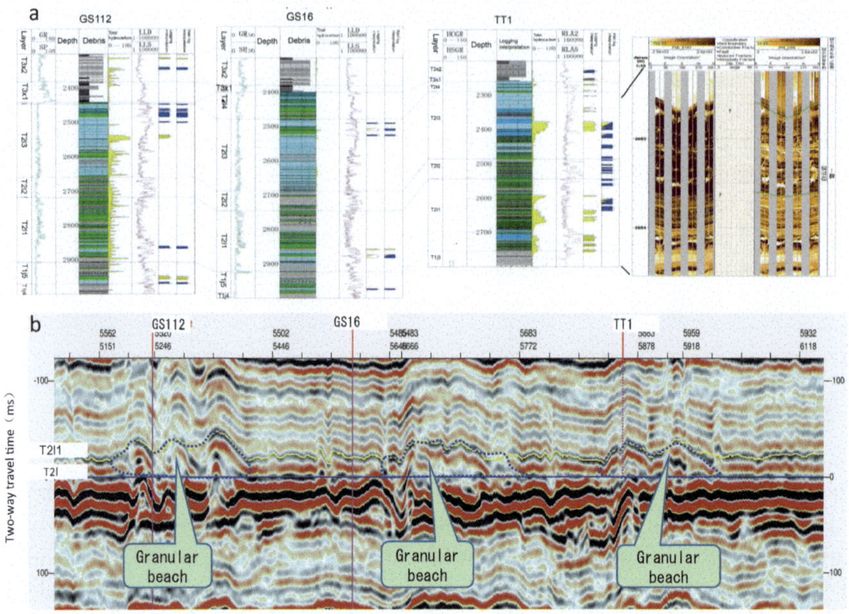

Fig. 111 Characteristics of granular beach reservoir in Lei1[1] Submember of Well Guogaoshi 112–Gaoshi 16–Tongtan 1

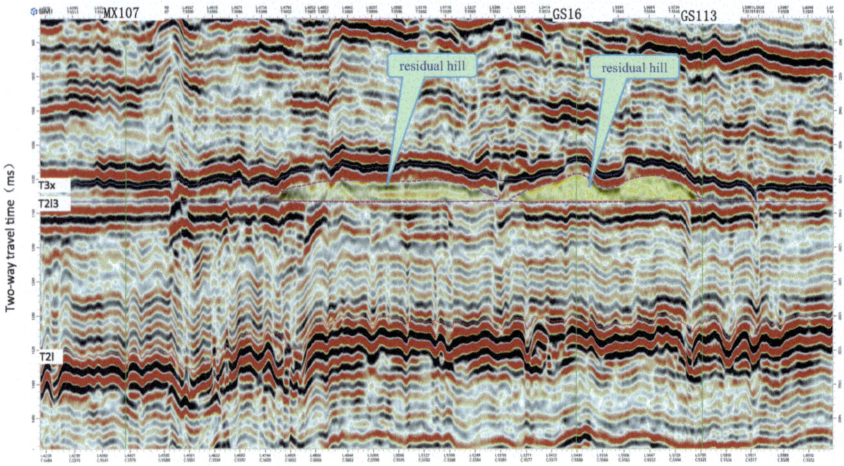

Fig. 112 Seismic profile across Well Gaoshi 16–Tongtan 1

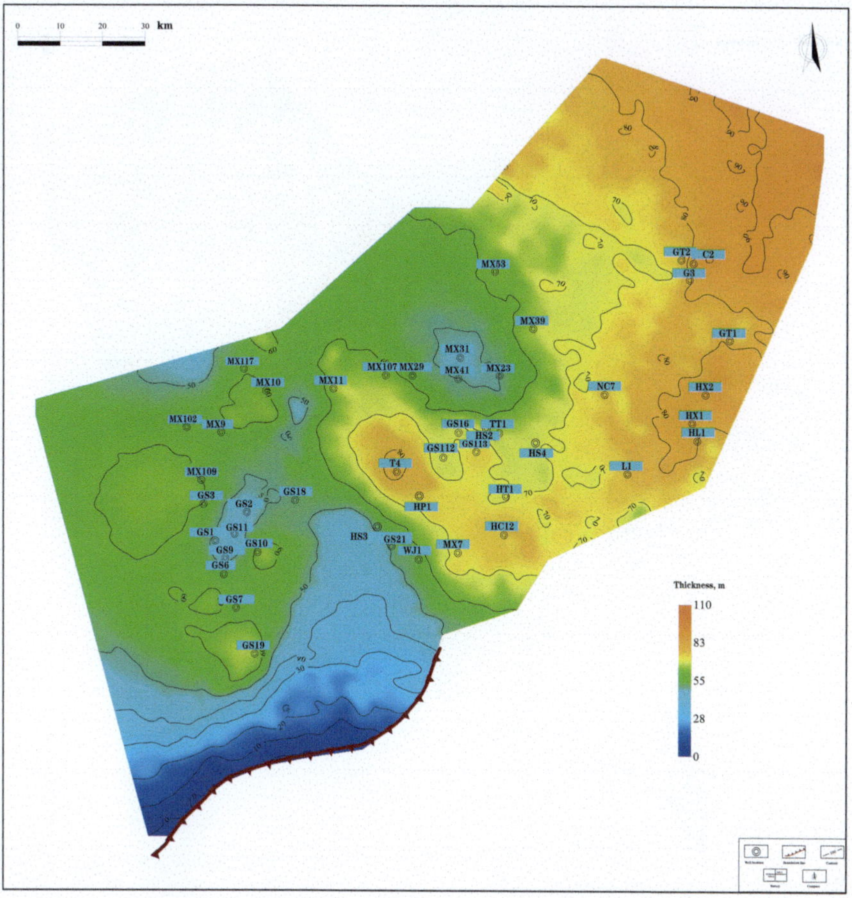

Fig. 113 Stratum thickness map of Lei1[1] Submember of Leikoupo Formation

section on the time section. It is considered that the reservoirs in granular beach of Lei1[1] Submember can be exactly descripted by f RMS amplitude (Fig. 114).

According to the analysis results of seismic characteristics of granular beach reservoir in Lei1[1] Submember, the facies belts for good reservoirs are further divided by the use of automatic seismic facies division technology. Based on 2D and 3D seismic data, waveform features are extracted in the time windows from the top of Lei1[1] Submember to the bottom of Leikoupo Formation. Various waveforms are classified according to the similarity and processed by Manhattan method. They are divided into seven types. Each color represents one type of waveform. The similarity of two adjacent waveforms is better than that of non-adjacent types. The analysis results are shown in Fig. 115. The seismic facies of granular beach are characterized by weak amplitude and weak continuity. Chaotic reflection represents the obvious increase of thickness, which is the warm-color area in the map. The seismic facies of

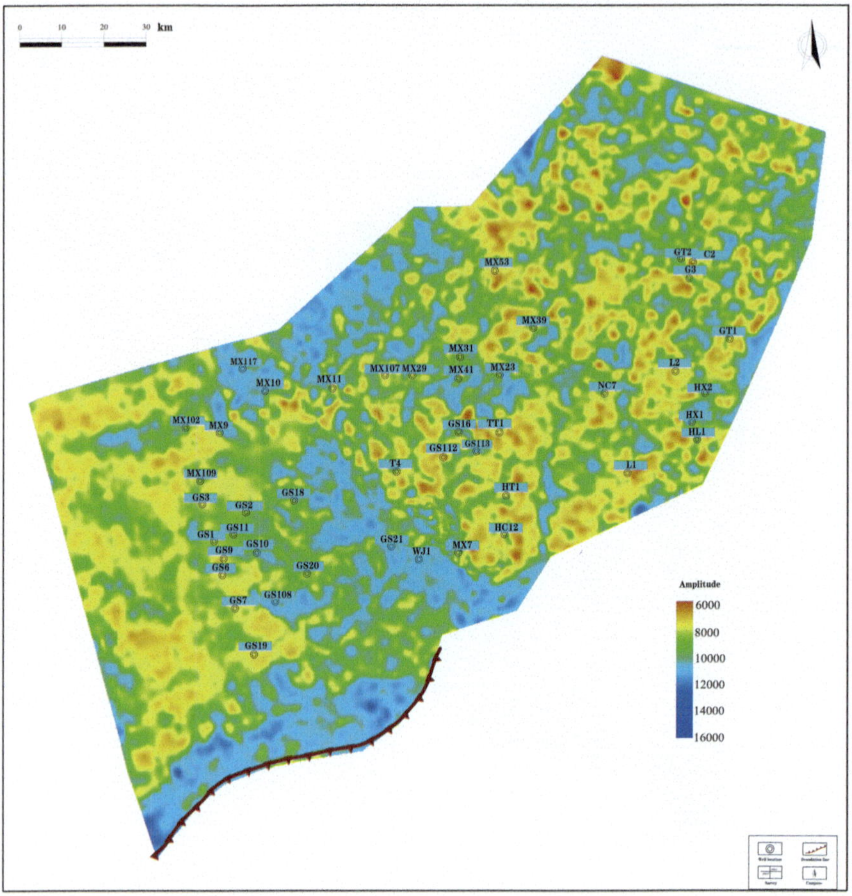

Fig. 114 Amplitude attribute map of Lei1[1] Submember of Leikoupo Formation in Central Sichuan

restricted platform are characterized by medium amplitude, medium continuity and parallel reflection, which are located in the cold-color area in the map. According to the analysis results of seismic facies, the development range of granular beach reservoir in Lei1[1] Submember can be well divided.

6.5 Lithofacies Paleogeographic Characteristics

The lithofacies paleogeography map of Lei1[1] Submember is drawn based on the thickness variation of the Maokou Formation, together with sedimentary facies and seismic facies (Fig. 116).

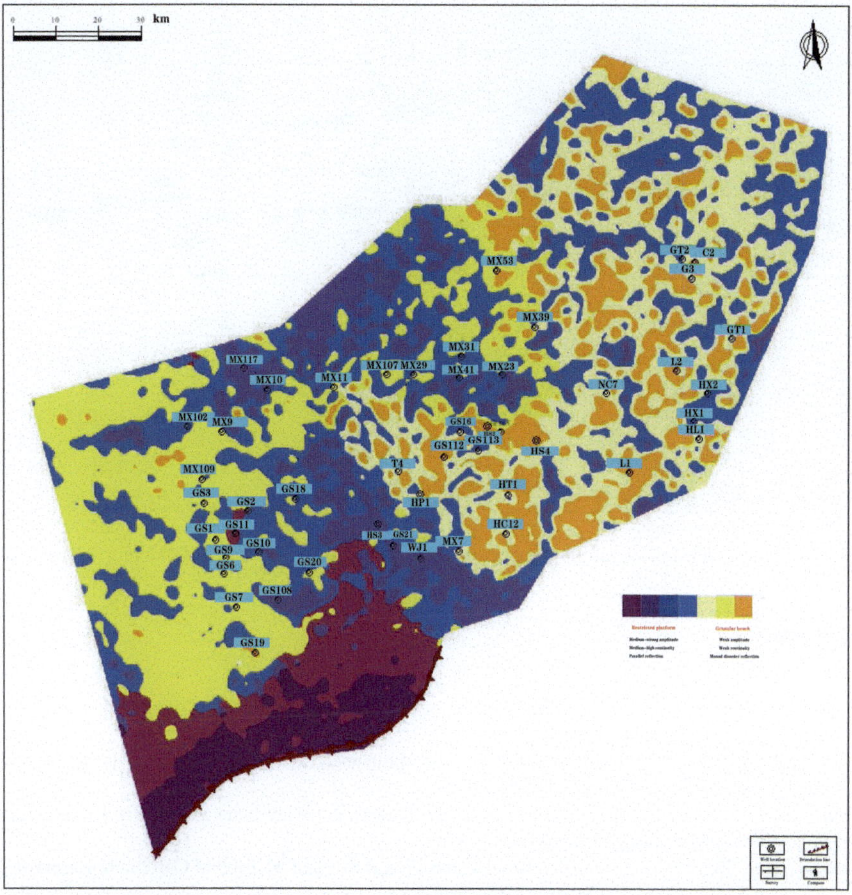

Fig. 115 Seismic facies of Lei1[1] Submember of Leikoupo Formation in Central Sichuan

It can be seen from the map that the overall Lei1[1] Submember is deposited as restricted platform, with the sedimentary trend of NW. The single beach body of granular beach is distributed with large scale, and mostly in isolation, which are mainly distributed along the NW direction.

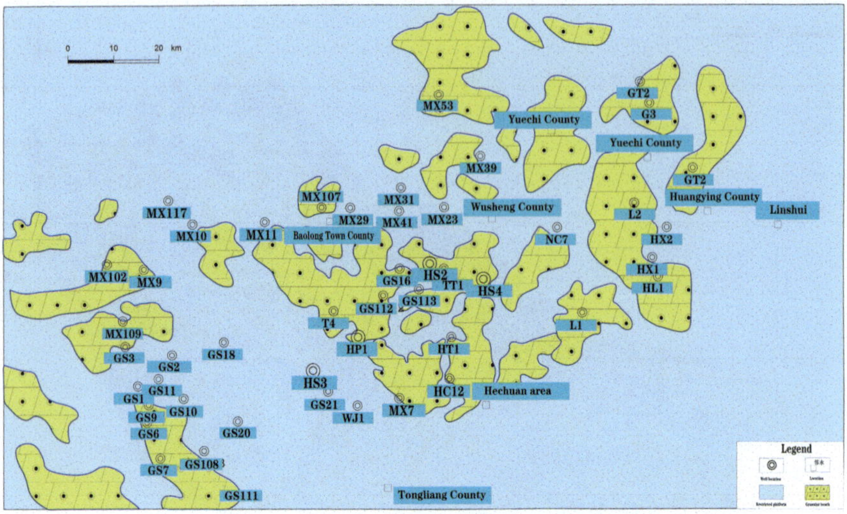

Fig. 116 Lithofacies paleogeography map of Lei1[1] Submember in Central Sichuan

References

Burne R V, Moore L S. 1987. Microbialites: organosedimentary deposits of benthic microbial communities [J]. Palaios, 2: 241–254.

Han Z Zh, Chen J T, Chi N J, et al. 2009. Microcarbonates:a review and perspecrives [J]. Marine Geology&Quatemary Geology, 29(4): 30–38.

Huang Zh Ch, Bian L Z. 1983. Discovery of the Ordovician Alga Nuia Siberica in China. Acta Palaeontologica Sinica, 22(1): 61–64.

Luo P, Wang, Sh, Li P W et al. 2013. Review and Prospectives Of M icrobial Carbonate Reservoirs [J]. Acta Sedimentologica Sinaca, 31(5): 807–823.

Mei M M. 2007. Discussion on advances of microbial carbonates from the terminological change of thrombolites [J]. Geological Science and Technology Information, 26(6): 1–9.

Riding R. 1991. Classification of microbial carbonates. In: Riding R. (Ed.), Calcareous Algae and Stromatolites [J]. Springer-Verlag, Berlin: 21–51.

Riding R. 2000. Microbial carbonates: the geological record of calcified bacterial-algal mats and biofilms [J]. Sedimentology, 47(Suppl. 1): 179–214.

Zhang B M, Shan X Q, Zhang J, et al. 2017. Geology and genesis of marine carbonate reservoirs in China [M]. Science Press, Beijing.

Chapter 4
Reservoir Characteristics and Distribution Prediction

1 Reservoir Characteristics and Distribution Prediction of the Sinian Dengying Formation

The results of drilling and research prove that the superimposition and transformation of weathering crust karstification on carbonate rocks of mound and beach is the main formation mechanism of high-quality reservoir in Dengying Formation. The related research is basically at a standstill as few wells are drilled in the Dengying Formation of the basin, mainly distributed in the Weiyuan structure. However, with the recent discovery of the Gaoshiti gas field in Sichuan Basin, and the data of boreholes and regional seismic data profiles added, it becomes possible to conduct a exhaustive study of the karst reservoirs of Dengying Formation in Sichuan Basin, and the distribution characteristics of favorable reservoirs in Dengying Formation can be predicted by the comprehensive application of paleogeomorphology, sedimentary facies, seismic facies and seismic inversion.

1.1 Genetic Types and Characteristics of Reservoirs

Based on the observation and analysis of cores in Well Gaoshi 16, Gaoshi 21, Gaoshi 1, Gaoshi 2 and Gaoshi 18 in Central Sichuan, the formation types of reservoir are defined with the summary of petrological characteristics and reservoir spaces.

1.1.1 Petrological Characteristics of Reservoirs in Deng4 Member

The effective reservoirs of Deng 4 Member in Central Sichuan are mostly concentrated in the Upper Deng 4 Member and the upper part of high-frequency sequences inside the Deng 4 Member. Through the cores' thin sections observation, the main

Q. Li and T. Wang, *Paleogeography and Gas Exploration of Marine Multiple Formations in Central Sichuan Basin*, https://doi.org/10.1007/978-981-99-2528-5_4

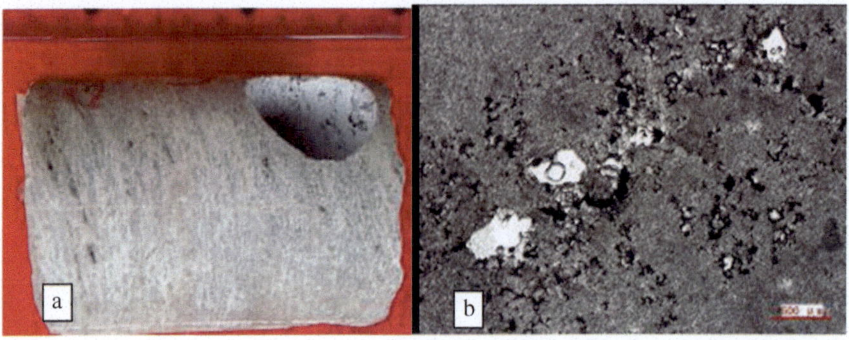

Fig. 1 Petrological characteristics of reservoirs in Deng 4 Member (Well Gaoshi 18, transition belt). **a** Algal granular dolomite, with a large number of needle-like dissolution pores and a small number of dissolution pores, Well Gaoshi 18, 5135.8 m. **b** Algal granular dolomite, with dissolution pores (dark areas), ordinary thin section, single polarization, Well Gaoshi 18, 5135.8 m

petrological characteristics of the reservoir can be defined as follows (Figs. 1, 2 and 3). (1) Microbial mound. Microbial mounds are usually composed of algal bindstones, and the reservoir pores are mostly developed in thrombolite dolomites and algal framework dolomites. (2) Granular beach. Granular beaches are usually composed of algal-bonding structures with different degrees of algal bonding, and the reservoir spaces are mostly developed in algal dolomites and sandy dolomites. In addition, a small amount of dissolution pore is also developed in a small amount of muddy dolomite, micritic silty dolomite and breccia dolomite.

1.1.2 Reservoir Spaces of Deng 4 Member

The reservoir pores in Deng 4 Member are developed with complex genesis and different types. Through the analysis of macro cores, conventional thin sections and cast thin sections, it is found that the reservoirs of Deng 4 Member are mainly composed of secondary pores, caves and fractures, mainly including the following types.

Pore

Intergranular Dissolved Pore

Intergranular dissolved pores are pores formed as a result of dissolution expansion or strong dissolution of intergranular dolomite cements based on residual intergranular pores. Intergranular dissolved pores in the study area are mainly developed in granular dolomites and algal dolomites, mainly in the size of 0.5–2 mm, locally associated with small karst caves, with the face rate of about 2–5%, mostly half-filled by asphalt, with a pore diameter of about 0.05–1 mm (Fig. 4). Algal debris, algal laminae and

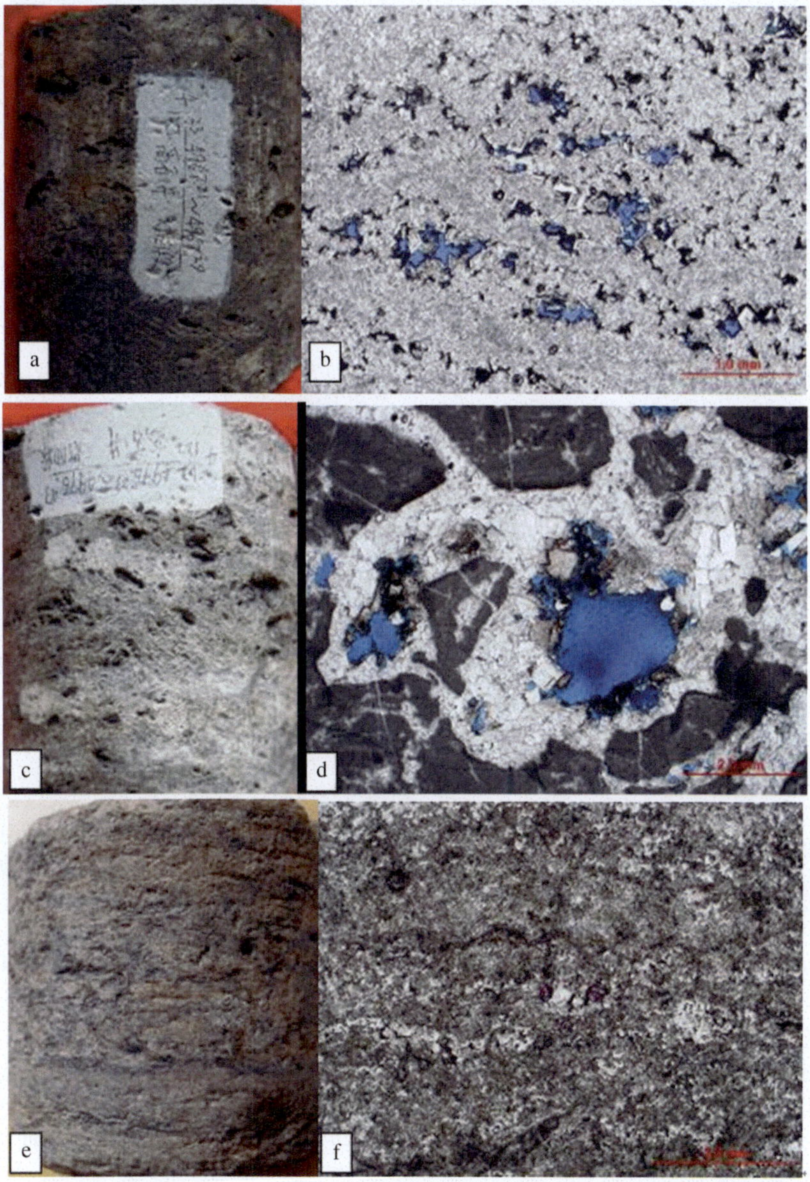

Fig. 2 Petrological characteristics of reservoirs in Deng4 Member (platform margin). **a** Algal-framework dolomite with irregular algal lamina, and needle-like dissolution pores in the algal lamina. Dissolved enlarged pores are developed in the framework pores (bird-eye), Well Gaoshi 1, 4967.3 m. **b** Algal laminar dolomite, with dissolved pores in the laminar, Well Gaoshi 1, 4767.3 m. **c** Thrombolite dolomite, with a large number of dissolved pores, Well Gaoshi 1, 4975.5 m. **d** Thrombolite dolomite, with dissolution pores in thrombolites, sparry cementation between particles, Well Gaoshi 1, 4975.5 m. **e** Algal laminar dolomite, a small number of dissolution pores, Well Gaoshi 2, 4014.25 m. **f** Algal laminar dolomite, with a small number of dissolution pores, Well Gaoshi 2, 4014.25 m

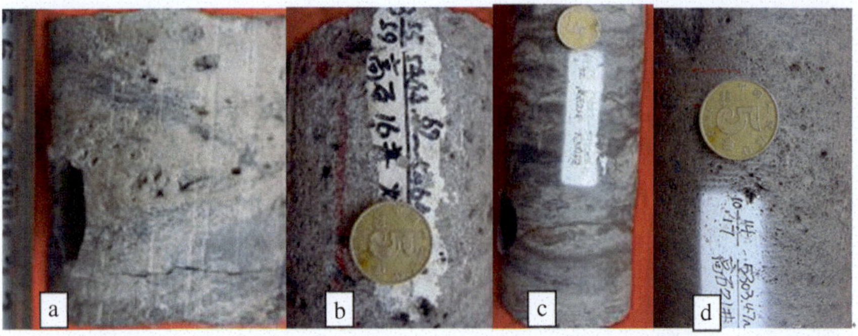

Fig. 3 Petrological characteristics of reservoirs in Deng4 Member (Well Gaoshi 16 and Gaoshi 21, intra-platform). **a** Algal-framework dolomite, Well Gaoshi 16, 5482.9 m, Deng 4 Member, with dense and tangled algal laminae, wave-resistant structures and framework dissolution pores. **b** Thrombolite dolomite, with many dissolution pores, Well Gaoshi 16, 5464.7 m. **c** Algal laminated dolomite, with a small number of dissolution pores, Well Gaoshi 21, 5322.8 m. **d** Algal granular dolomite, with many needle-like dissolution pores, 5303.5 m

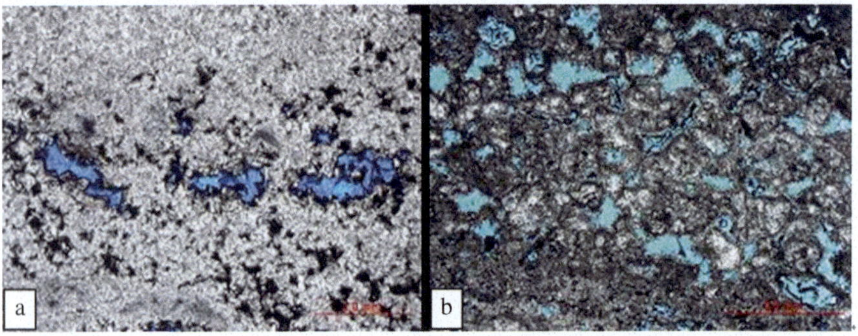

Fig. 4 Characteristics of intergranular dissolution pores. **a** Algal granular dolomite with enlarged intergranular dissolution pores filling by dark asphalt, Well Gaoshi 1, 4959.4 m. **b** Algal-laminated dolomite, with many dissolution pores between and inside algal bodies, Well Gaoshi 2, 5393.65 m

related residual structures are commonly developed in dolomites in Central Sichuan, indicating the uncompleted dolomitization; however traces of algae and algal particles affecting pore formation can be seen, which further supports the idea that the sedimentary environment can affect the formation of dissolution pores.

Intercrystal Dissolved Pore

Intercrystal pores are mostly readjusted by the fine intercrystal pores in the process of recrystallization from microcrystalline dolomite into coarse-grained dolomite. The pores are generally developed with large diameter of mostly 0.1–2 mm. They are developed between euhedral and semi-euhedral dolomite crystals, showing irregular polygon, irregular distribution and poor connectivity (Fig. 5). Intercrystal dissolved

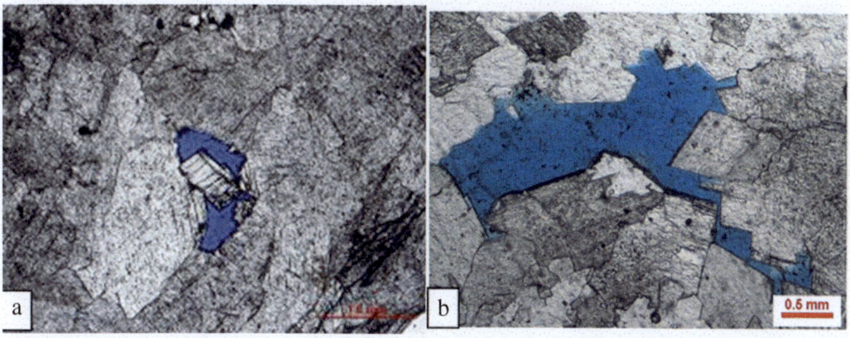

Fig. 5 Characteristics of intercrystalline dissolution pores. **a** Coarse-grained dolomite with intercrystalline dissolution pores. The dolomite crystals, large in size, are of saddle dolomites, which may be of hydrothermal origin. Well Gaoshi 1, 4958.6 m. **b** Coarse-grained dolomite, with intercyrstal dissolution pores. The pores may be formed in such a way that the existing dissolution pores are filled with coarse-grained dolomites, form residual intergranular dissolution pores. Well Gaoshi 101, 5505.9 m

pores are primarily developed in coarse-grained dolomites that may be related to burial or hydrothermal dolomitization. The constructive and destructive effects of intercrystal dissolved pores on the process of reservoir formation are still under exploration.

Dissolution Cave

Karst cave refers to various dissolution spaces with a diameter greater than 2 mm. The formation mechanism of the karst cave is similar to that of karst pores, but karst cave is developed with larger size and scale compared with karst pores. The main characteristics of karst caves in Deng 4 Member of the study area are shown as follows (Fig. 6). (1) The karst caves are developed in algal lamina, algal lamination and algal-framework dolomites, generally along algal lamina, and the long axis is consistent with the direction of structural fractures. (2) The karst caves are also developed in thrombolite dolomites, clot dolomites and micritic dolomites. (3) The karst caves are less developed in algal dolomites. (4) The karst caves are constantly filled with dolomite, quartz and other minerals.

Fracture

Carbonate rock is brittle and is easy to fracture, and the development of fracture is a common geological phenomenon. Fractures in carbonate reservoirs are developed as not only reservoir pores, but also important infiltration channels. According to the origin of the fractures, the fractures in Deng4 Member of the study area can be divided into structural fracture, dissolution fracture, and pressure solution fracture.

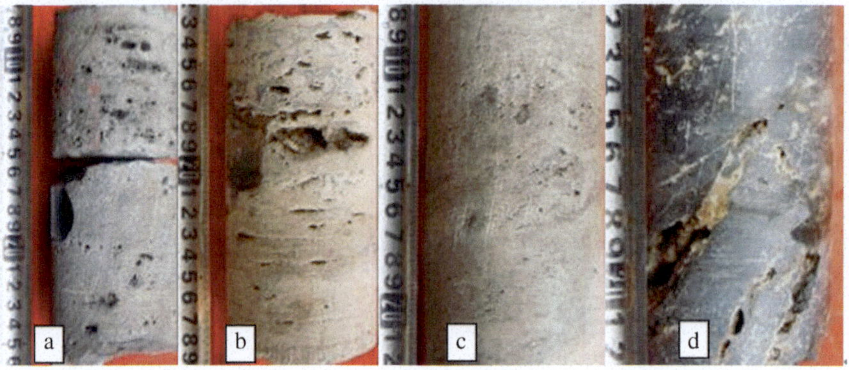

Fig. 6 Core characteristics of dissolution caves. **a** Algal lamina dolomite, with many dissolution caves along the lamina, Well Gaoshi 16, 5453.2 m. **b** Clot dolomite, with dissolution caves characterized by bedding development. The cave is filled with dolomites. Well Gaoshi 21, 5248.8 m. **c** Algal granular dolomite, mainly composed of dissolution pores, with a small number of dissolution caves, Well Gaoshi 21, 5303.6 m. **d** Mecritic dolomite, with dissolution cave developed along fractures, filled with dolomites, Well Moxi 105, 5323.9 m

Structural Fracture

Structural fracture refers to the fracture formed as a result of the fracturing of rock when the structural stress exceeds the elastic limit of rock under the action of tectonic stress. Based on the occurrence angle between the fracture and the stratum, the structural fractures of Deng4 Member of Dengying Formation in the study area are mostly high-angle fractures and vertical fractures, unfilled or filled with asphalt and dolomite (Fig. 7).

Dissolution Fracture

Dissolution fracture refers to the fracture filled or half-filled by dolomite, organic matter or pyrite with irregular shape and multiple-stage development, and is formed by the dissolution or transformation of acid diagenetic fluid. The dissolution fracture is developed with uneven fracture walls and an obvious dissolution phenomenon. Network dissolution fractures filled with asphalt and dolomite can be seen in Deng4 Member of the study area. There are pores and caves formed by expanding dissolution along the fractures (Fig. 8). The causes of dissolution fractures are complex. Firstly, the causes of fractures are complex, which can be formed by structural fractures, karst fracture of weathering crust, cave collapse or compaction. Secondly, the dissolution is complex, and might be associated with atmospheric freshwater leaching or deep-buried dissolution.

Fig. 7 Characteristics of structural fractures. **a** Algal laminar dolomite, with high-angle fractures, Well Gaoshi 18, 5138.9 m. **b** Algal laminar dolomite, with high-angle fractures, Well Gaoshi 16, 5468.9 m

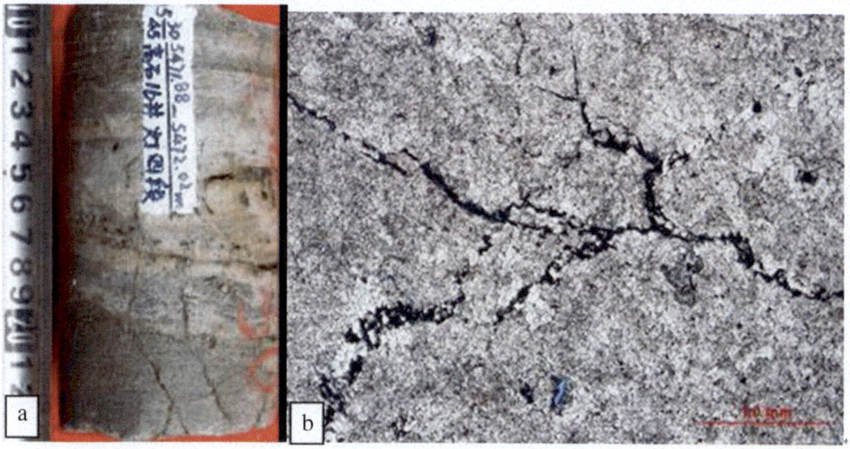

Fig. 8 Characteristics of dissolution fractures. **a** Laminar dolomite, with symbiosis of irregular dissolution fractures and caves, forming fracture-cavity bodies filled with asphalt, Well Gaoshi 16, 5471.9 m. **b** Residual algal fine-grained dolomite, with irregular dissolution fractures filling by asphalt, Well Gaoshi 1, 4956.3 m

1.1.3 Formation Characteristics of Reservoir

During a long and complex diagenetic evolution process, complex reservoir spaces including pores, caves, fractures and complexes are formed in Dengying Formation of ancient Sinian in a complex sedimentary environment. As a result, it is self-evident that the formation types of the reservoir are complex. Based on description of the differences of reservoir types in cores and thin sections and the analysis of sedimentary and reservoir characteristics, the formation types of reservoirs in Deng 4 Member can be divided into the following three: early karst reservoir of mound-beach facies, late supergene karst reservoir superimposed by early karst, and reservoir superimposed and transformed by other factors.

Early Karst Reservoir of Mound-Beach Facies

Basic Features

This kind of reservoir is controlled by sedimentation and is mainly developed in mound beach facies. It is mainly composed of pinholes, small cave, intercrystal dissolution pores and bedding dissolution pores (caves). The development of dissolution pores and caves is closely related to sedimentary laminae. The karst cave has a diameter of 5 mm. Bedding pinholes are widely developed in granular dolomite, and bedding pores are also widely developed in algal dolomite. Under the microscope, the dissolved pores between particles filled with dolomite and asphalt can be clearly observed, as well as the bedding pores in algal dolomite.

The Dissolution Pores of Algal-Framework Dolomite, Algal Thrombolite Dolomite and Algal Debris Dolomite Related to Microbial Mound Are Highly Developed

From the relationship between lithology and the development of reservoir space, it can be seen that the dissolution pores are well developed in algal-framework dolomite, algal-thrombolite dolomite, and algal-laminated dolomite; fairly well developed in in algal debris dolomite; while poorly developed in micritic silty/brecciated dolomite. From the statistical analysis of the development of effective reservoirs of different lithologies, it can be seen that the cumulative coring thickness of Well Gaoshi 16 and Gaoshi 21 is about 115 m, and the cumulative thickness of effective reservoir (according to the standard of whether there are obvious reservoir spaces identified by hand specimen) is 98 m. The proportion relationship of cumulative thickness in effective reservoir and total core thickness of different lithologies is established to form a statistical histogram in order to analyze the occurrence probability of effective reservoir of different lithologies. It can be seen that the effective reservoirs are mostly developed in algal-framework dolomite and algal thrombolite dolomite related to microbial mound, and the effective reservoir of algal debris dolomite related to algal granular beach is most developed. The development degree of effective reservoirs in algal laminar dolomite is low, so is the case with the pores in micritic dolomite and brecciated dolomite. Thus, it is certain that sedimentary facies unquestionably influence the formation of reservoirs, and the high-energy facies belts are the material basis for the high-quality reservoirs. At the same time, the formation of facies-controlled dissolution pores is generally related to sedimentation and quasi contemporaneous karstification.

Dissolution Pores Highly Developed in the Upper Part of High-Frequency Sequence (Parasequence Set)

Sea level changes generally control the high-frequency sequences (parasequences/parasequence sets) of marine carbonates. In the parasequences or parasequence sets, during the slow decline of sea level, the sedimentary water body becomes shallower and high-energy facies belts are deposited above the parasequences or parasequence set, which is conducive to the development of primary pores and the dissolution of quasi-syngenetic atmospheric water–rock, laying a material foundation for the formation of dissolution pores in the later stage. According to the core observation of Well Gaoshi 16 and Gaoshi 21 (Figs. 9 and 10), multiple high-frequency cycles related to parasequences and parasequence sets are clearly identified. According to the comprehensive histogram of cores in sedimentary reservoirs with parasequence set as a unit, it can be seen that, (1) the parasequence set is generally vertically represented by the lithologic combination of thin-layered micritic dolomite and thick-layered algal dolomite, that is, the microfacies combination of inter-mound sea and microbial mound or granular beach. And the natural gamma-ray has the trend of reverse rhyme, such as the parasequence sets corresponding to the 9th and 8th coring of Well Gaoshi 16 (Fig. 10). (2) The porosity of the parasequence sets tends to increase vertically, which can be clearly identified in the parasequence sets of the two wells, and the corresponding dissolution pores and caves are mainly developed in the upper part of the parasequence sets.

Dissolution Pores and Caves Are also Well Developed in Intra-Platform Microbial Mounds and Beaches

Gaoshiti-Moxi gas field is located in the platform margin, while the research area of Hechuan-Tongnan is located in the intra-platform area. The presence of well-developed mound-beach reservoirs inside the intra-platform region has long been a concern for exploration. Based on the sedimentary correlation profiles from platform margin to intra-platform area, it is clear that relatively isolated microbial mound-beach combinations are developed in the intra-platform area of Well Gaoshi 16 and Gaoshi 21, with a certain vertical cumulative thickness; dissolution pores and caves are mostly developed in algal frameworks and algal laminated dolomites; and needle-like dissolution pores are also developed in algal granular beaches. The scales of mound-beach body and mound-beach reservoir in intra-platform area are smaller than that in platform margin, but the thickness of mound-beach body in single wells

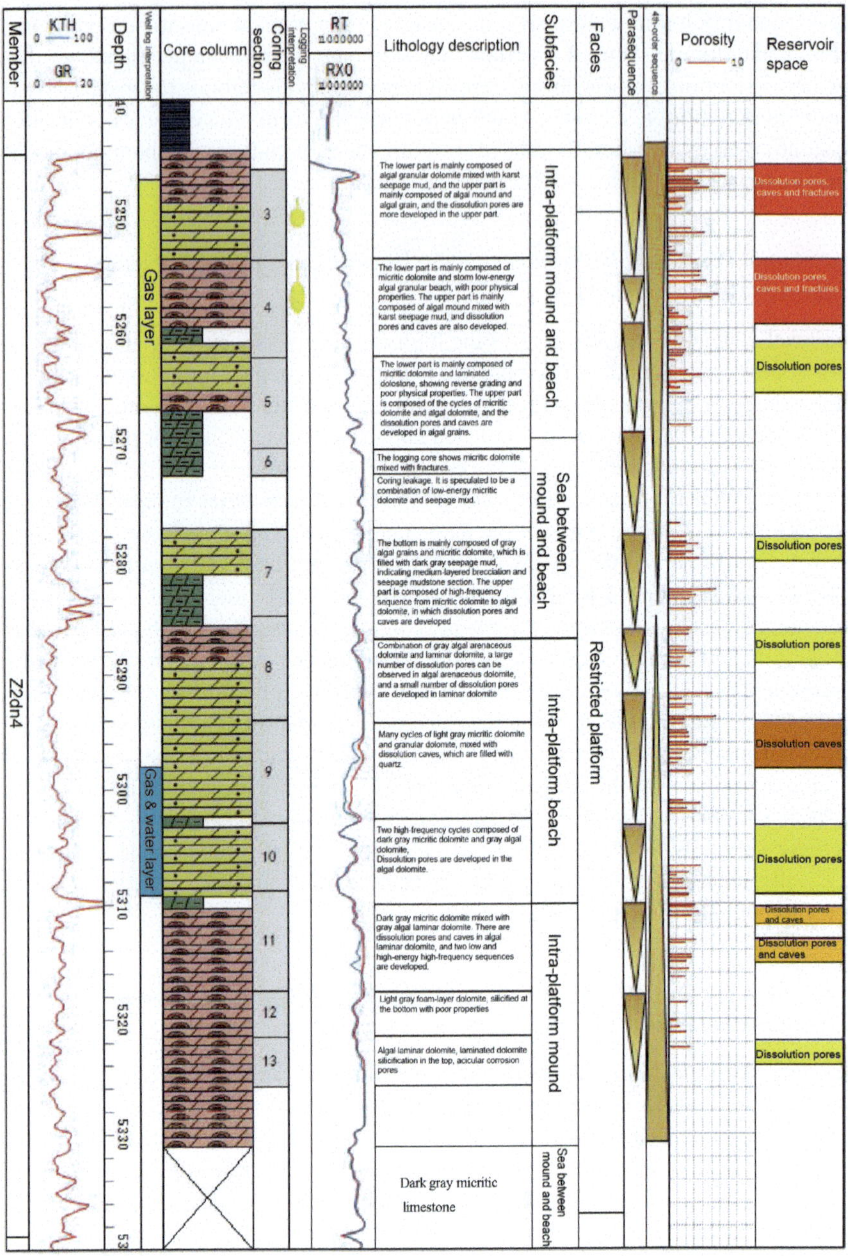

Fig. 9 Comprehensive histogram of reservoirs in cores of Well Gaoshi 21

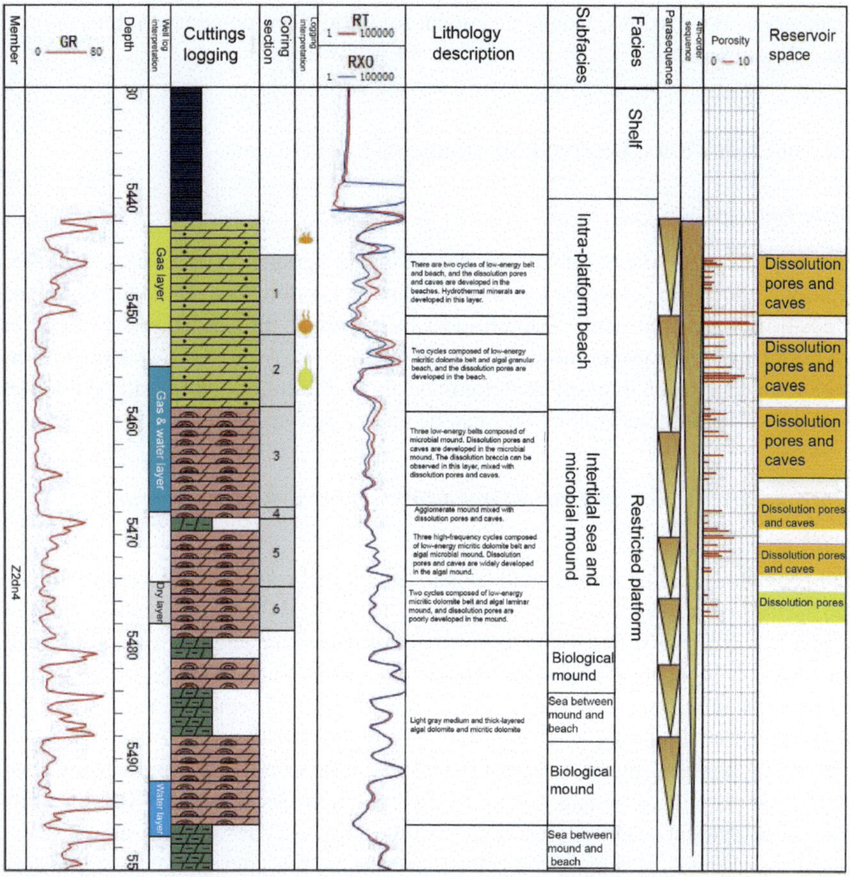

Fig. 10 Comprehensive histogram of reservoirs in cores of Well Gaoshi 16

of intra-platform area still accounts for more than 65%, and the thickness of reservoir accounts for 40%. Based on the formation mechanism, the possible reasons for this phenomenon are as follows. (1) There is little difference in sedimentary paleogeomorphology between the platform margin and the intra-platform area of Sinian, and the change of sea level mainly controls the sedimentary sequences. Under the same sea-level conditions, slight difference paleogeomorphology may lead to the difference in deposition rates between carbonate sediments. The micro paleogeomorphology at the platform margin is higher and the deposition rate is faster, which finally shows the characteristics of great prosperity and large thickness of mound-beach body. While the intra-platform mound-beach body is developed with relatively low deposition rate and small deposition scale. (2) The intra-platform area may be developed with heterogeneity, and there may be intra-platform depressions. The locations with relatively slightly higher paleogeomorphology may be developed in the periphery of depressions, which are conducive to the development of intra-platform mounds and beaches.

Late Supergene Karst Reservoir Superimposed by Early Karst

Basic Features

This kind of reservoir is mostly developed in the upper part of Deng 4 Member, usually within 40 m below the top unconformity. It is often associated with dissolved deposition breccias, irregular fractures and karst seepage fillings. The reservoir pores are dominated by honeycomb caves, pores, and karst cave system, and such reservoirs are widely seen in granular dolomites, algal dolomites, and micritic dolomites (Fig. 11). The diameter of the pores can be up to a few centimeters.

Superimposed Transformation of Karstification in Supergene Weathering Crust

The Tongwan movement uplifted the Sinain Dengying Formation at the end of Sinian, and the top layers were weathered, denuded and dissolved. The identification marks of weathering crust karstification in terms of petrological characteristics are as follows. (1) Ablation breccia related to paleokarst caves. (2) Karst seepage materials. (3) Irregular fractures and karst seepage materials filling the irregular fractures. Similar features can be found by observing the core characteristics in the Gaoshiti-Moxi area, which confirms the existence of paleokarst weathering. In general, paleoweathering crust karstification is a favorable superposition based on early karstification and has no evident selectivity for reservoir lithology. As a result, favorable dissolution fractures and caves can be developed in various lithologies. However, the core observation in the work area shows that the development degree and probability of dissolution fractures and caves in the biological mound are significantly higher than those in the low-energy sedimentary facies belt, indicating that the supergene weathering crust karst superposition based on early quasi-syngenetic karst is conducive to the formation of dissolution fractures and caves.

Reservoir Superimposed and Modifies by Other Factors

After a long process of burial and evolution, the diagenesis in Dengying Formation is very complex. The superposition factors affecting reservoir formation are mainly structure and hydrothermal factors, and can be summarized as follows: (1) reservoirs with pore-cave-fracture dominated by structure (fracture)-hydrothermal activity; (2) fractured reservoirs dominated by high-angle structural fracture system.

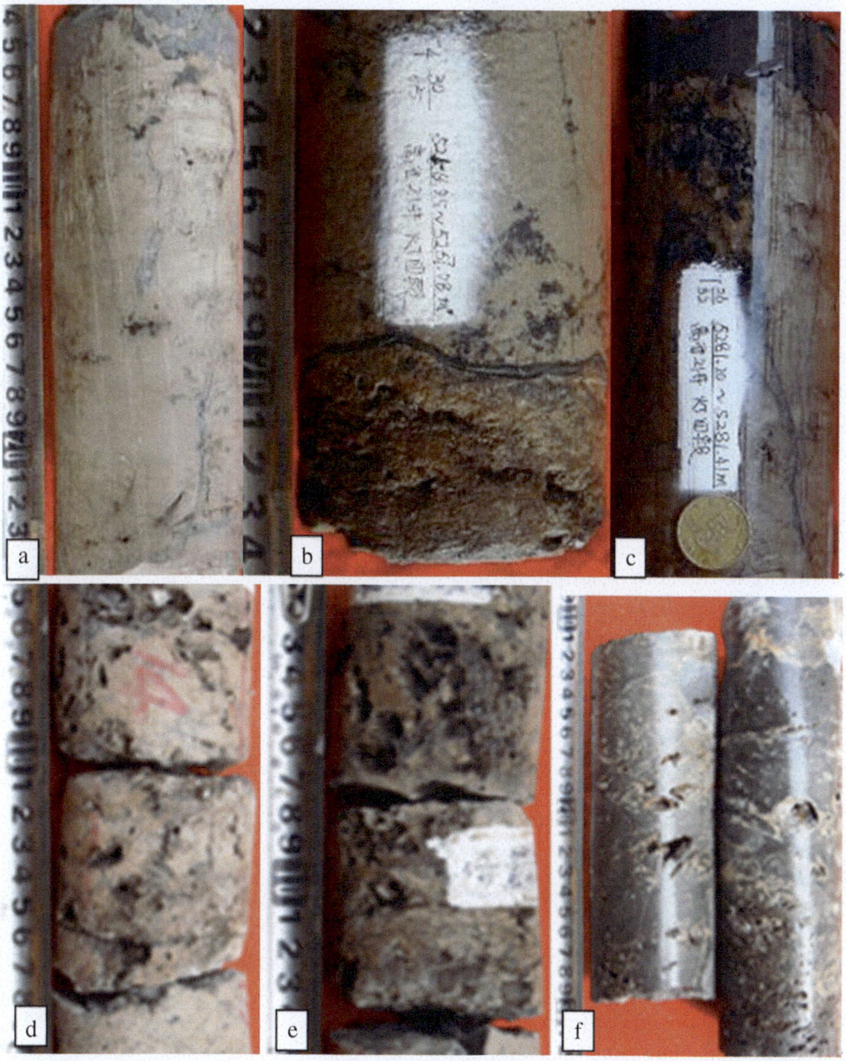

Fig. 11 Characteristics of supergene karst reservoir. **a** Clot dolomite, filled with dark-gray spoted karst seepage materials and irregular fractures filled with karst seepage materials, and a small number of irregular dissolution fractures and caves, Well Gaoshi 21, 5280.8 m. **b** Clot dolomite, filled with horizontal dark-gray karst seepage materials, and irregular dissolution caves under it. Well Gaoshi 21, 5257.1 m. **c** Clot dolomite, filled with dark-gray spoted karst seepage materials, irregular fractures filled by karst seepage materials, a small amount of irregular dissolution fractures and caves, Well Gaoshi 21,5281.4 m. **d** Algal granular dolomite, with irregular dissolution caves, Well Moxi 105, 5356.3 m. **e** Granular dolomite, with honeycomb karst caves, Well Moxi 105, 5356.3 m. **f** Micritic dolomite with karst caves, Well Moxi 105, 5307.7 m

Reservoir of Pore-Cave-Fracture Dominated by Structure (Fracture)-Hydrothermal Action

This reservoir can be mainly represented as fractures and related expanded dissolution fractures, dissolution pores and caves which are mostly filled with quartz, coarse-grained saddle dolomite and other minerals. Based on the output characteristics of the fracture-cave bodies of Dengying Formation, it can be seen that the structures and hydrothermal actions have positive transformation significance for the reservoir, but they may also have negative effects at the same time. For example, the filled hydrothermal minerals related to structural activities will lead to a small reservoir space. In ancient carbonate strata, structural fractures and hydrothermal actions in the burial stage commonly occur. The influence of these two actions on reservoir formation has always been a complex scientific problem. The study on these two actions involves many uncertain factors, such as action period, fluid properties and sources, diagenetic environment, etc.

Fractured Reservoir Dominated by High-Angle Structural Fracture System

According to the core description, high-angle structural fractures are mostly developed in Well Gaoshi 16 and Gaoshi 21, and most of the high-angle fractures are open. Dissolutions are developed in the fractures, mostly filled with asphalt. It shows that the existence of high-angle fractures has obvious positive significance for reservoir reconstruction.

1.2 Prediction of Reservoir Distribution

1.2.1 Petrophysical Characteristics of Reservoirs

The main high-quality gas-producing reservoirs of Dengying Formation in Central Sichuan are the reservoirs of pore-cave-fractures, which is characterized by logging at relatively low speed and low density. But further petrophysical analysis is needed as lithology change has great interference on reservoir logging response. Most gas tests in the study area are combined tests with different gas test thicknesses. According to the cross plot of porosity curves of dry layer-poor gas reservoir-gas reservoir and AC in Well Gaoshi 16, Gaoshi 21, Hetan 1, Moxi 23, Moxi 39, Moxi 107, Gaoshi 108 and Gaoshi 10 (Fig. 12), the reservoir porosity distribution range is greater than 3%, and most of the reservoirs with porosity less than 3% are tight layers. Therefore, the effective reservoir of Dengying Formation can be identified by porosity inversion.

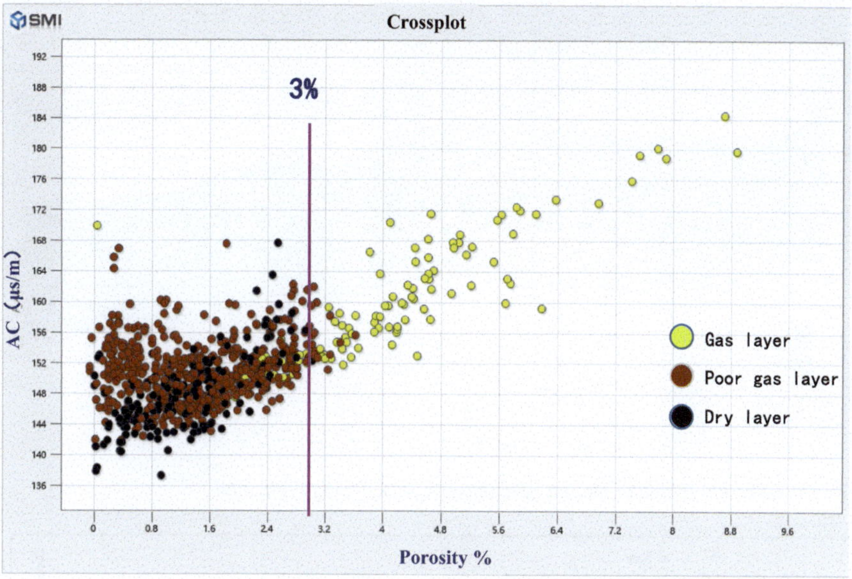

Fig. 12 Cross plot of AC curve and porosity curve in Dengying Formation

1.2.2 Reservoir Predicted by Waveform Indication Inversion

The main high-quality gas-producing reservoir in Dengying Formation is the reservoir of pore-cave-fracture. The petrophysical chart analysis shows that the reservoir is characterized by logging of relatively low speed and low density, which is the basis of wave impedance inversion to predict reservoir. In this study, the inversion method of seismic waveform indication is adopted, and 3D seismic data and logging curves are used to carry out the inversion. The reliability of the inversion results needs to be comprehensively considered in combination with other prediction methods as the mound-beach reservoirs in Dengying Formation are developed with great lateral change. Figure 13 is the inversion profile of the mound-beach bodies in Deng2 and Deng4 Member. It can be seen that the characteristics of lateral boundary of the mound-beach body can be clearly described by the inversion results, as a result, the porosity inversion results can better reflect the reservoir characteristics of mound-beach body.

By further extracting the average porosity and reservoir thickness of Deng2 and Deng4 Member, it can be seen that the distribution characteristics of mound-beach bodies in the Dengying Formation can be clearly described by waveform indication inversion. A 3D reservoir with an area of 559 km^2 of Deng 2 Member in Central Sichuan (Figs. 14 and 15) is depicted, as well as a 3D reservoir with an area of 643 km^2 in Deng 4 Member (Figs. 16 and 17).

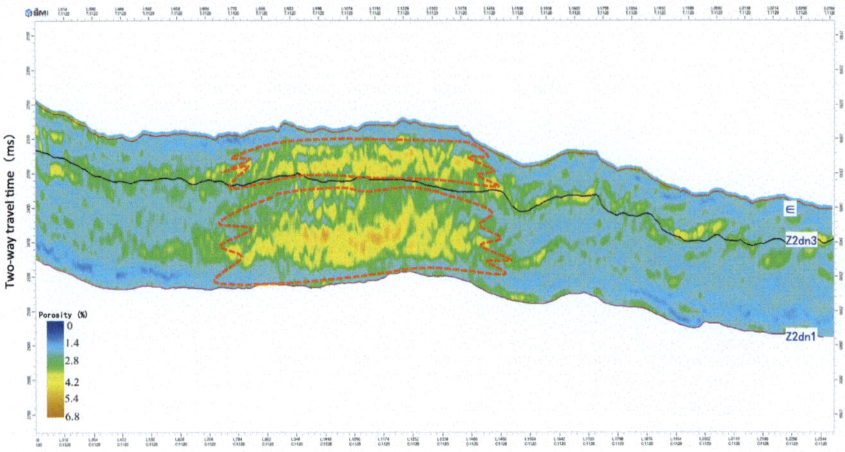

Fig. 13 Inversion profile of mound-beach bodies in Dengying Formation

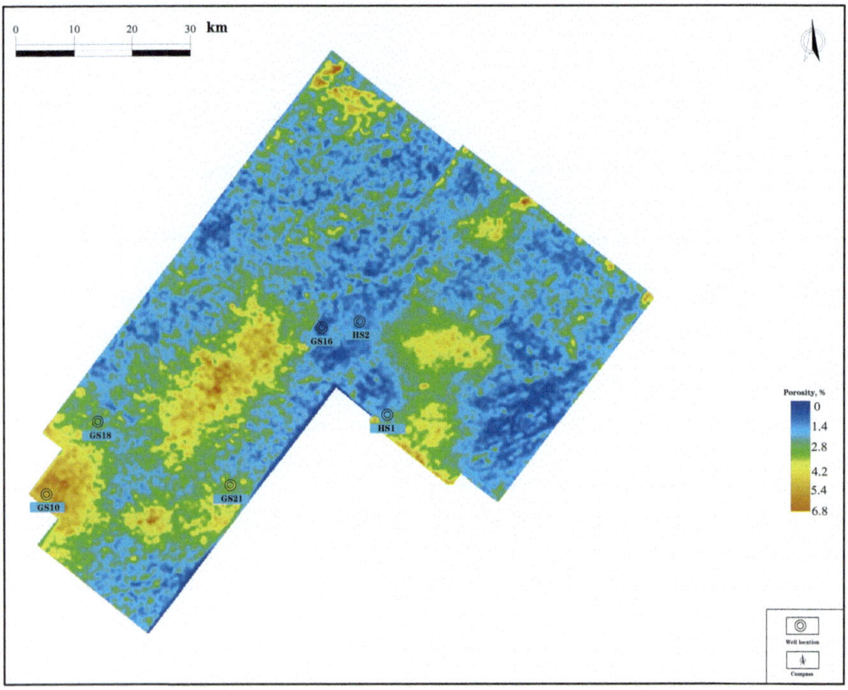

Fig. 14 Average porosity of Deng 2 Member in 3D area of Central Sichuan

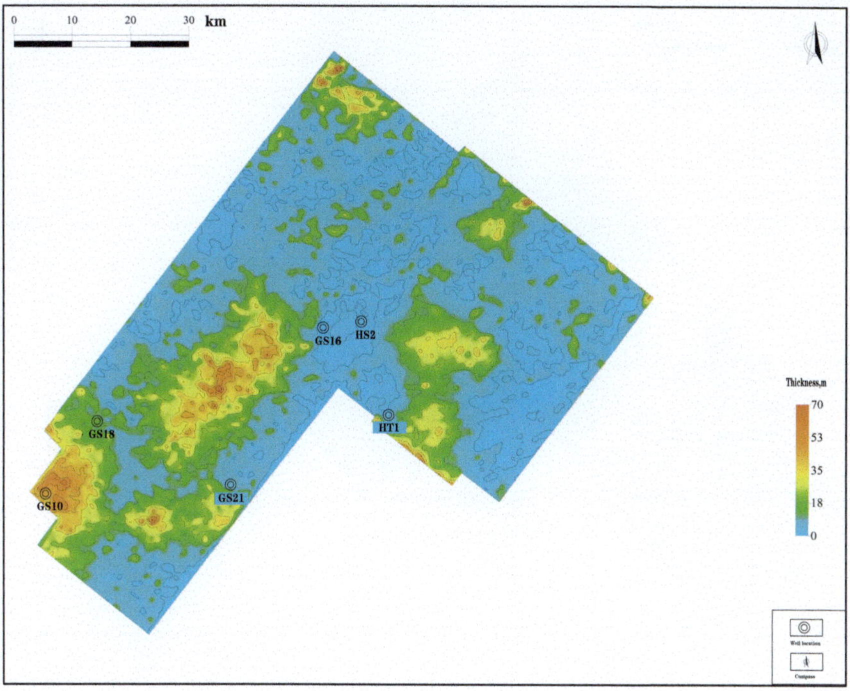

Fig. 15 Reservoir thickness of Deng 2 Member in 3D area of Central Sichuan

1.2.3 Oil and Gas Detection

In CM oil and gas detection, the characteristic information such as amplitude and frequency of seismic data can be analyzed by cumulative energy normal probability analysis. The attribute parameters such as cumulative energy of low-frequency and high-frequency bands and their ratios to the global energy can be calculated by low-frequency and high-frequency bands determined by sensitive frequency points.

The drilling frequency amplitude spectrum analysis in Well Moxi 129H shows the typical bimodal characteristics (Fig. 18), and the daily gas production of Well Moxi 129H is 1.41×10^6 m^3. The relative amplitude peak in low-frequency band is 0.6–0.8, which is distributed in low-frequency band within 15 Hz, and the drop from the sub-peak in the high-frequency band is very small. The high-frequency band is distributed at 20–50 Hz. Well Gaoshi 8 is drilled with the daily gas production of 7.672×10^5 m^3, and shows the similar characteristics to Well Moxi 129H. Well Gaoshi 16 is drilled with the daily gas production of 1.059×10^5 m^3, and shows the similar characteristics to Well Moxi 129H. Well Gaoshi 21 is tested as a dry layer with no obvious bimodal characteristics, that is, the energy of low-frequency band is weak, and the high-frequency band is distributed at 40–80 Hz, which is 20–30 Hz higher than that of Well Moxi 39 and Tongtan 1.

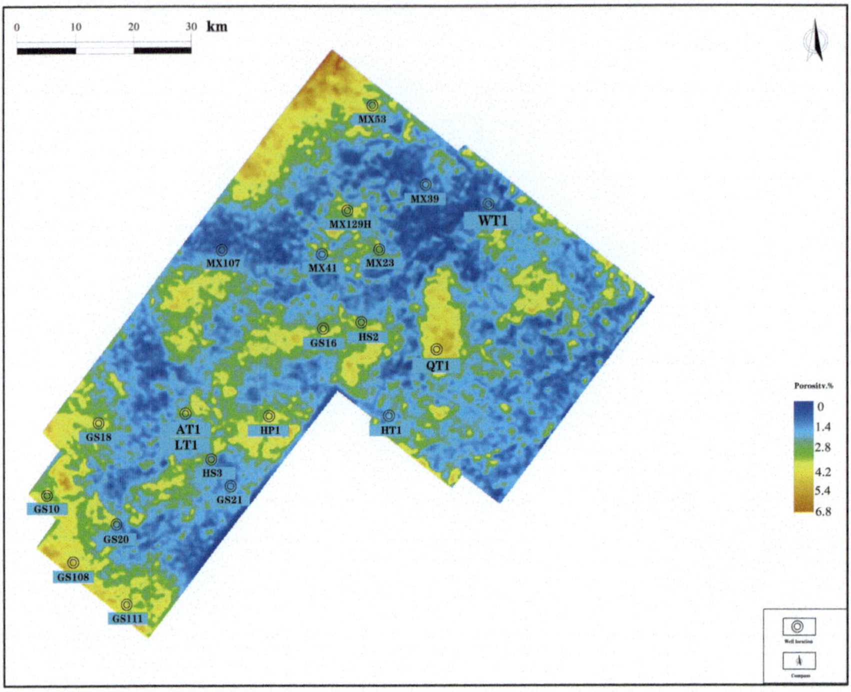

Fig. 16 Average porosity of Deng 4 Member in 3D area of Central Sichuan

It can be seen from the oil and gas detection attribute map of Well Gaoshi 16 and Gaoshi 21 in Fig. 19 that the gas layers in Well Gaoshi 16 show prominent characteristics of low high-frequency energy attribute, high low-frequency energy attribute and increased low/high frequency energy ratio attribute (DIV), and the dry layers in Well Gaoshi 21 show obvious characteristics of high high-frequency energy attribute, low low-frequency energy attribute and reduced low/high frequency energy ratio attribute (DIV).

From Div attribute profile of Well Gaoshi 16 and Gaoshi 21 in Fig. 20, it can be seen that the gas layers of Well Gaoshi 16 show the attribute characteristics of obvious increase of low/high frequency energy ratio (Div), and the dry layers of Well Gaoshi 21 show the attribute characteristics of obvious decrease of low/high frequency energy ratio (Div). By further extracting the Div attribute plane distribution of Deng2 (Fig. 21) and Deng4 Member (Fig. 22), and comparing with the test and logging interpretation results, it can be seen that the coincidence rate of Deng2 Member and Deng4 Member is 84.3% and 85% respectively.

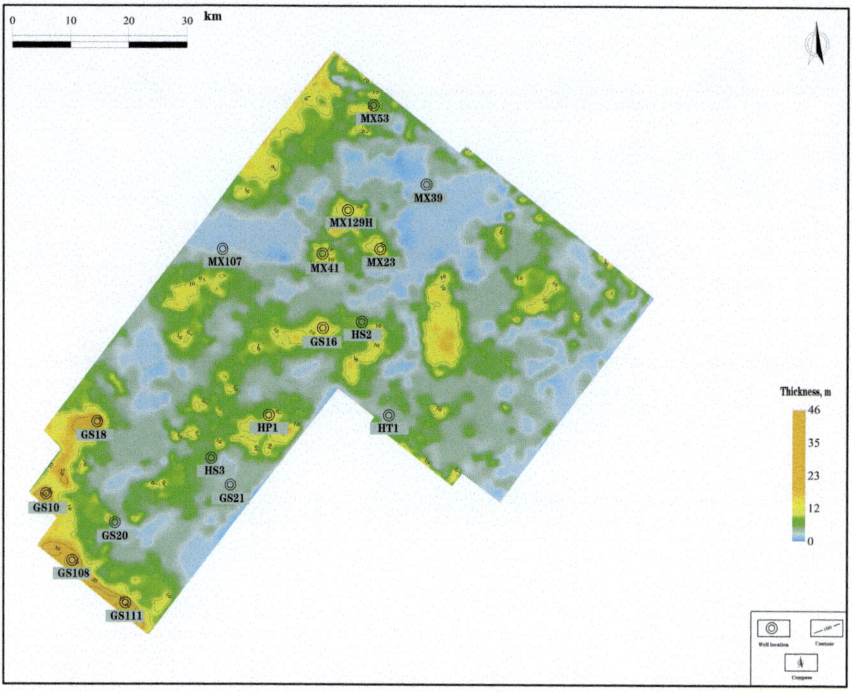

Fig. 17 Reservoir thickness of Deng 4 Member in 3D area of Central Sichuan

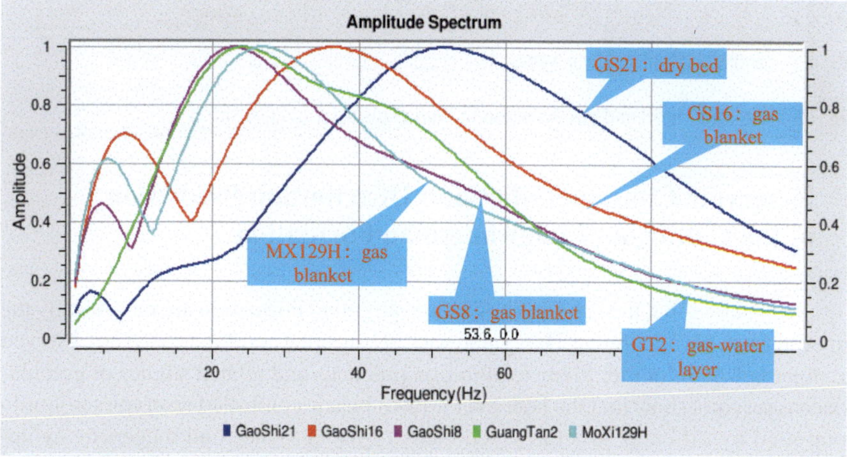

Fig. 18 Analysis of frequency amplitude spectrum in Dengying Formation

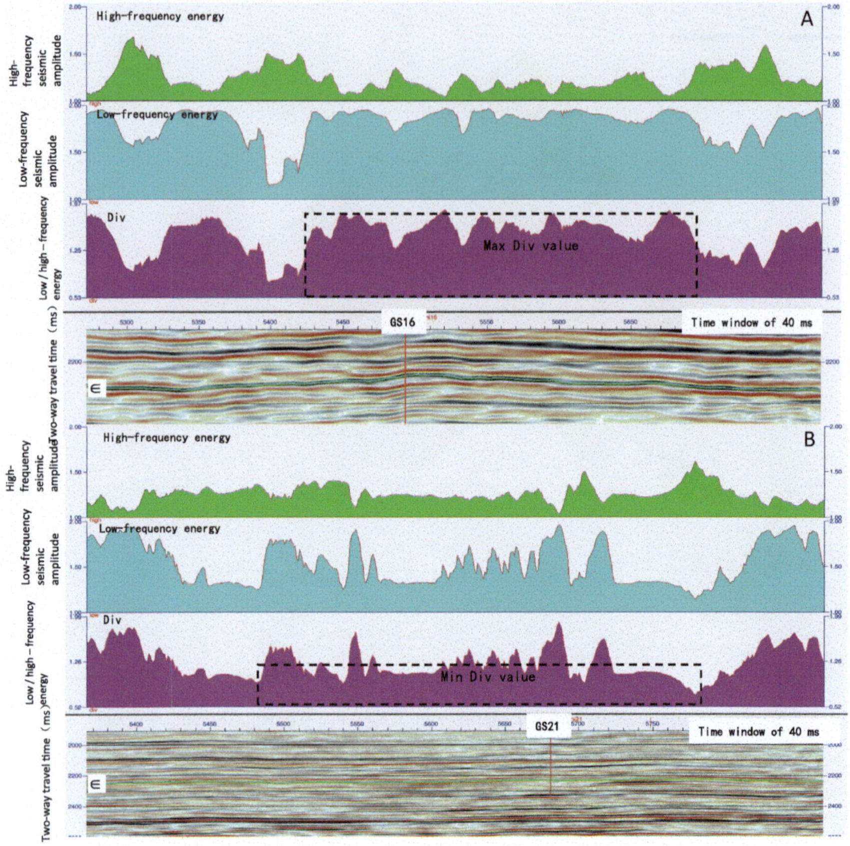

Fig. 19 Attributes of oil and gas detection in Well Gaoshi 16 and Gaoshi 21

2 Reservoir Characteristics and Distribution Prediction of the Cambrian Longwangmiao Formation

The study shows that the reservoirs of Longwangmiao Formation are modified by the superposition of Caledonian-Hercynian weathering crust and karst based on observed granular beach dolomite. Many exploration practices and related studies of granular beach reservoirs show that the pore evolution of beach-controlled reservoirs is jointly controlled by different structures, sedimentary backgrounds and diagenetic evolutions, and there are clear differences among different types of granular beach reservoirs. Therefore, based on the study of the geological characteristics developed in the granular beach of Longwangmiao Formation, the micro amplitude structure and reservoir thickness developed in the granular beach are identified by the fine seismic interpretation, and the plane distribution characteristics of reservoirs are predicted with seismic attributes and reservoir inversion technologies.

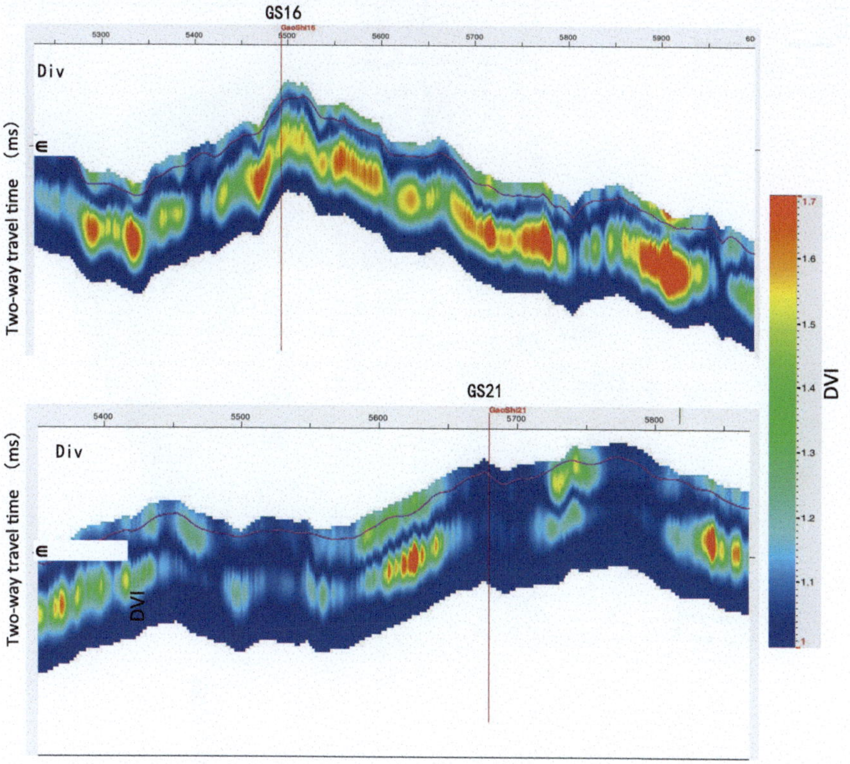

Fig. 20 Div attribute profile of Well Gaoshi 16 and Gaoshi 21

2.1 Genetic Types and Characteristics of Reservoirs

2.1.1 Petrological Characteristics of Reservoirs

The petrological types of the primary reservoirs in Central Sichuan's Longwangmiao Formation include granular (sandy debris) dolomite and crystal dolomite, as shown by cores and thin sections (Figs. 23 and 24). In the granular dolomite, the doloarenite is the main favorable reservoir, with the main reservoir spaces of dissolution pores and caves. Oolitic dolomite reservoir is underdeveloped in the research area. Crystal dolomite refers to dolomite mainly composed of dolomite grains, and the residual sedimentary fabrics are difficult to identify. According to the grain size, it can be further divided into micritic microcrystal dolomite, silty dolomite and fine-grained dolomite. Effective reservoir spaces are less developed in micritic microcrystal and silty dolomites. A small amount of intercrystal pores can be seen in fine-medium crystal dolomites, and intergranular dissolved pores and small caves are formed by local dissolution expansion. Dense bedding dissolved pores and small caves are also observed on the cores, which can form a better reservoir. The doloarenite and grain

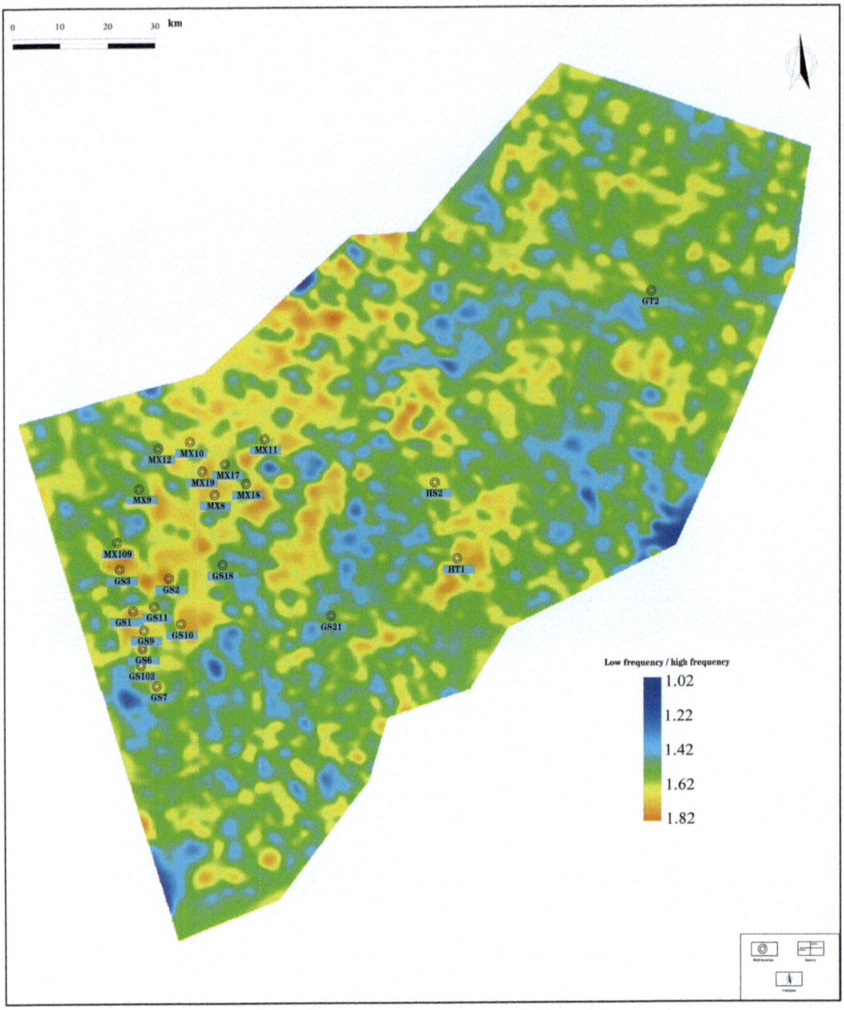

Fig. 21 Oil and gas detection (DIV attribute) in Deng2 Member of Central Sichuan

dolomite of Longwangmiao Formation in the research area form well-developed dissolution pores and caves. The doloarenite accounts for the main body and its pore development degree is relatively higher than that of crystal dolomite. While the reservoir space of Longwangmiao Formation in Well Tongtan 1 is highly developed, and the dissolution pores and caves in Well Guangtan 2 at the northeast margin of the research area are almost not developed.

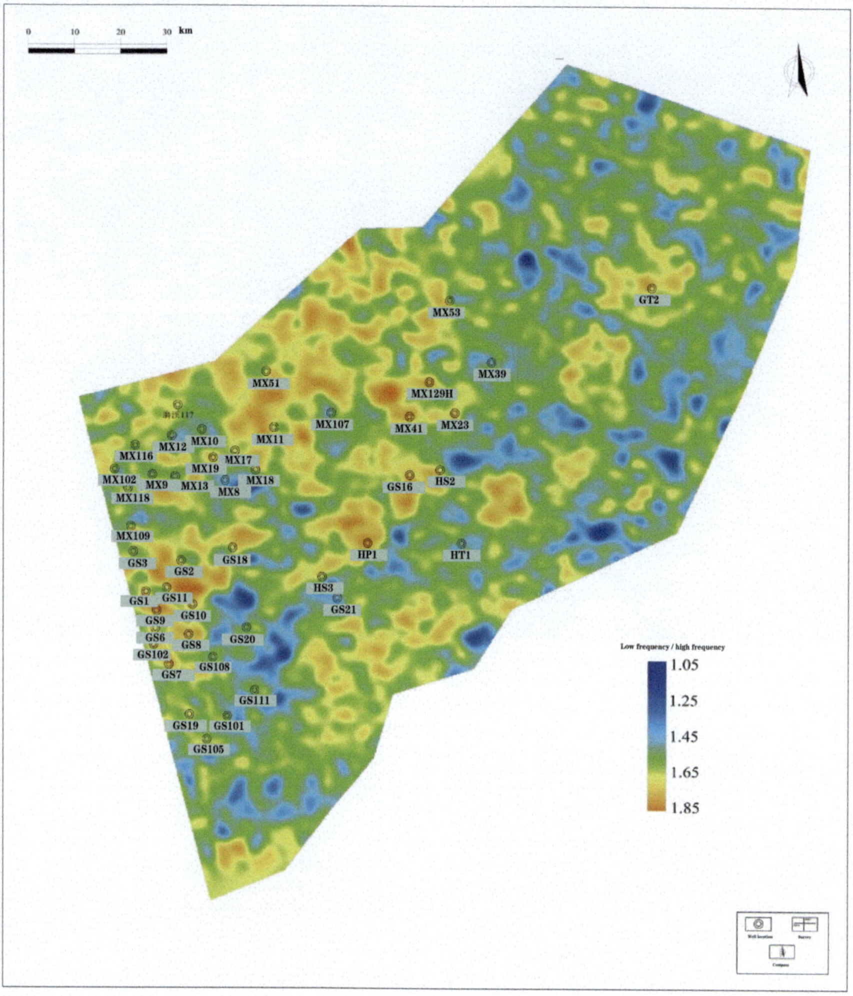

Fig. 22 Oil and gas detection (DIV attribute) in Deng4 Member of Central Sichuan

2.1.2 Reservoir Space Type

The types of reservoir pores in the Cambrian Longwangmiao Formation of Central Sichuan are relatively simple, mainly including dissolution pores and caves, and the fracture-related reservoir space is underdeveloped.

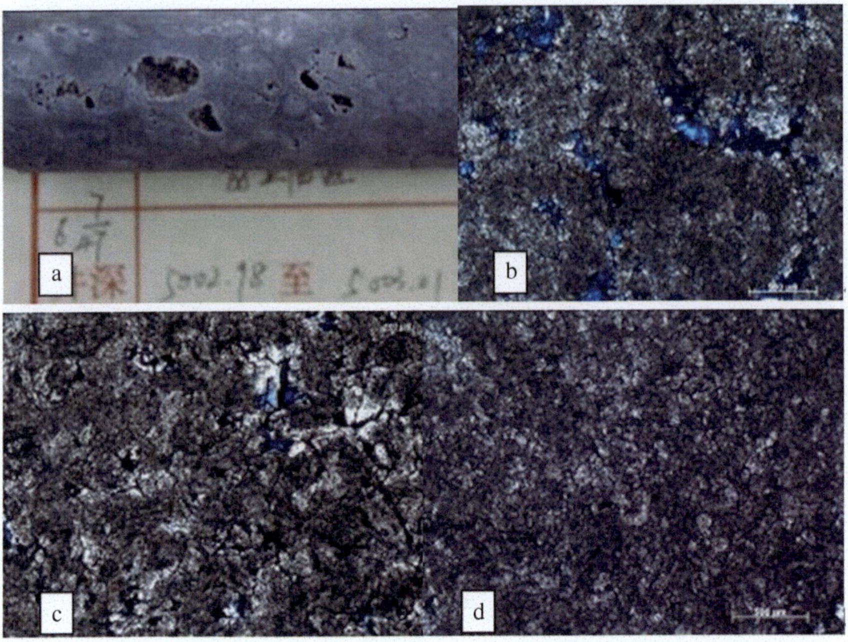

Fig. 23 Petrological characteristics of Longwangmiao Formation in Well Tongtan 1. **a** Gray doloarenite, with dissolution pores and caves, Well Tongtan 1, 5002.98 m. **b** Doloarenite, with residual sandy structure and dissolution pores, Well Tongtan 1, 4995.87 m. **c** Medium-crystal dolomite, with high degree of dolomite crystallization and dissolving intergranular pores, Well Tongtan 1, 5007.7 m. **d** Silty dolomite, with no dissolution pore, Well Tongtan 1, 4989.1 m

Pore

Intergranular Dissolved Pore

Intergranular dissolved pores are pores produced by dissolution expansion or strong dissolution of intergranular dolomite cement on the basis of residual intergranular pores. Intergranular dissolved pores in the study area are mainly developed in granular dolomite, mainly in the size of 0.5–2 mm, and locally associated with small karst caves. Algal debris and related residual structures are common in dolomites in the research area, indicating that dolomitization is incomplete, but traces of algae and algal particles affecting pore formation can be seen, which further supports the idea that the sedimentary environment can affect the formation of dissolution pores.

Intercrystal Dissolved Pore

The fine intergranular pores readjust intergranular pores in the process of microcrystalline dolomite are transformed into coarse-grained dolomite through recrystallization. The pores are generally developed with a large diameter of mostly 0.1–2 mm. They are developed between euhedral and semi-euhedral dolomite crystals, showing

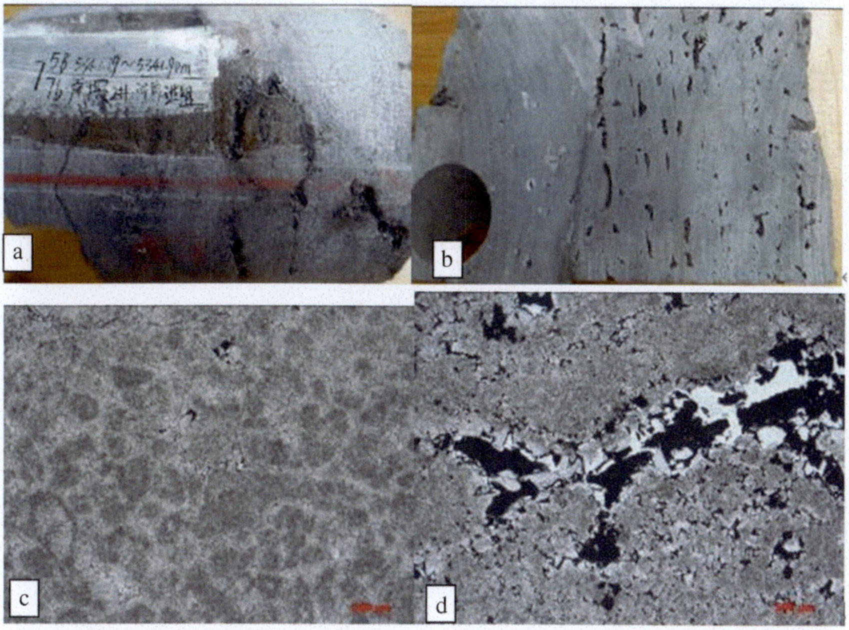

Fig. 24 Petrological characteristics of Longwangmiao Formation in Well Guangtan 2. **a** Doloarenite, with dissolution pores and caves, Well Guangtan 2, 5341.9 m. **b** Fine-crystal dolomite, with bedding dissolution pores and caves, Well Guangtan 2, 5343.37 m. **c** Doloarenite, with a small number of dissolution pores, Well Guangtan 2, 5341.9 m. **d** Fine-crystal dolomite, with dissolution pores, Well Guangtan 2, 5343.37 m

irregular polygon, irregular distribution and poor connectivity (Fig. 24d). Intergranular dissolution pores are mostly developed in coarse-grained dolomites, which may be related to burial or hydrothermal dolomitization. The constructive and destructive effects of intergranular dissolved pores on reservoir formation are still under exploration.

Dissolution Cave

Dissolution cave refers to various dissolution spaces with a diameter greater than 2 mm. The formation mechanism of dissolution caves is similar to that of various karst pores, but the size and scale of dissolution caves are larger than that of dissolution pores. Main characteristics of dissolution caves in the study area are as follows (Fig. 24d). There are isolated caves with different sizes (Fig. 23a) in doloarenite, which are mostly developed along the lamina, and there is the case that the long axis of dissolution cave is consistent with the direction of structural fracture (Fig. 24b).

2.1.3 Genetic Characteristics of Reservoir

According to the comprehensive analysis of reservoir space and sedimentary characteristics revealed by cores, the basic reservoir characteristics of Longwangmiao Formation are summarized into three types. (1) Doloarenite reservoir dominated by dense and bedding dissolution pores (Fig. 25). (2) Crystal dolomite reservoir (Class III) with "spotted" structure, dominated by dissolved pores and sporadic small caves (Fig. 26a, b). (3) Spotted dolomite reservoir with needle-like dissolution pores (Fig. 26c, d).

High-Energy Doloarenite Intra-Platform Beach is the Material Basis for High-Quality Reservoirs, the Reservoir Spaces of Storm Beach and Oolitic Beach Are Underdeveloped

By analyzing the core data of Longwangmiao and Xixiangchi Formation in Well Tongtan 1, Guangtan 2 and He 12, it can be clearly found that a large amount of dissolution pores and caves are mainly developed in spotted doloarenite and residual

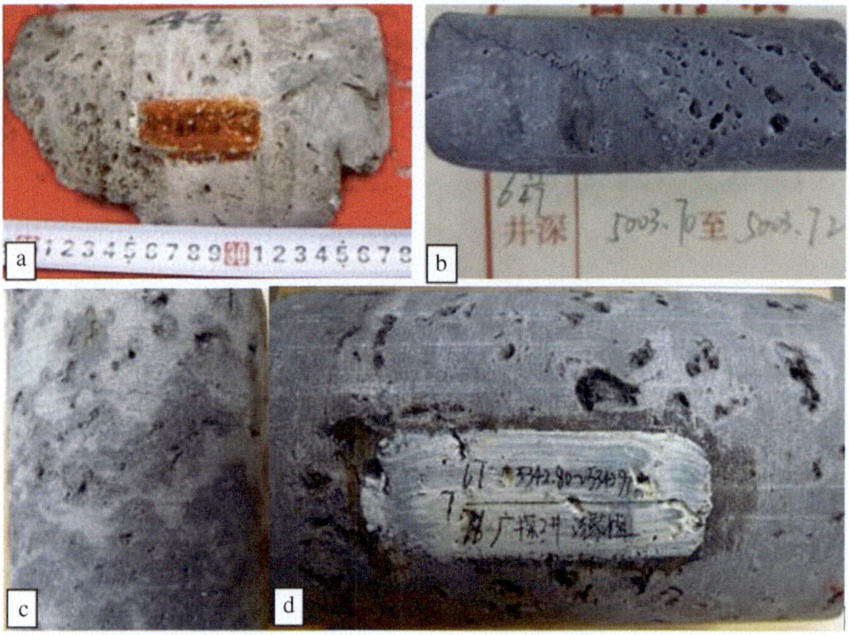

Fig. 25 Dense and bedding dissolution pores and caves in Longwangmiao Formation. **a** Fine-grained doloarenite, with small honeycomb pores, Well Moxi 20, 4608.1 m. **b** Doloarenite with dense bedding pores, Well Tongtan 1, 5003.7 m. **c** Spotted doloarenite, with bedding dissolution pores and caves, Well Guangtan 2, 5345.2 m. **d** Spotted doloarenite, with dissolution caves, Well Guangtan 2, 5342.8 m

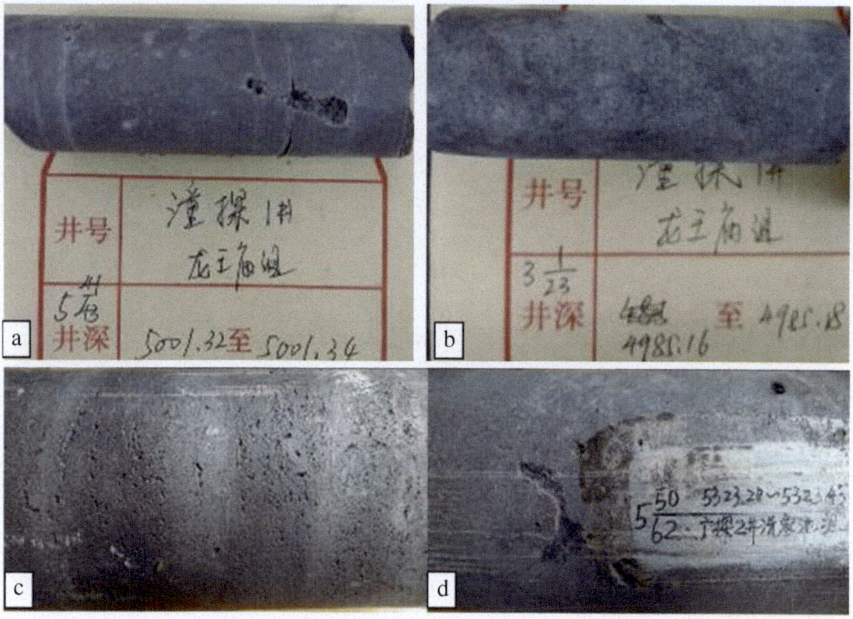

Fig. 26 Dissolution pores in spotted doloarenite of Longwangmiao Formation. **a** Spotted doloarenite, mainly composed of needle-like dissolution pores, and a small number of dissolution caves, Well Tongtan 1, 5001.3 m. **b** Spotted dolomite with needle-like dissolution pores, Well Tongtan 1, 4985.18 m. **c** Doloarenite, needle-like dissolution pores, Well Guangtan 2, 5320.1 m. **d** Doloarenite, needle-like dissolution pores, Well Guangtan 2, 5323.2 m

sandy crystal dolomite, while there is almost no dissolution pore in oolitic dolomite and storm dolomite. This phenomenon can well support the understanding that the high-energy facies belt is the material basis for the high-quality reservoirs. Since the high-quality reservoir is controlled by high-energy facies belt, it is also affected by the high-frequency sequence related to high-frequency sea-level change vertically, and the dissolution pores and caves are mostly developed in the upper part of the sequence.

Evidence of Weathering Crust in Longwangmiao Formation

Generally speaking, the superposition and transformation of weathering crust karstification is the key to the formation of carbonate reservoirs. Previous studies have shown that the top Longwangmiao Formation had experienced erosion which is related to short-term high-frequency sea-level decline, and there may be weathering crust karstification in quasi-contemporaneous period (Tan et al., 2015). In this research, the evidence of karst breccia in the upper part of Longwangmiao Formation (Fig. 27) have been found, which further confirmed the existence of weathering crust karst. However, the lithologic observation of several wells confirmed that there is

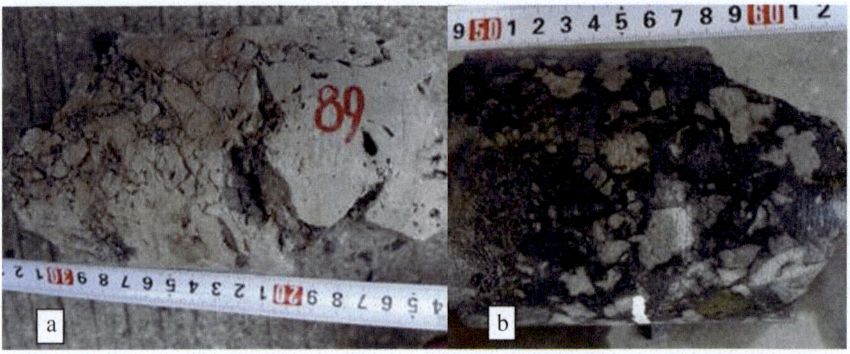

Fig. 27 Evidence of weathering crust karstification in Longwangmiao Formation. **a** Light-gray fine-grained doloarenite, with dissolution pores, irregular brecciation in the surface of rocks, gray seepage material filled in caves, Well Moxi 203, 4769.3 m. **b** Spotted breccia with irregular light-gray breccias of single composition, and the breccias are filled with dark-gray karst materials

no evidence of weathering crust karstification in the Xixiangchi Formation, which further confirmed that the Xixiangchi Formation and the bottom of Ordovician were in successive sedimentation.

2.1.4 Reservoir Correlation and Distribution

Similar to the idea of comparative analysis of reservoirs in Dengying Formation, the reservoirs in Longwangmiao Formation are mainly controlled by high-energy facies belt. Therefore, the idea of sedimentary facies correlation is mainly inherited in the reservoir correlation, and it is considered that the high-energy intra-platform beach is the main part of reservoir. From the reservoir correlation profile across Well Moxi 16-Moxi 46-Gaoshi 16-Gaoshi 113-Hetan 1, and Gaoshi 112-Gaoshi 16-Moxi 23-Moxi 39-Guangtan 2, it can be seen that two favorable reservoirs are developed vertically in Longwangmiao Formation and mostly in the highstand system of the sequence. Laterally, these reservoirs gradually thin eastward.

2.2 Reservoir Prediction

2.2.1 Petrophysical Characteristics of Reservoir

The velocity and density of granular beach reservoirs in Longwangmiao Formation may decrease significantly due to the reservoir space, and the density is a better identifier of the granular beach reservoirs of Longwangmiao Formation. Therefore, the linear relationship between the density curve and porosity can be established (left of Fig. 28). The expression is: porosity = 96.991–33.814 * density, and the correlation

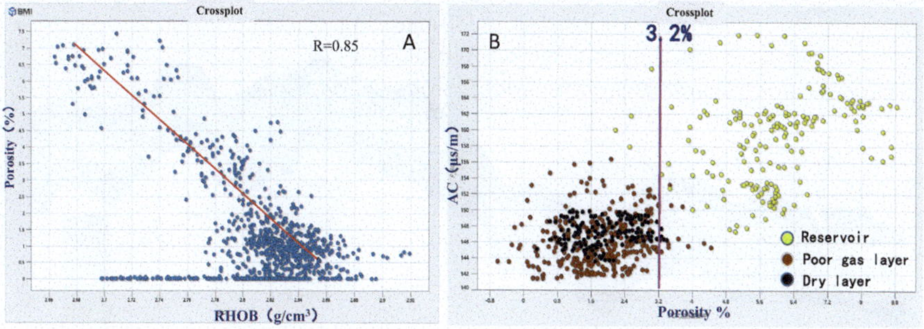

Fig. 28 Cross plot of porosity and density curve (left), cross plot of porosity and AC (right)

R = 0.85. The porosity curve can be obtained according to the expression given above (sample wells: Gaoshi 16, Gaoshi 21, Hetan 1, Moxi 23, Moxi 39, Gaoshi 112 and Gaoshi 113). The reservoir interpretation results obtained through logging interpretation are the basis for reservoir zonatation with threshold value of reservoir porosity being 3.2%. Therefore, the granular beach reservoir of Longwangmiao Formation can be identified through porosity curve (Fig. 28 right).

2.2.2 Reservoir Predicted by Waveform Indication Inversion

The velocity of granular beach reservoir will be significantly lower, as a result, a large wave impedance difference will form between beach reservoir and surrounding rocks. Petrophysical chart analysis shows that there are some differences in porosity between granular beach reservoir and nonreservoir of Longwangmiao Formation, which acts as the basis for porosity prediction. In this study, the seismic waveform indication inversion method is adopted, and the inversion work is carried out by using 3D seismic data and logging curves. Based on the great horizontal change of carbonate reef reservoir, the reliability of the inversion results should be comprehensively considered in combination with other prediction methods. Figure 29 shows the porosity inversion results of Well Gaoshi 112—Gaoshi 16—Moxi 23—Moxi 39. In Fig. 29 it can be seen that the inversion results are consistent with the actual drilling test results (the daily gas production of Well Gaoshi 16 is 2.044×10^5 m^3, the daily gas production of Well Moxi 23 is 1.108×10^6 m^3, the testing layer in Well Gaoshi 112 is water layer and in Well Moxi 39 is the dry layer). Therefore, the porosity inversion results can better reflect the characteristics of granular beach reservoir of Longwangmiao Formation.

By further extracting the average porosity plan (Fig. 30) and reservoir thickness map (Fig. 31) of Longwangmiao Formation, it can be seen that the distribution characteristics of granular beach in Longwangmiao Formation can be clearly described by waveform indication inversion. The area of favorable reservoir predicted by inversion of Longwangmiao Formation described in 3D area is 1120.5 km^2. By comparing

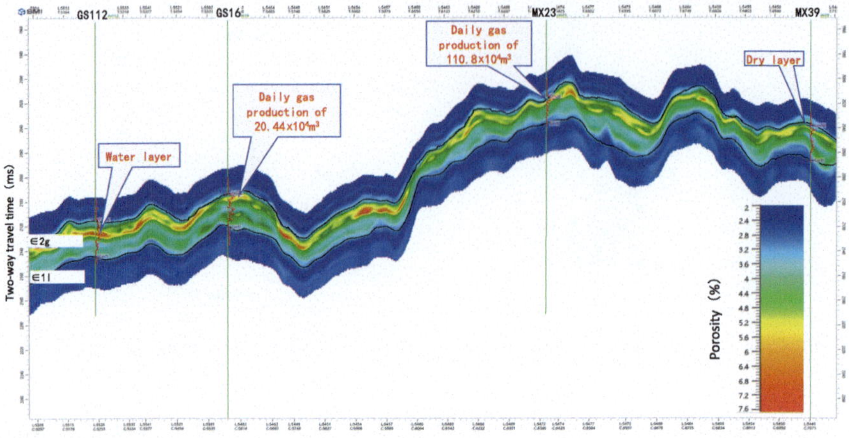

Fig. 29 Porosity inversion profile of Well Gaoshi 112-Gaoshi 16-Moxi 23-Moxi 39

the logging interpretation, the reservoir prediction coincidence rate is 95%, and the thickness error is less than 2 m.

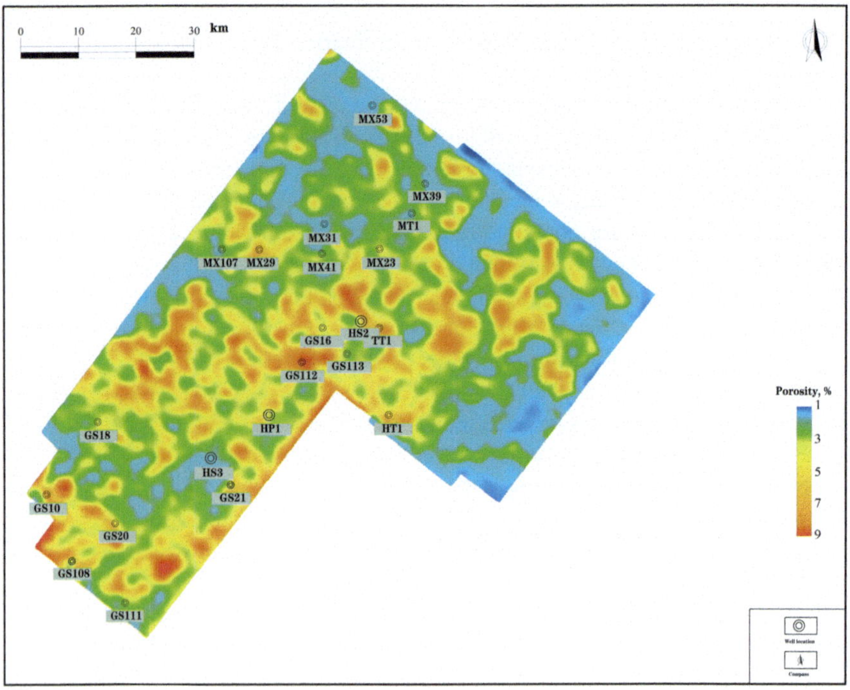

Fig. 30 Porosity inversion plan of Longwangmiao Formation in 3D area of Central Sichuan

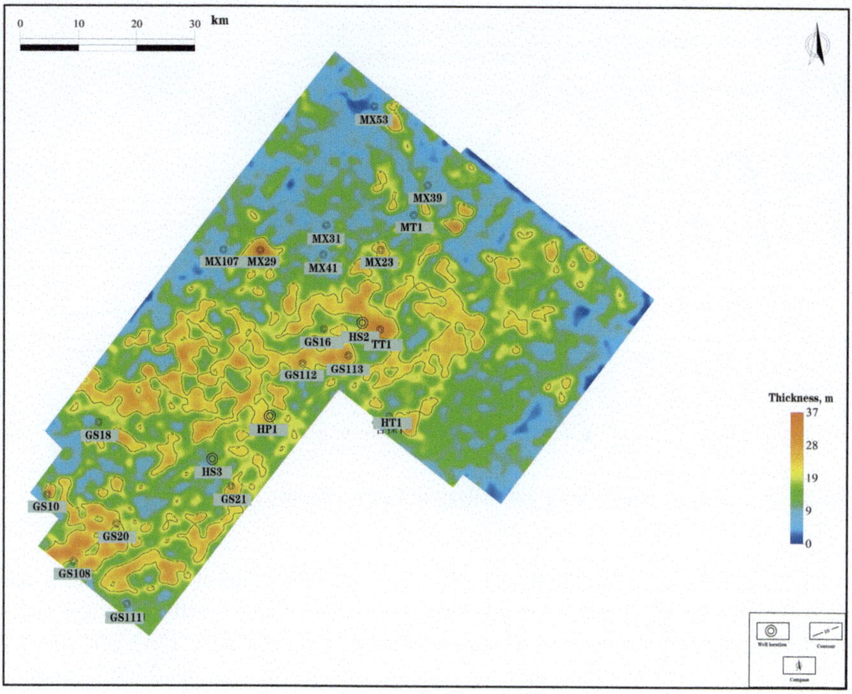

Fig. 31 Reservoir thickness of Longwangmiao Formation in 3D area of Central Sichuan

2.3 Oil and Gas Detection

In the process of CM oil and gas detection, the characteristic informations such as amplitude and frequency of seismic data can be analyzed by the method of cumulative energy normal probability analysis. The attribute parameters such as cumulative energy of low-frequency and high-frequency bands and its ratio to the global energy can be calculated based on low-frequency and high-frequency bands determined by sensitive frequency points.

The analysis results of drilling frequency amplitude spectrum show that (Fig. 32), the daily gas production of Well Moxi 23 is 1.108×10^6 m^3, showing a typical bimodal characteristic. The relative amplitude peak in low frequency band is 0.6–0.8, which is distributed in low frequency band within 15 Hz, and the drop from the sub peak in the high frequency band is very small. The high frequency band is distributed at 20–40 Hz. Well Gaoshi 16 has a daily gas production of 2.044×10^5 m^3, showing similar characteristics to Well Moxi 23, and the high frequency band is distributed at 20–50 Hz. Well Moxi 39 is tested as a dry layer and Well Gaoshi 112 is tested as a water layer. The two wells show no obvious bimodal characteristics, that is, the low frequency-band energy is weak, and the high frequency-band is distributed at 40–80 Hz, which is 20–40 Hz higher than Well Moxi 39 and Tongtan 1.

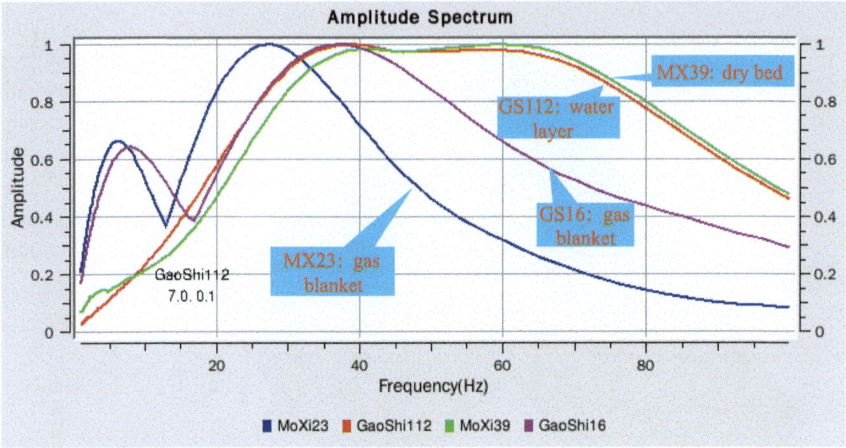

Fig. 32 Frequency amplitude spectrum analysis of Longwangmiao Formation

It can be seen from the oil and gas detection attribute map of Well Moxi 23 and Moxi 39 in Fig. 33 that the gas layer of Well Moxi 23 shows the characteristics of obvious increase of low/high frequency energy ratio attribute (Div), and the dry layer of Well Moxi 39 shows the characteristics of decrease of low/high frequency energy ratio attribute (Div). From the oil and gas detection profile across Well Gaoshi 112-Gaoshi 16, it can be seen that the test gas layer of Well Gaoshi 16 shows the characteristics of obvious increase of low/high frequency energy ratio attribute (Div), and the water layer of Well Gaoshi 112 shows the characteristics of decrease of low/high frequency energy ratio attribute (Div). The oil and gas detection attribute can well detect the gas-bearing property of Longwangmiao Formation.

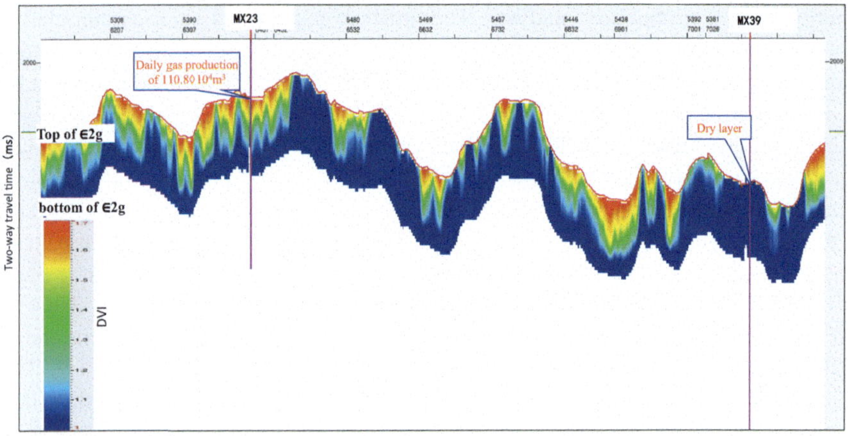

Fig. 33 Profile of oil and gas detection across Well Moxi 23-Moxi 39

The plane distribution of Div attribute in Longwangmiao Formation (Fig. 34) is further extracted. Through comparison with test results and logging interpretations, the coincidence rate of Div is 84.6%.

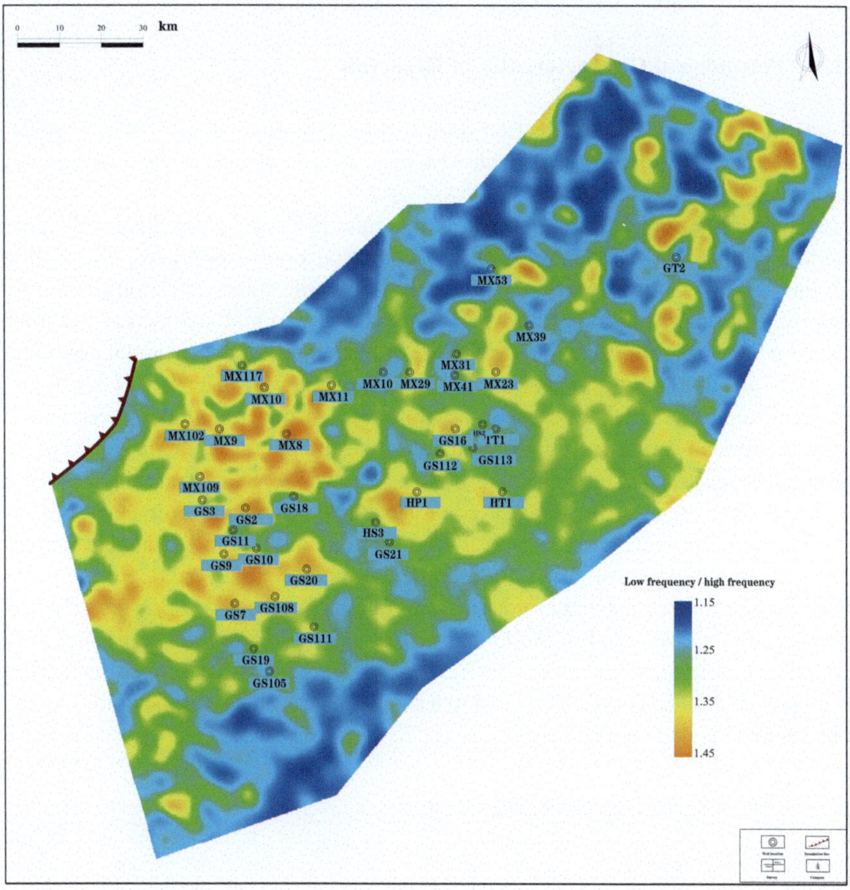

Fig. 34 Oil and gas detection plan of Longwangmiao Formation in Central Sichuan

3 Reservoir Characteristics and Distribution Prediction of the Cambrian Xixiangchi Formation

3.1 Genetic Types and Characteristics of Reservoirs

3.1.1 Petrological Characteristics of Reservoir

Observations of core and thin section show that the petrological types of main reservoirs in Xixiangchi Formation of Central Sichuan are granular dolomite (doloarenite) and crystal dolomite (Fig. 35).

Among the granular dolomites, the doloarenite reservoir is the main favorable reservoir with main reservoir spaces of dissolution pores and caves, and oolitic dolomite reservoir is underdeveloped. Crystal dolomite refers to dolomite mainly composed of dolomite grains, and indistinguishable residual sedimentary fabrics. Based on the grain size, it can be further divided into micritic microcrystal dolomite, silty dolomite and fine-grained dolomite. Effective reservoir space is less developed in micritic microcrystal dolomite and silty dolomite. A small amount of intergranular pores can be seen in fine-medium crystal dolomite, and intergranular dissolved pores and small caves are formed by local dissolution expansion. Dense bedding dissolved pores and small caves are also observed on the cores, which can form a better reservoir. The development degree (proportion) of crystal dolomite in Xixiangchi Formation is greater than that of granular dolomite, and the development degree of pores in crystal dolomite is also better than that of granular dolomite (doloarenite). Reservoirs are highly developed in Well Guangtan 2, which are dominiated by medium-crystalline dolomite. While reservoirs in Well He 12 are poorly developed and underdeveloped dissolution caves are found in a small amount of crystalline dolomite and doloarenite.

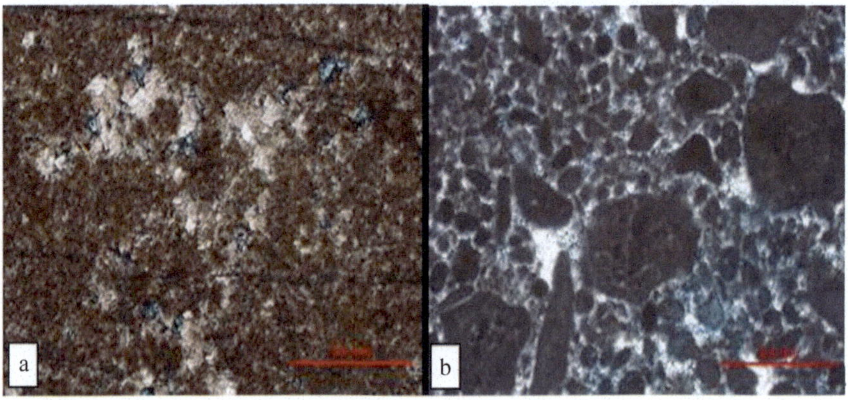

Fig. 35 Petrological characteristics of Xixiangchi Formation in Well He 12. **a** Residual doloarenite, with dissolution pores, Well He 12, Block 2-81. **b** Doloarenite, Well HE 12, Block 7-1

3.1.2 Types of Reservoir Space

The reservoir space types of Cambrian Xixiangchi Formation in Central Sichuan are relatively simple, mainly including dissolution pores and caves, and the fracture-related reservoir space is underdeveloped.

Pore

Intergranular Dissolved Pore

Intergranular dissolved pores are pores produced by dissolution expansion or strong dissolution of intergranular dolomite cement based on residual intergranular pores. In the study area, intergranular dissolved pores are predominantly formed in granular dolomite, usually with a diameter of 0.5–2 mm, and they are associated with small karst caves. Algal debris and related residual structures are common in dolomites in the research area, indicating that dolomitization is incomplete, but traces of algae and algal particles affecting pore formation can be seen, which further supports the idea that the sedimentary environment can affect the formation of dissolution pores.

Intercrystal Dissolved Pore

The fine intergranular pores mostly readjust intergranular pores in the process of microcrystal dolomite recrystallization into coarse-grained dolomite. The pores are generally developed with large diameter of mostly 0.1–2 mm. They are developed between euhedral and semi-euhedral dolomite crystals, showing irregular polygon, irregular distribution and poor connectivity. Intercrystal dissolved pores are mostly developed in coarse-grained dolomites that may be related to burial or hydrothermal dolomitization. The constructive and destructive effects of intercrystal dissolved pores on reservoir formation are still under exploration.

Dissolution Cave

Dissolution cave refers to various dissolution spaces with a diameter greater than 2 mm. The formation mechanism of dissolution caves is similar to that of various karst pores, but the size and scale of dissolution caves are larger than that of dissolution pores. Main characteristics of dissolution caves in the study area are as follows. There are isolated caves with different sizes in doloarenite, which are often developed along the lamina, and the long axis of dissolution cave is consistent with the direction of structural fracture.

3.1.3 Reservoir Correlation and Distribution

Similar to the idea of comparative analysis of reservoirs in Dengying Formation, the reservoirs in Xixiangchi Formation are mainly controlled by high-energy facies belt. Therefore, the idea of sedimentary facies correlation is mainly inherited in the reservoir correlation, and it is considered that the high-energy intra-platform beach is the main part of reservoir. From the reservoir correlation profiles across Well Moxi 16-Moxi 46-Gaoshi 16-Gaoshi 113-Hetan 1, and Well Gaoshi 112-Gaoshi 16-Moxi 23-Moxi 39-Guangtan 2, it can be seen that two favorable reservoir zones are developed vertically and horizontally distributed in a large area in Xixiangchi Formation. The reservoir area of Longwangmiao Formation is larger than that of Xixiangchi Formation.

3.2 *Reservoir Prediction*

3.2.1 Petrophysical Characteristics of Reservoir

The velocity and density of granular beach reservoirs in Xixiangchi Formation may decrease significantly due to the reservoir space, and the density is the better identifier of the granular beach reservoirs of Xixiangchi Formation. Therefore, the linear relationship between density curve and porosity can be established, and the porosity curve is obtained (sample wells: Gaoshi 16, Gaoshi 21, Hetian 1, Moxi 23, Gaoshi 112, Gaoshi 113 and Tongtan 1). Based on the reservoir classification obtained by reservoir interpretation results through logging interpretation, the threshold value of reservoir porosity is 4.0%. Therefore, the granular beach reservoir of Xixiangchi Formation can be identified through porosity curve (Fig. 36).

3.2.2 Reservoir Predicted by Waveform Indication Inversion

Petrophysical chart analysis shows that there are obvious differences in porosity between granular beach reservoir and nonreservoir of Xixiangchi Formation, which is the basis of porosity prediction. In this study, the seismic waveform indication inversion method is adopted, and the inversion work is carried out by using seismic data and logging curves. Based on the great horizontal change of carbonate reef reservoir, the reliability of the inversion results should be comprehensively considered in combination with other prediction methods. Figure 31 shows the porosity inversion results of Well Gaoshi 112-Gaoshi 16-Moxi 23-Moxi 39. It can be seen from Fig. 37 that the inversion results are consistent with the actual drilling test results (the daily gas production of Well Gaoshi 16 is 7.82×10^4 m^3, the daily gas production of Well Moxi 23 is 2.11×10^4 m^3. Therefore, the porosity inversion results can better reflect the characteristics of granular beach reservoir of Xixiangchi Formation.

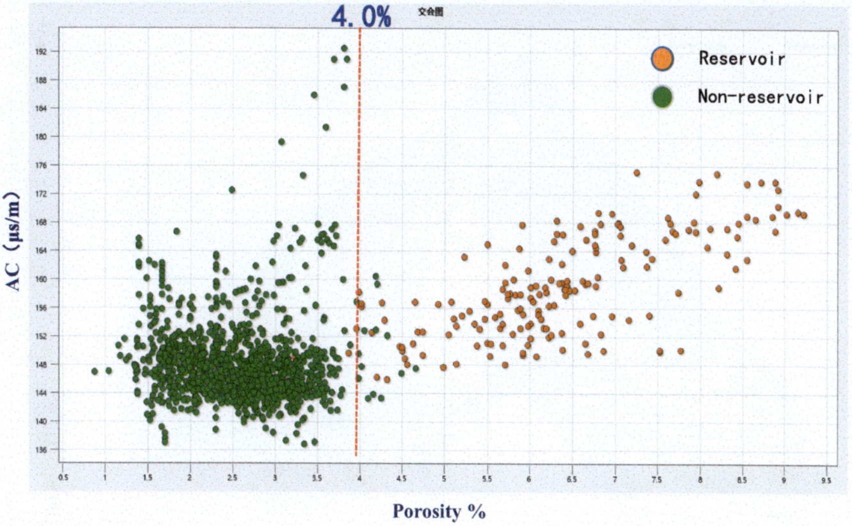

Fig. 36 Cross plot of porosity and AC of Xixiangchi Formation

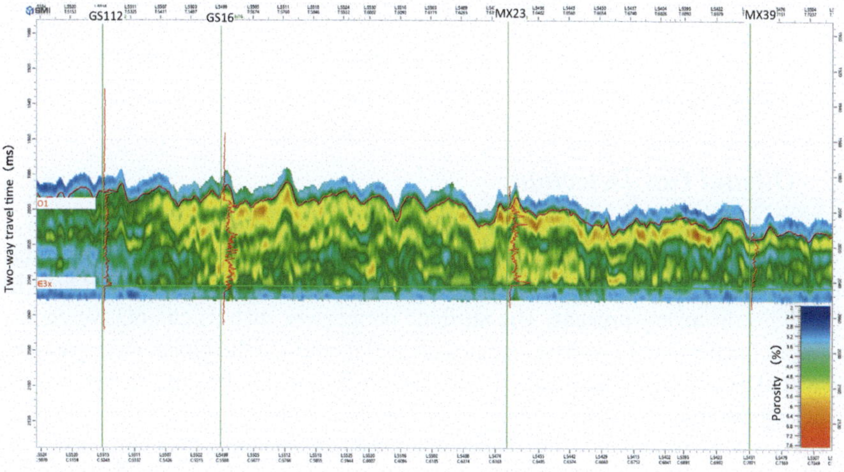

Fig. 37 Porosity inversion profile of Well Gaoshi 112-Gaoshi 16-Moxi 23-Moxi 39 (Cm2x flattened)

By further extracting the average porosity plan (Fig. 38) and reservoir thickness map (Fig. 39) of Xixiangchi Formation, it can be seen that the distribution characteristics of granular beach in Xixiangchi Formation can be clearly described by waveform indication inversion. The area of favorable reservoir predicted by inversion of Xixiangchi Formation described in 3D area is 518 km^2 that coincides with logging interpretation to an extent of 94%.

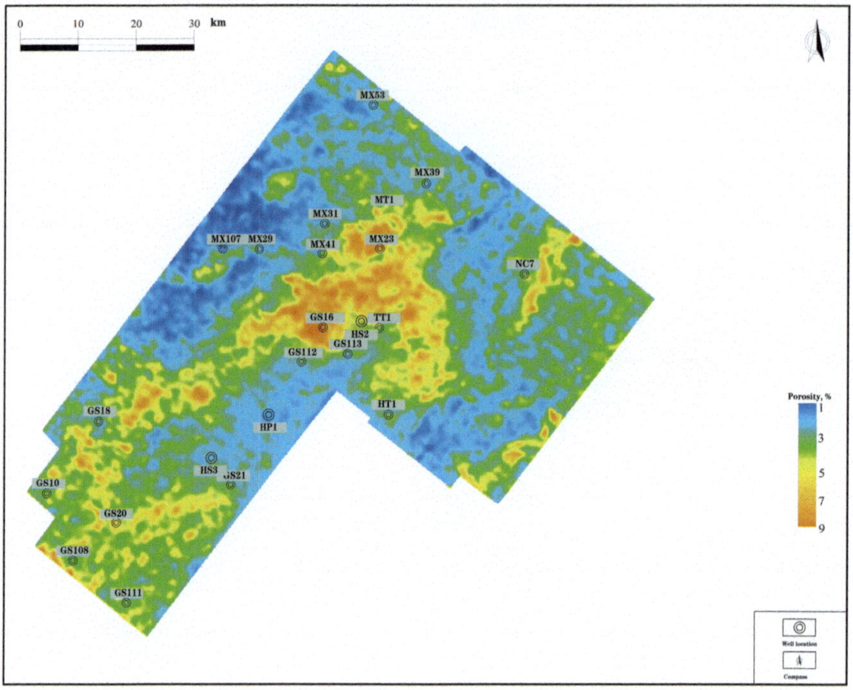

Fig. 38 Average porosity plan of Xixiangchi Formation in 3D area of Central Sichuan

3.3 Oil and Gas Detection

The cumulative energy normal probability analysis technique may be used to examine the information in seismic data such as amplitude and frequency throughout the CM oil and gas discovery process. The attribute parameters such as cumulative energy of low-frequency and high-frequency bands and its ratio to the global energy can be calculated based on low-frequency and high-frequency bands determined by sensitive frequency points.

The analysis results of drilling frequency amplitude spectrum show that (Fig. 40), the daily gas production of Well Gaoshi 16 is 7.82×10^4 m^3, showing typical bimodal characteristics. The relative amplitude peak in low frequency band is 0.6–0.8, which is distributed in low frequency band within 15 Hz, and the drop from the sub peak in the high frequency band is very small. The high frequency-band is distributed at 20–60 Hz. Well Moxi 23 has a daily gas production of 2.11×10^4 m^3, showing similar characteristics to Well Gaoshi 16, and the high frequency band is distributed at 20–60 Hz. Well Moxi 41 is tested as a dry layer, a bimodal characteristics of Well Moxi 41 is no obvious, that is, the low frequency-band energy is weak, and the high frequency-band is distributed at 30–80 Hz, which is 10–20 Hz higher than that of Well Gaoshi 16 and Well Moxi 23.

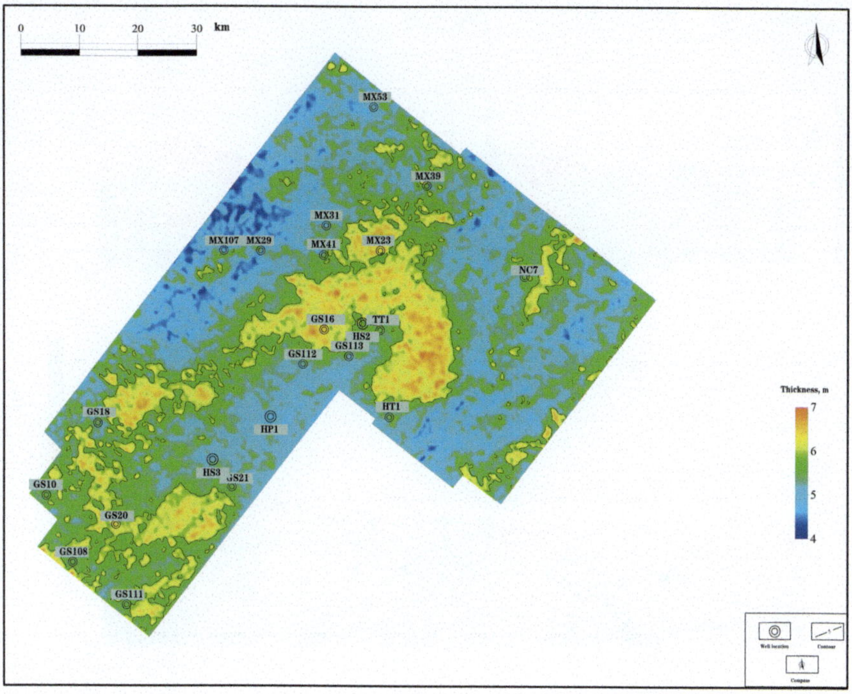

Fig. 39 Reservoir thickness of Xixiangchi Formation in 3D area of Central Sichuan

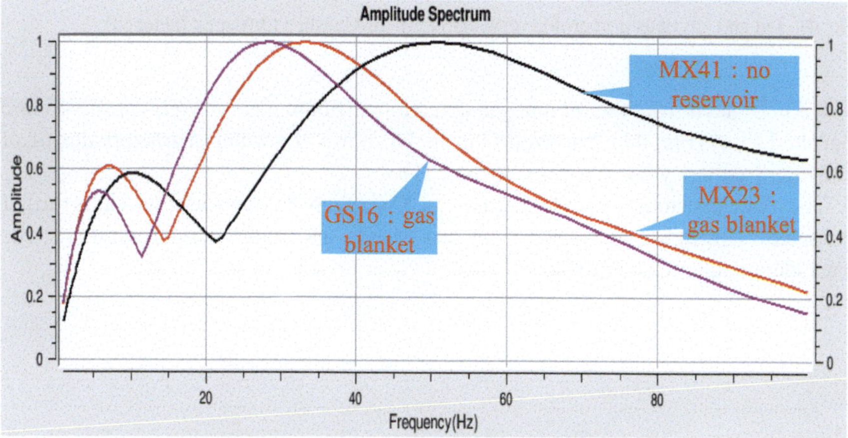

Fig. 40 Frequency amplitude spectrum analysis of Xixiangchi Formation

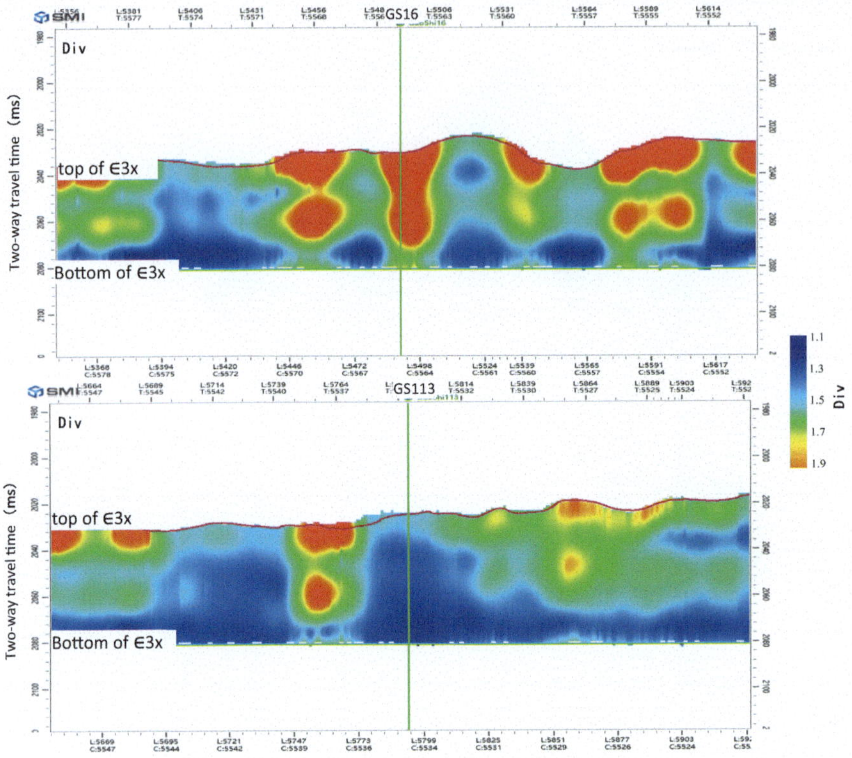

Fig. 41 Oil and gas detection profile of Well Gaoshi 16-Gaoshi 113 (Cm2x flattened)

It can be seen from the oil and gas detection attribute map of Well Gaoshi 15 and Gaoshi 113 in (Fig. 41) that Well Gaoshi 16 shows abnormal characteristics of of low/high frequency energy ratio attribute (Div).

The plane distribution of Div attribute in Xixiangchi Formation (Fig. 42) is further extracted. Through comparison with test results and logging interpretations, the coincidence rate of Div is 84.6%.

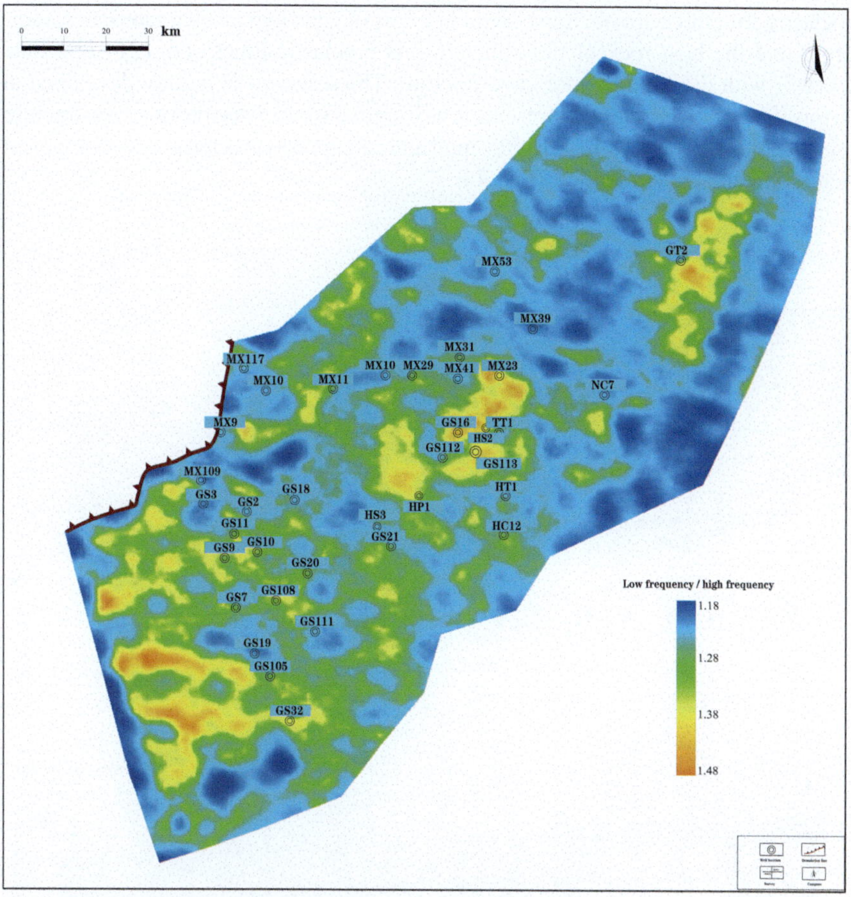

Fig. 42 Oil and gas detection plan of Xixiangchi Formation in Central Sichuan

4 Reservoir Characteristics and Distribution Prediction of the Permian Qixia-Maokou Formation

4.1 Genetic Types and Characteristics of Reservoirs

4.1.1 Petrological Characteristics of Reservoir

The main petrological characteristics of reservoir in Qixia-Maokou Formation can be defined as follows. (1) Bioclastic dolomite reservoir with dissolution pore-cave-fracture (Fig. 43). This kind of reservoir is mainly characterized by dolomite, with well-developed dissolution pores and caves, which are usually associated with fractures, showing the characteristics of pore-cave-fracture reservoir. The residual bioclastic structures are obviously seen in terms of petrological characteristics, indicating that the host rock of dolomitization is bioclastic limestone. (2) Limestone reservoir with dissolution pore-cave-fracture. The reservoir is mainly developed at the top of Maokou Formation, presenting as irregular (brecciated) pore –cave-fracture reservoir, mostly filled with dark-gray mud and calcite, which is the product of typical weathering crust karstification (Fig. 44).

Fig. 43 Dolomite reservoir with dissolution pore-cave-fracture in Qixia-Maokou Formation. **a** Residual bioclastic dolomite, with needle-like dissolution pores, a small amount of dissolution caves and irregular dissolution fractures, constitutes a pore-cave-fracture reservoir, Well Moxi 42, 4657 m, Qixia Formation. **b** Residual bioclastic dolomite, with fractures and expanded dissolution fractures, filled with coarse-grained dolomite, Well Guangtan 2, Maokou Formation, 4717.9 m. **c** Gray bioclastic dolomite, with a large number of dissolution pores, caves and fractures, forming a pore-cave-fracture reservoir, Well Guangcan 2, 4–61/68, Maokou Formation

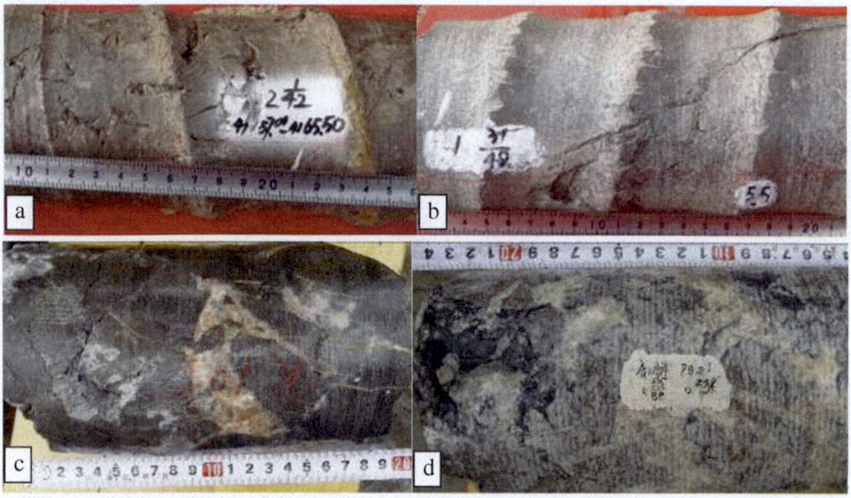

Fig. 44 Characteristics of limestone reservoir with pore-cave-fracture in Maokou Formation. **a** Micritic bioclastic limestone, with irregular reticular fractures and dissolution pores, filled with mud and calcite, Well Tongtan 1, 4157 m. **b** Gray micritic bioclastic limestone, with high-angle fractures filled by mud, Well Tongtan 1, 4154.1 m. **c** Gray micritic bioclastic limestone, with irregular network fractures and a small amount of dissolution pores. The fractures are filled with calcite and dark-gray mud. Well Lai 1, 4078.8 m **d** Brecciated limestone with irregular fractures, calcite and dark-gray mud, Well He 12, Maokou Formation, 3929 m

4.1.2 Geological Model of Hydrothermal Dolomite Reservoir with Dissolution Pore-Cave-Fracture

Petrological Characteristics

The analysis of the coring section (4615–4574 m) in Maokou Formation of Well Guangcan 2 in Central Sichuan found that residual bioclastic medium-coarse crystal dolomites with residual bioclastic medium-coarse crystal dolomites with the thickness of 24 m and the color of dark-gray to gray are developed. The dissolution pores, caves and fractures of the core are widely developed, and most of them are filled with dolomite clusters. In addition, coarse-grained dolomites are also found in the dissolution fractures of dolomites in Well Guangtan 2. Thin section analysis shows that the dolomites are medium crystal dolomite of subhedral-anhedral with some curved crystal planes, residual bioclastics, and residual sand debris. The original rocks are bioclastic limestones and some bright coarse-grained dolomites can be seen. The dolomite crystals in the dissolution caves of cores are typical coarse-megacrystalline dolomites. Under single polarization, the average crystal diameter is greater than 1 mm, which is typical saddle dolomite. Under orthogonal polarization, wavy extinction can be seen.

Isotope Analysis

Carbon and oxygen isotopic characteristics of dolomites of Well Guangcan 2 in Central Sichuan are as follows (Fig. 45). (1) $\delta^{13}C$ (PDB) of medium-coarse grained dolomite is 3.44–3.7‰, and of saddle dolomite is 3.56‰, which are within the range of medium-coarse grained dolomite, indicating that the genetic relationship between saddle dolomite and medium-coarse grained dolomite, and indecating that the formation of the two dolomites should had been affected by the same factors. (2) $\delta^{18}O$ (PDB) of medium-coarse grained dolomite is – 7.64 to 7.77‰, and of saddle dolomite is – 7.95 ‰. The values in these two wells are similar, and the average value is – 7.78‰. The $\delta^{18}O$ (PDB) of marine micritic limestone ranges from – 2.95 to – 3.61‰, and the average value is – 3.31‰ which is significantly higher than that of marine micrite limestone and buried dolomite in Northwest Sichuan (– 6.38 ‰), but lower than that in Southwest Sichuan (– 10.73 ‰), indicating that dolomite in this area has experienced thermal fractionation, but the influence degree of high-temperature event is lower than that in Southwest Sichuan. (3) By projecting the value of ^{13}C–^{18}O in medium-coarse grained dolomite in Central Sichuan on the carbon–oxygen isotope discrimination chart of carbonate sediments and cements by Chaqutte & James (1990), it can be seen that the value of ^{13}C–^{18}O is projected in the range of saddle dolomite crystal, indicating the deep buried hydrothermal dolomitization.

Rare Earth Element Analysis

According to the distribution of rare earth elements in dolomites and limestones of Well Guangcan 2 and Tong 4 in Central Sichuan, the average ΣREE in bioclastic limestones is 4.7×10^{-6}. When EU < 1, it indicates negative anomaly, and when CE > 1, it indicates positive anomaly. The two values indicate that the sedimentary environment is of weak reducibility. What is special is that the EU of crystal dolomite in Well Guangcan 2 is obviously greater than 1, indicating a high-temperature environment and obvious hydrothermal genesis. The REE distribution pattern is characterized by a slight enrichment of light rare earths and a slight loss of heavy rare earths, the value of Σ LREE/ Σ HREE is similar to that of marine normal limestones (Fig. 46).

Homogenization Temperature of Inclusion

Based on the analysis of inclusion homogenization temperature in Southwest Sichuan, it can also be inferred that the upper limit of the formation time of dolomite reservoirs in Well Guangcan 2 is Jurassic. By analyzing the burial-thermal history curve of Well Guangcan 2 in Central Sichuan, it can be seen that the maximum burial depth is about 6500 m, the burial depth of Jurassic is 3200–4700 m, and the variation range of ground temperature is about 105–145 °C. The homogenization temperature of inclusions in saddle dolomites of Maokou Formation is 125–160 °C, and the main

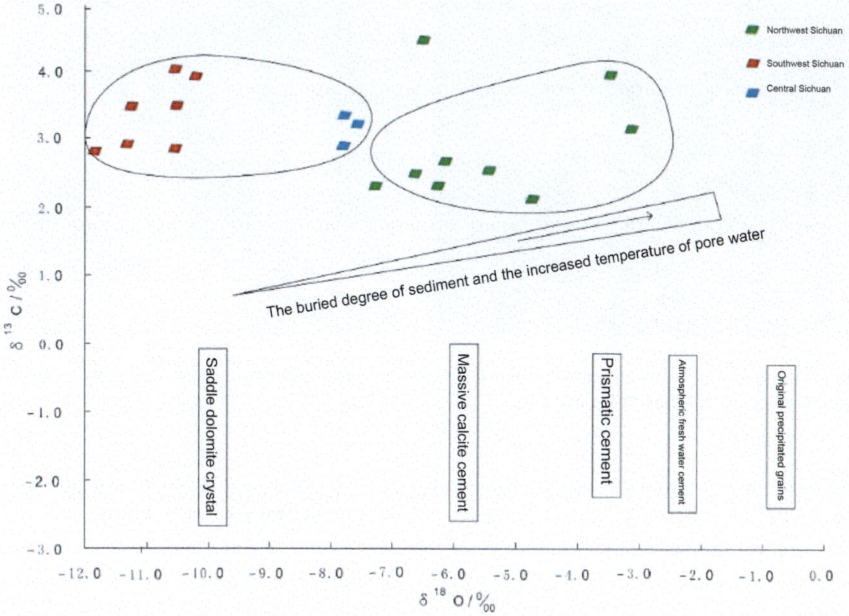

Fig. 45 Discrimination chart between carbon and oxygen isotopes of dolomites in Qixia-Maokou Formation of Northwest, Southwest and Central Sichuan and foreign carbonate sediments and cements (according to Chaqutte and James 1990)

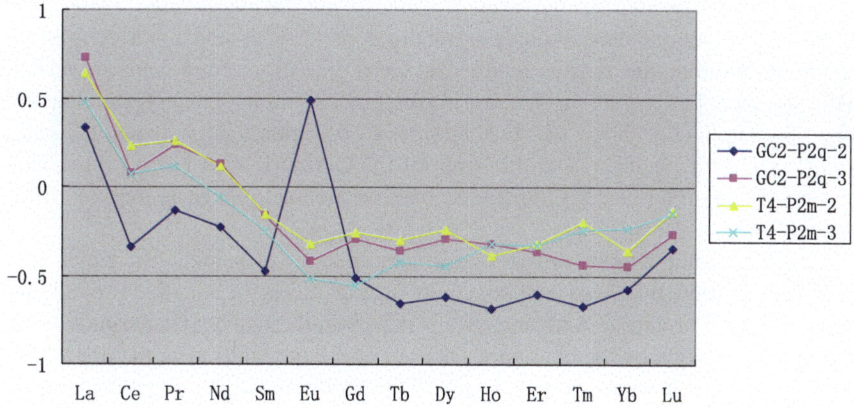

Fig. 46 REE distribution pattern of dolomites in Central Sichuan

range of homogenization temperature is about 155 °C (Fig. 47), which is significantly higher than the normal ground temperature in Jurassic. Davies and Smith (2006) counted the homogenization temperature of a large number of hydrothermal dolomite inclusions and pointed out that the formation temperature is generally 60–250 °C, and most of them are ranged between 90 and 120 °C. Based on this feature, it

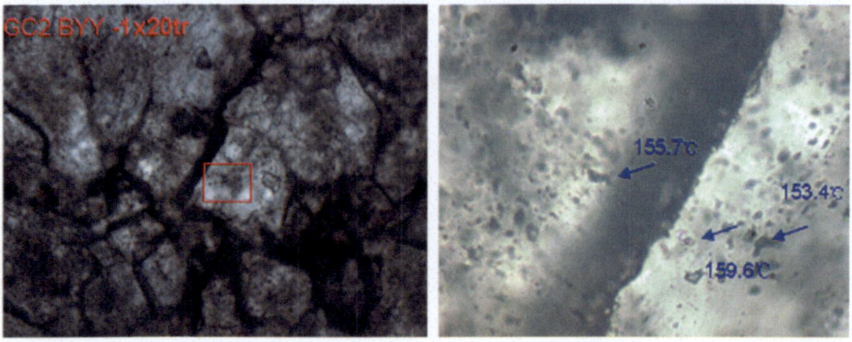

Fig. 47 Homogenization temperature of brine inclusions near fractures of saddle dolomite particles (Well Guangcan 2, Maokou Formation)

can be clearly pointed out that the dolomites in Well Guangcan 2 of Central Sichuan are affected by thermal events and are developed as hydrothermal dolomite.

Based on the above evidence of petrology, isotope, trace elements and inclusion homogenization temperature, it shows that hydrothermal dolomites of Qixia-Maokou Formation in Central Sichuan are represented by Well Guangcan 2, Guangtan 2 and Moxi 42.

Geological Model of Hydrothermal Dolomite

Based on the comprehensive analysis of the genetic types and main controlling factors of dolomite, the genetic geological model of hydrothermal dolomite related to the basement faults of Qixia-Maokou Formation in Central Sichuan is summarized (Fig. 48). The model shows that the activities of basement faults created the paleouplift in Central Sichuan, and the strata of the Qixia-Maokou Formation eroded the underlying strata in the Central and Eastern Sichuan, which is the margin of paleouplift.

A high paleotopography was formed in the margin of paleouplift, which is a favorable position for the development of intra-platform beach of the overlying Qixia-Maokou Formation. And magnesium-rich hydrothermal fluid in the underlying stratum was brought by the continuous activities of large faults after the deposition of Qixia-Maokou Formation. As a result, the hydrothermal dolomitization reaction occurred between magnesium rich hydrothermal fluid and limestones with good permeability in intra-platform beaches, faults and fractures, and the hydrothermal dolomites formed. It can be seen that the activity of basement faults is the internal cause, and the developments of paleouplift margin, intra-platform beach, faults and fractures are the external cause. The hydrothermal dolomites in Central Sichuan are widely developed by the organic combination of internal and external causes. As

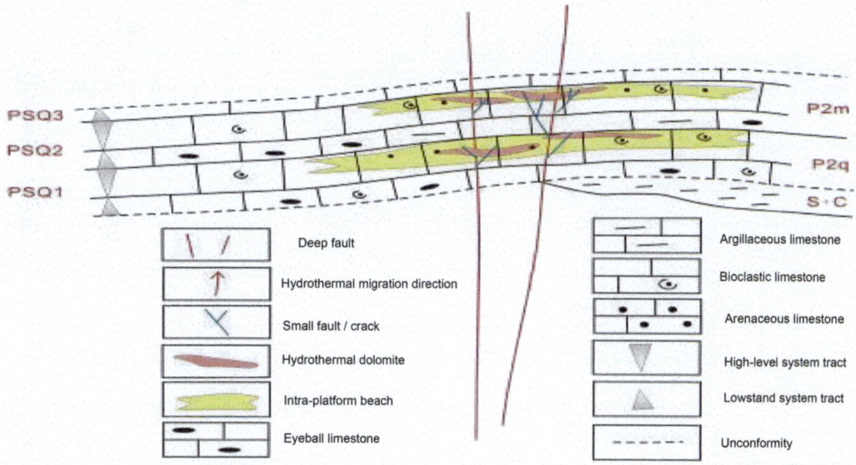

Fig. 48 Genetic geological model of hydrothermal dolomite related to basement faults in Central Sichuan

shown above, the deep magnesium-rich hydrothermal fluid was brought by the activities of tensional-shear basement faults, which is easy to react with limestones in high-energy facies belt or fault-fracture development belt to form hydrothermal dolomites. This is hydrothermal dolomitization. As a result, the seismic response characteristics of dolomite reservoirs and the genetic geological model of hydrothermal dolomite in Central Sichuan may be used to predict dolomite reservoirs to a large extent.

Based on the genetic geological model, the seismic response characteristics of hydrothermal dolomite are summarized by the detailed dissection of seismic profile across Well Guangcan 2 (Fig. 49). The dolomites with a thickness of about 40 m are developed in the upper part of Maokou Formation in Well Guangcan 2, with the host limestone of intra-platform bioclastic beach. From the seismic profile, it can be seen that two basement faults are developed near Well Guangcan 2. These faults are tensional-shear basement faults. The dolomite section at the top of Maokou Formation is characterized by the seismic response of weak amplitude on the whole and relatively strong amplitude in thin layers with poor continuity and worm shape. While the dolomite section just corresponds to the strong amplitude (the dominant frequency of the seismic profile in this section is 45 Hz, and the minimum stratum thickness that can be distinguished is 25 m, therefore, a wave impedance interface can be occurred at the location of a dolomite stratum with the thickness of 40 m and relatively good physical properties). At the same time, the macro seismic response is characterized by depression.

To sum up, the seismic response characteristics of Qixia-Maokou Formation in Central Sichuan are analyzed and the seismic identification laws of dolomite reservoir are summarized as follows. (1) The dolomite reservoirs are mainly developed in the intra-platform beach in highstand system of Upper Qixia Formation and Middle-Upper Maokou Formation. (2) Basement faults are developed. (3) The characteristics

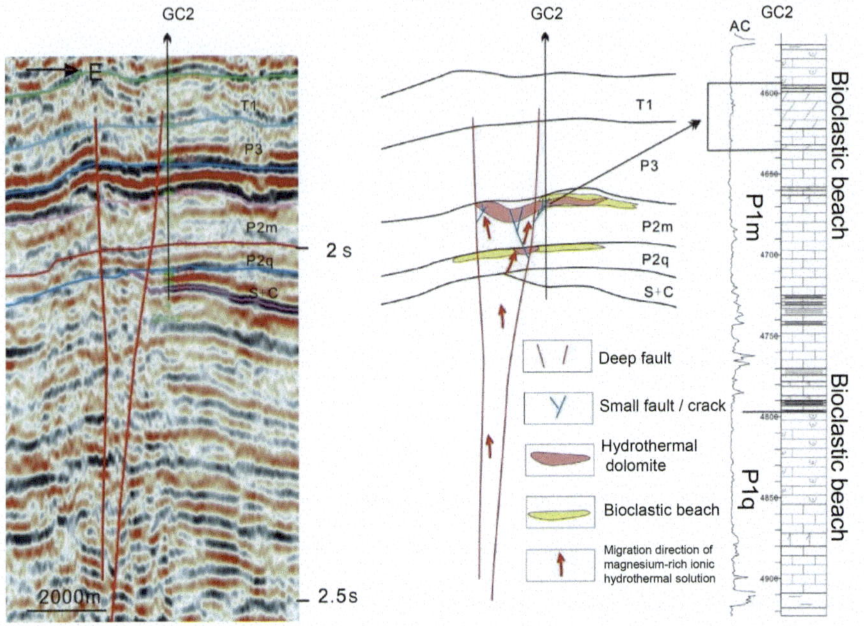

Fig. 49 Genetic interpretation profile of hydrothermal dolomites of Qixia-Maokou Formation in Central Sichuan

of depression in seismic events. (4) Strong amplitude occurs under the background of weak amplitude, and the reflection has the characteristics of worm shape and poor continuity.

4.1.3 Pore-Cave-Fracture Limestone Reservoirs Related to Weathering Crust Karstification

The positive transformation of carbonate rocks can form Pore-cave-fracture reservoirs through weathering crust karstification. The weathering crust karstification forms the large-scale pore-cave-fracture reservoirs in the regional unconformity of Top Maokou Formation, which has long been confirmed by the exploration results of Sichuan Basin, and this study will not detail here.

4.1.4 Reservoir Correlation and Distribution

The reservoirs of Qixia Formation are mainly controlled by high-energy facies belt and hydrothermal dolomitization, and the reservoirs of Maokou Formation are mainly controlled by hydrothermal dolomitization and weathering crust karstification. From the reservoir correlation profile across Well Moxi 16-Moxi 46-Gaoshi 16-Gaoshi

113-Hetan 1 and Well Gaoshi 21-Gaoshi 112-Gaoshi 113-Tongtan 1-Nanchong 7-Guangtan 2, it can be seen that hydrothermal dolomite reservoirs of Qixia Formation are developed with large thickness difference and poor horizontal continuity. They are mainly developed in the middle and upper parts of Qixia Formation and may be controlled by high-energy facies belt. From the reservoir correlation profile across Well Moxi 16-Moxi 46-Gaoshi 16-Gaoshi 113-Hetan 1 and Well Gaoshi 21-Gaoshi 112-Gaoshi 113-Tongtan 1-Nanchong 7-Guangtan 2, it can be seen that the fracture-cave reservoirs in karst weathering crust are intensively developed at the Top Maokou Formation with quasi-layers and a thickness of 10–30 m. Hydrothermal dolomite reservoirs are developed with large thickness variation and poor horizontal continuity, which are developed in any well section of Maokou Formation.

4.2 Reservoir Prediction

4.2.1 Reservoir Prediction of Qixia Formation

Petrophysical Characteristics of Reservoir

The velocity and density of granular beach reservoirs in Qixia Formation may decrease significantly due to the reservoir space, and the density is the better identifier for the granular beach reservoirs of Qixia Formation. Therefore, the linear relationship between density curve and porosity can be established, and the porosity curve is obtained (sample wells: Gaoshi 16, Gaoshi 112, Hetan 1, Moxi 23, Gaoshi 18, Moxi 41, Nanchong 7, Tongtan 1 and Lai 1). Based on the reservoir classification obtained by reservoir interpretation results through logging interpretation, the threshold value of reservoir porosity is 4.0%. As a result, the porosity curve may be used to identify the Qixia Formation's granular beach reservoir (Fig. 50).

Reservoir Predicted by Waveform Indication Inversion

The velocity of granular beach reservoir will be significantly lower and form a large wave impedance difference with the surrounding rocks. Petrophysical chart analysis shows that there are obvious differences in porosity between granular beach reservoir and nonreservoir of Qixia Formation, which is the basis of porosity prediction. In this study, the seismic waveform indication inversion method is adopted, and the inversion work is carried out by using 3D seismic data and logging curves. Based on the great horizontal change of carbonate reef reservoir, the reliability of the inversion results should be comprehensively considered in combination with other prediction methods. Figure 45 shows the porosity inversion results of Well Gaoshi 112—Gaoshi 16—Moxi 23—Moxi 39. It can be seen from Fig. 51 that the porosity inversion results can better present the reservoir characteristics of granular beach in Qixia Formation.

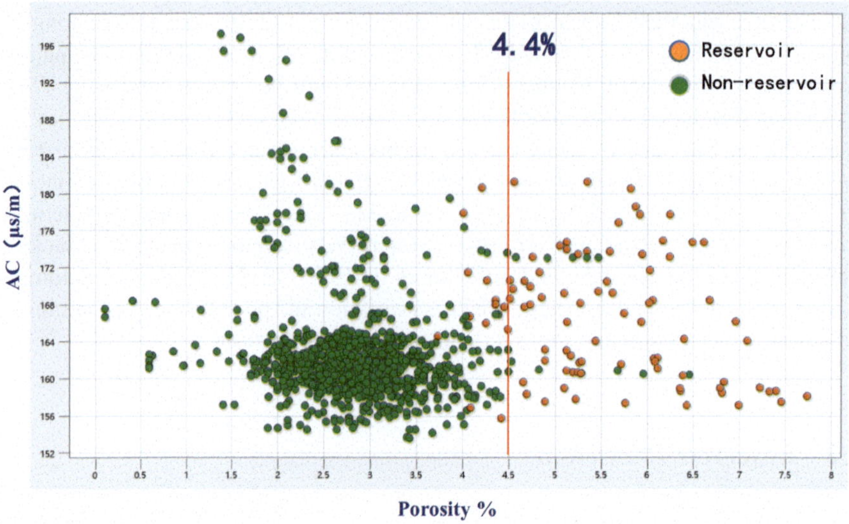

Fig. 50 Cross plot of porosity and AC of Qixia Formation

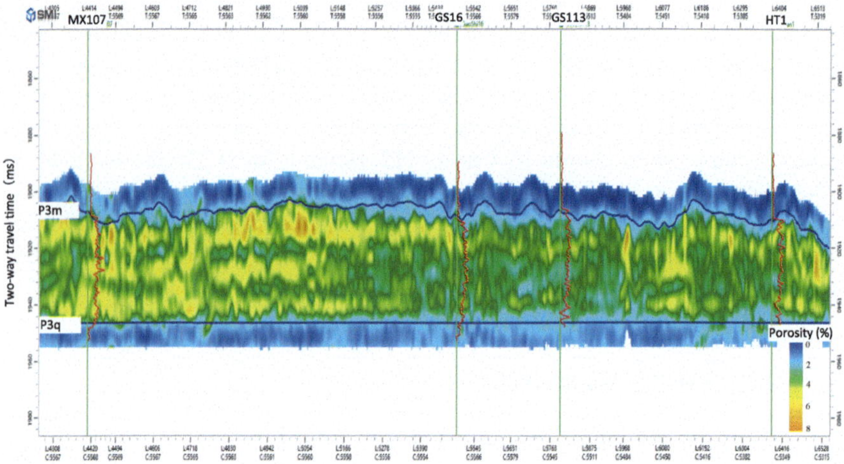

Fig. 51 Porosity inversion profile of Well Moxi 107-Gaoshi 16-Gaoshi 113-Hetan 1

By further extracting the average porosity plan (Fig. 52) and reservoir thickness map (Fig. 53) of Qixia Formation, it can be seen that the distribution characteristics of granular beach in Qixia Formation can be clearly described by waveform indication inversion. The area of favorable reservoir predicted by inversion of Qixia Formation described in 3D area is 839 km^2 that coincides with logging interpretation to an extent of 91%.

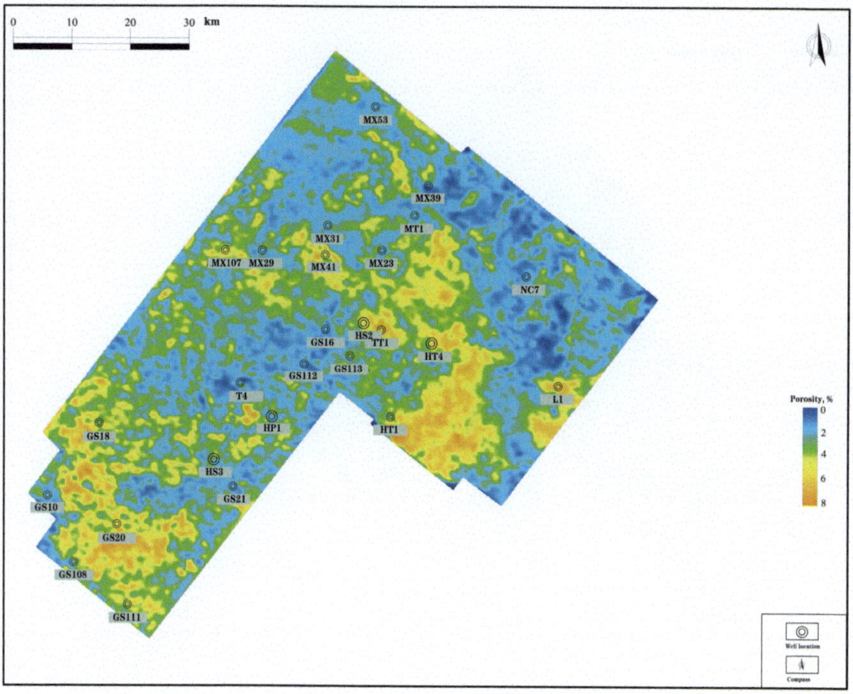

Fig. 52 Porosity plan of Qixia Formation in 3D area of Central Sichuan

4.2.2 Reservoir Prediction of Maokou Formation

Reservoir Petrophysical Characteristics

The velocity and density of granular beach reservoirs and fault-karst reservoirs in Maokou Formation may decrease significantly due to the reservoir space, and the density is a better identifier of the fault-karst reservoirs of Maokou Formation. Therefore, the linear relationship between density curve and porosity can be established, and the porosity curve is obtained (sample wells: Gaoshi 16, Gaoshi 112, Hetian 1, Moxi 23, Moxi 39, Gaoshi 112, Nanchong 7, Tongtan 1 and Lai 1). The porosity curve can be used to identify reservoirs in Maokou Formation as the threshold value of reservoir porosity is 3.6% (Fig. 54).

Reservoir Predicted by Waveform Indication Inversion

Pores, caves and fractures are developed in fault-krast reservoirs and hydrothermal dolomite reservoirs in Maokou Formation. Petrophysical chart analysis shows that there are obvious differences in porosity between granular beach reservoir and non-reservoir, which is the basis of porosity prediction. In this study, the seismic waveform

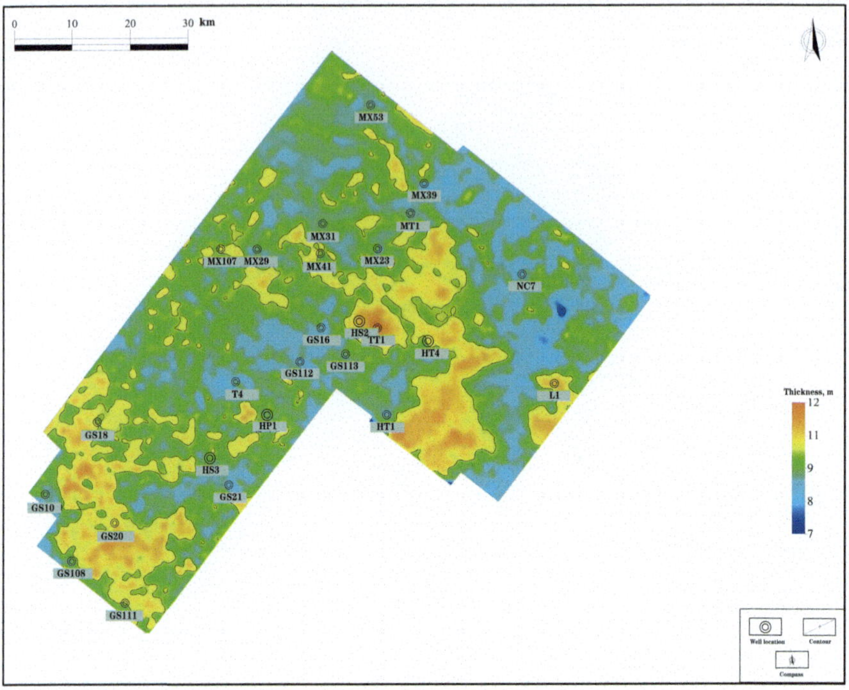

Fig. 53 Reservoir thickness of Qixia Formation in 3D area of Central Sichuan

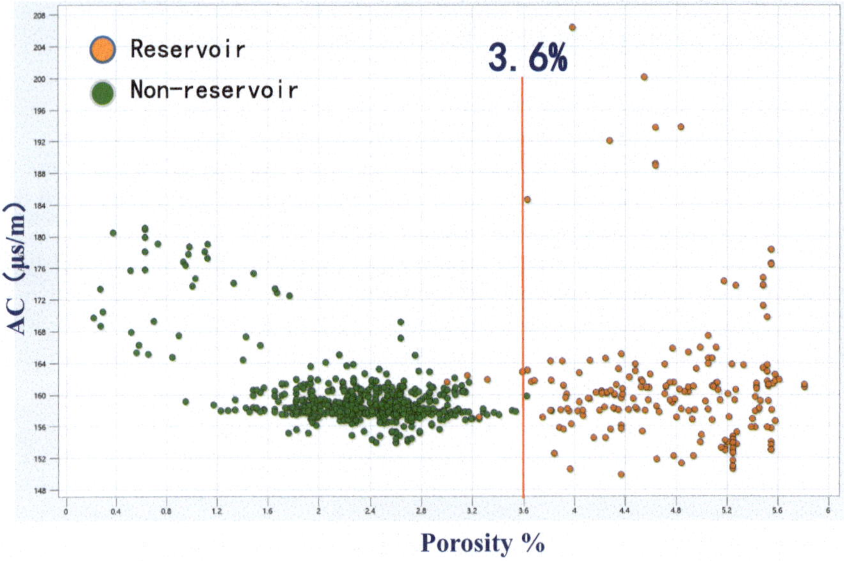

Fig. 54 Cross plot of acoustic curve and porosity curve of Maokou Formation

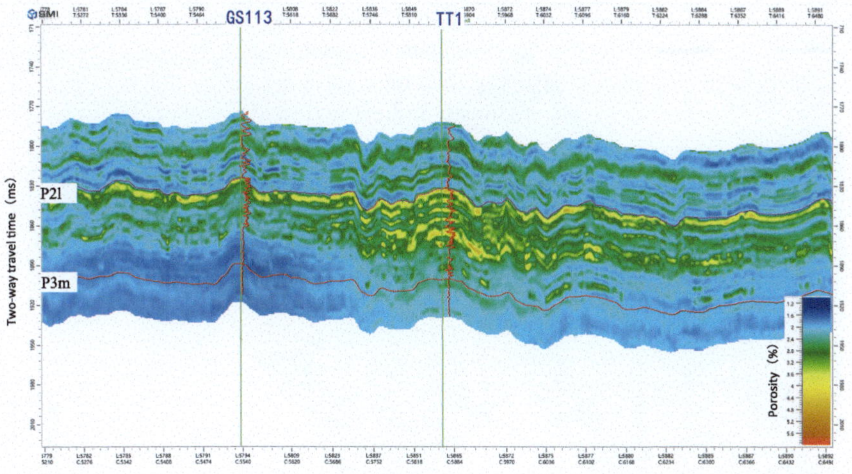

Fig. 55 Porosity inversion profile across Well Gaoshi 113-Tongtan 1

indication inversion method is adopted, and the inversion work is carried out by using 3D seismic data and logging curves. Based on the great horizontal change of reservoir, the reliability of the inversion results should be comprehensively considered in combination with other prediction methods. Figure 55 shows the porosity inversion results across Well Gaoshi 113-Tongtan 1, and it can be seen that the superposition relationship between reservoirs and interlayers vertically is clear in inversion results. As a result, the characteristics of fault-karst reservoirs and granular beach reservoirs in Maokou Formation can be better reflected by the wave impedance inversion results.

By further extracting the porosity plan of Maokou Formation (Fig. 56), it can be seen that the high-porosity areas are mainly distributed in well area of Tong 4, Tongtan 1, Moxi 39, Nanchong 7, and southern, western and eastern parts of well area Lai 1, and the favorable reservoir area predicted by the waveform indication inversion is 915 km^2.

4.3 Oil and Gas Detection

In the process of CM oil and gas detection, the characteristic informations such as amplitude and frequency of seismic data can be analyzed by the method of cumulative energy normal probability analysis. The attribute parameters such as cumulative energy of low-frequency and high-frequency bands and its ratio to the global energy can be calculated based on low-frequency and high-frequency bands determined by sensitive frequency points.

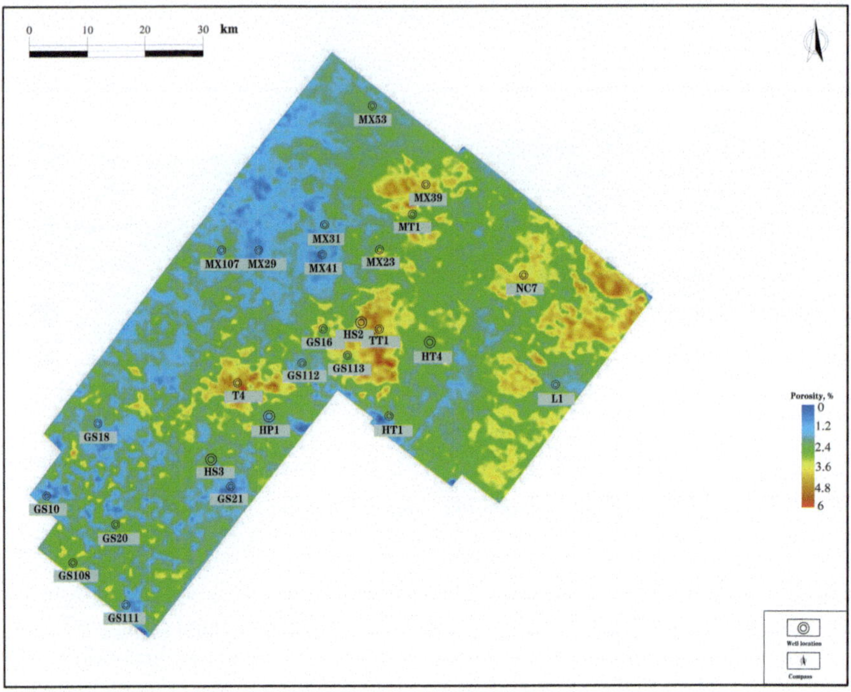

Fig. 56 Porosity plan of Maokou Formation in 3D area

4.3.1 Oil and Gas Detection of Qixia Formation

The analysis results of drilling frequency amplitude spectrum show that (Fig. 57), the daily gas production of Well Gaoshi 18 is 4.174×10^5 m^3, showing typical bimodal characteristics. The relative amplitude peak in low frequency band is 0.6–0.8, which is distributed in low frequency-band within 15 Hz, and the drop from the sub peak in the high frequency band is very small. The high frequency-band is distributed at 20–60 Hz. Well Gaoshi 113 is tested as a dry layer with week low frequency-band energy, showing no obvious bimodal characteristics, and the high frequency-band is distributed at 30–80 Hz, which is 10–20 Hz higher than Well Gaoshi 18.

It can be seen from the oil and gas detection profiles of Well Gaoshi 16 and Gaoshi 113 in Fig. 58 that the low/high frequency energy ratio attribute (Div) in Well Gaoshi 18 is obviously abnormal.

The plane distribution of Div attribute in Qixia Formation (Fig. 59) is further extracted. Through comparison with test results and logging interpretations, the coincidence rate of Div is 84.6%.

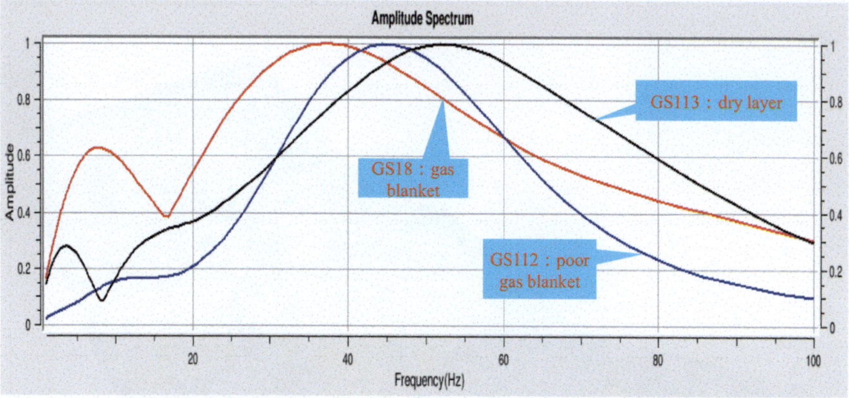

Fig. 57 Analysis of frequency amplitude spectrum in Qixia Formation

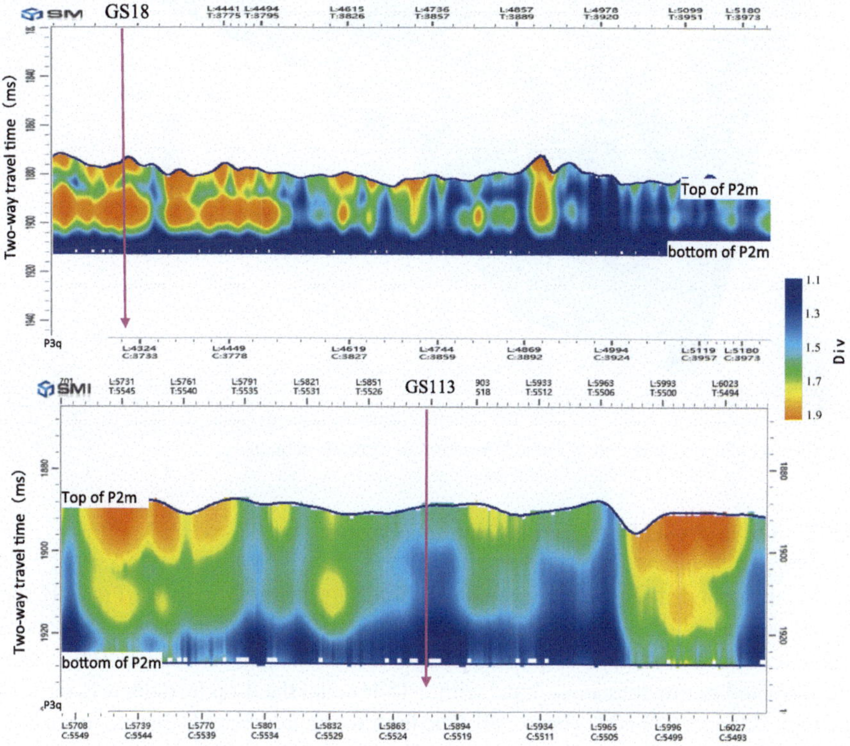

Fig. 58 Oil and gas detection profile across Well Gaoshi 18-Gaoshi 113

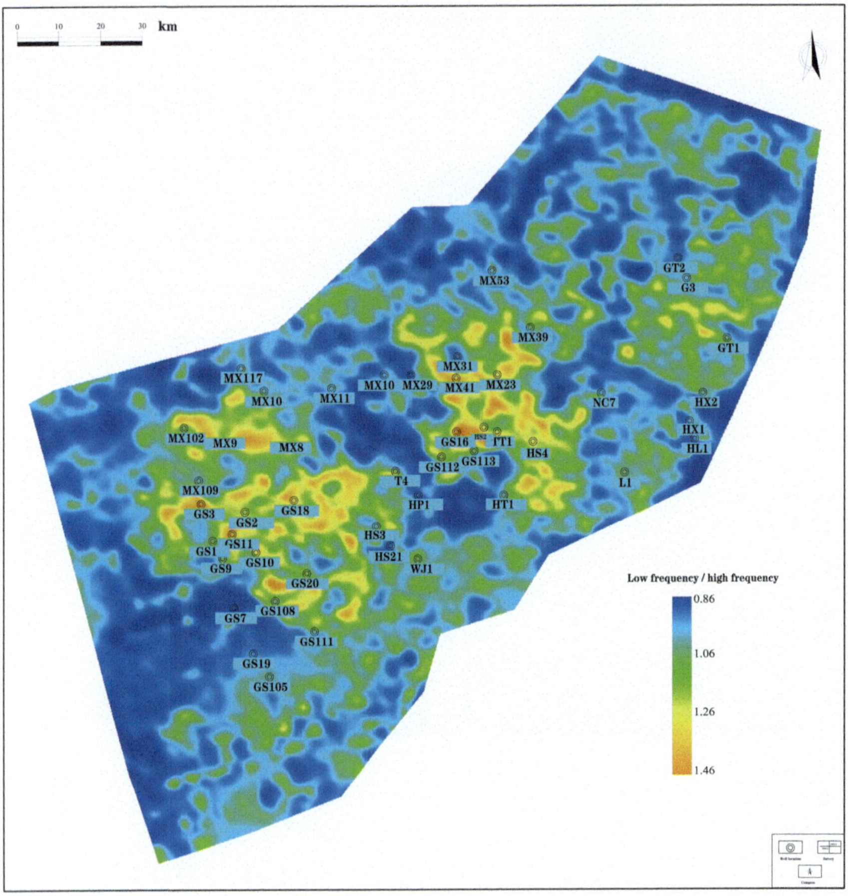

Fig. 59 Oil and gas detection of Qixia Formation in Central Sichuan

4.3.2 Oil and Gas Detection in Maokou Formation

The analysis of drilling frequency and amplitude spectrum shows that (Fig. 60), the daily gas production of Well Moxi 39 is 2.4679×10^5 m^3, showing a typical bimodal characteristic. The relative amplitude peak in low frequency band is 0.6–0.8, which is distributed in low frequency-band within 15 Hz, and the drop from the sub peak in the high frequency band is very small. The high frequency-band is distributed at 20–50 Hz. Well Tongtan 1 has a daily gas production of 3.1×10^5 m^3, showing similar characteristics to that in Well Moxi 39. The logging interpretation of Well Gaoshi 16 is poor gas reservoir with week low-frequency energy, and the high-frequency band is 20–50 Hz. Moxi 41 is tested as a dry layer with no obvious bimodal characteristics, that is, the energy of low-frequency band is weak, and the high frequency band is

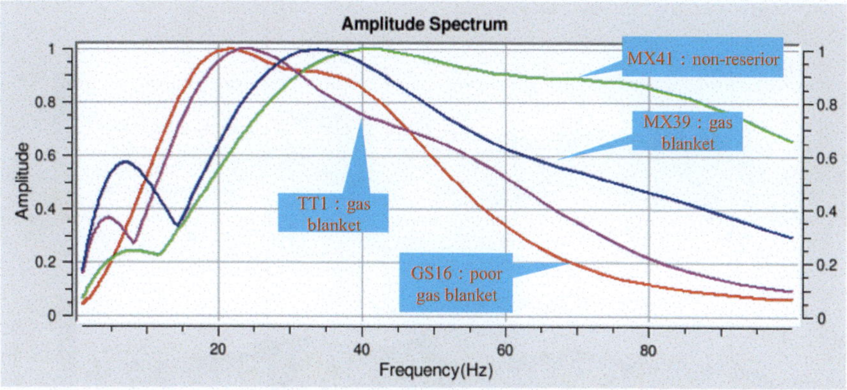

Fig. 60 Analysis of frequency amplitude spectrum in Maokou Formation

distributed at 40–80 Hz, which is 20–30 Hz higher than that of Moxi 39 well and Tongtan 1 well.

The low/high frequency energy ratio (upper), the high-frequency band (middle) and the low-frequency band (lower) energy profiles as seen in Fig. 61 across Well Gaoshi 16-Tongtan 1 indicate that the gas-bearing property of the Maokou Formation can be well predicted based on these three attributes in water layers. By comparing the test results and logging interpretation results of 42 wells, 37 wells are coincidence, and the coincidence rate is 85.7% (Fig. 62).

5 Reservoir Characteristics and Distribution Prediction of the Permian Changxing Formation

5.1 Genetic Types and Characteristics of Reservoirs

5.1.1 Reservoir Petrological Characteristics

Based on the analysis of drilling outcrop data from Well Lai 1 and the adjacent areas, the main lithology of favorable reservoirs in Changxing Formation is dolomite. On this basis, the dolomite is further identified as residual structure dolomite and crystal dolomite. The pore type of Changxing Formation is relatively single, and is mainly developed as dissolution pores.

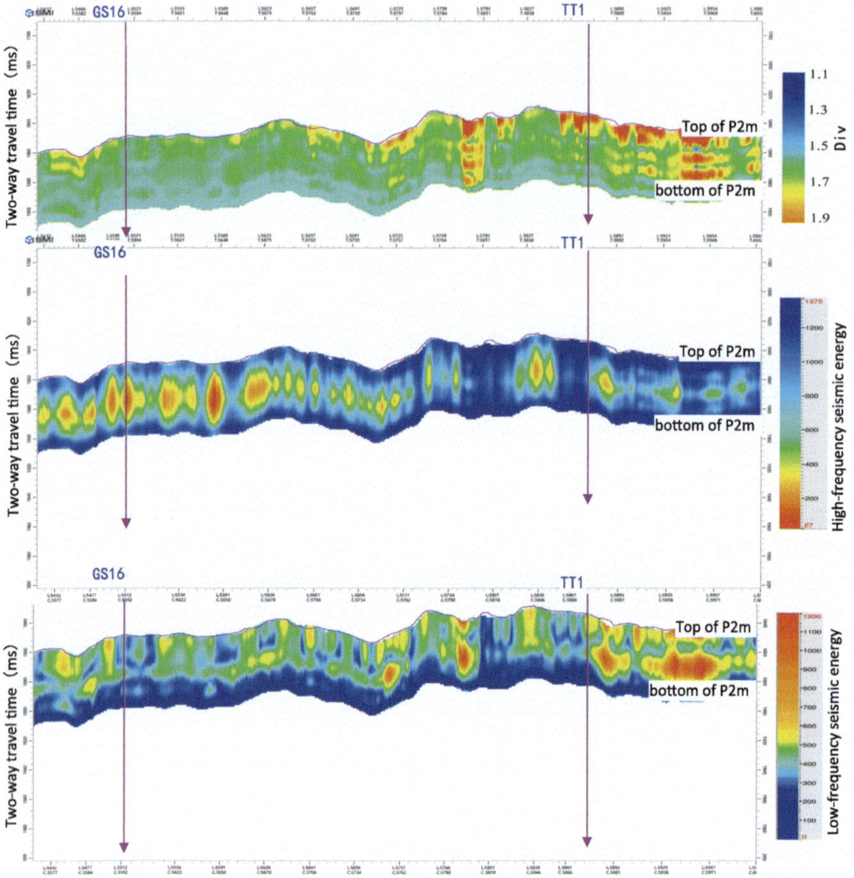

Fig. 61 Low/high frequency energy ratio (upper), high-frequency band (middle) and low-frequency band (lower) energy profiles across Well Gaoshi 16-Tongtan1

The residual reef-structure dolomites are mainly developed with the color of gray-light gray, thick-massive layers, and dissolution pores. Residual bioclastic dolomites are developed with a small amount of dissolution pores. Crystal dolomite are developed with no residual structure, mainly silly-fine crystals and intercrystalline dissolution pores. It can be seen that the development degree of dissolution pores in Changxing Formation is directly proportional to dolomitization. The dissolution pores in reef dolomite and grain dolomite with high degree of dolomitization are highly developed, and the bioclastic limestone is incompletely dolomitized with low degree of pore development (Fig. 63).

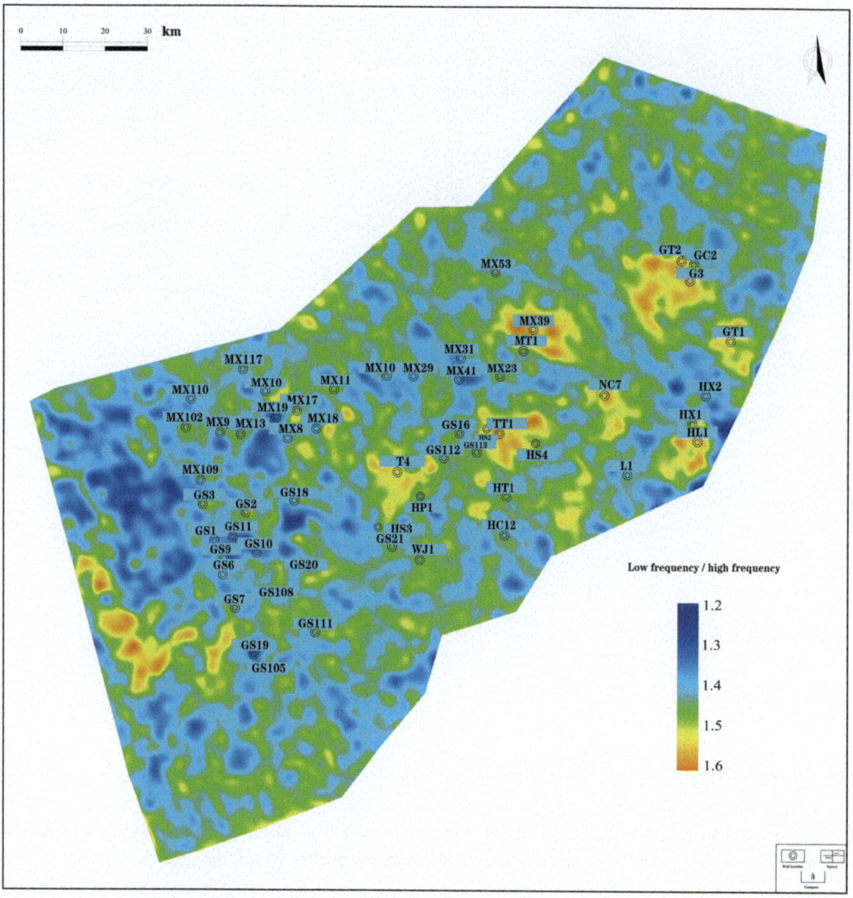

Fig. 62 Oil and gas detection of Maokou Formation in Central Sichuan

5.1.2 Reef Beach is the Basis for the Development of Favorable Dissolution Pores

By analyzing the drilling data in the research area and adjacent areas, it can be found that the dissolution pores are mainly developed in the facies belts related to reef beach, for example, high-quality reservoirs with dissolution pores which were developed in the reef beach of platform margin shown by reef beach gas reservoirs in Changxing Formation, Puguang gas field and Panlongdong outcrop profile. High-quality dissolution pores are developed in the reefs of Well Guang 3. In carbonate rocks, high-energy facies belt is usually the material basis for the porous reservoirs, and the understandings above also further support this universal law.

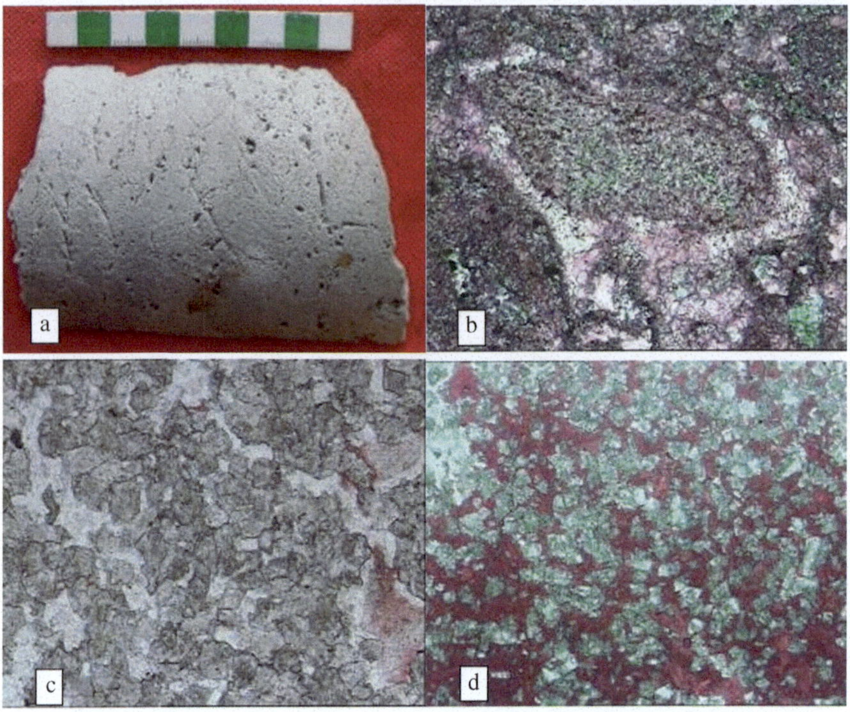

Fig. 63 Petrological characteristics of reservoirs in Changxing Formation. **a** Light-gray residual biological dolomite, with a large number of dissolution pores, Well Guang 3, 4211 m. **b** Bioclastic limestone, with a small amount of dolomites and dissolution pores, Well Moxi 1, 3919 m. **c** Fine-crystalline dolomite with intercrystalline dissolution pores, Well Moxi 1, 3962 m. **d** Silty dolomite, with intercrystalline pores and intercrystalline dissolution pores, Well Guangtan 1, 4107 m

5.1.3 Dolomitization Associated with Faults is the Key to the Formation of High-Quality Reservoirs

By analyzing the cross-well seismic interpretation profile and characteristics of reservoir and gas test in many wells drilled in Changxing Formation of Central Sichuan, it can be effectively summarized that the formation of reservoirs in Changxing Formation must be associated with faults and dolomitization. Well Guang 3 was tested at the depth between 4183.2 and 4209.2 m (equivalent to the upper part of the coring section), the total hydrocarbon increased to 77%, gas invasion and kick occurred, perforation test was performed with gas production of 1539 m^3/d, water of 112 m^3/d, formation pressure of 71.4 MPa and pressure coefficient of 1.74. Coring reveals the development of reef dolomites and the dissolution pores. Based on the seismic interpretation profile of Well Guang 3, it can be seen that the well is close to a deep and large fault. Well Moxi 1 was perforated and tested for gas at the depth range of 3891–3894 m and 3896–3922 m, with water of 240 m^3/d and gas of 53.7 m^3/d. The well-developed reservoirs can be confirmed, and a deep and large fault is seen on the seismic profile across Well Moxi 1. Gas logging abnormal is only seen in the upper part of Changxing Formation in Well Guangshen 1, and no obvious fault is found on the seismic profile across this well.

Based on the above geological and seismic facts, it can be summarized and put forward that dolomitization associated with faults is the key to the formation of high-quality reservoirs. Based on the understanding of Qixia-Maokou Formation, it can be further inferred that the formation of dolomites and dolomite reservoirs of Changxing Formation is largely related to deep and large faults, and should also be formed by hydrothermal dolomitization. This finding can provide valuable support for natural gas exploration in Permian.

5.2 *Reservoir Prediction*

5.2.1 Petrophysical Characteristics of Reservoir

Reefs grow and develop in the clean waters of shallow sea with high energy and high light transmittance. Therefore, reefs are developed where there is less terrigenous materials and very low content of mud, which is represented as a low value on the gamma ray curve; it is an extremely low value for the reef core section in particular. Flint lumps or siliceous layers are often deposited in non-reef facies in Changxing Formation. Therefore, in the low-gamma section of Changxing Formation, when the flint siliceous interlayers or lumps are developed, it indicates that Changxing Formation is developed as non-reef facies, while the reef facies is developed with pure limestones and dolomites. At the same time, due to the increase of pores, the reef facies is characterized by high AC and low density, with the acoustic wave usually greater than 170 us/m, the density less than 2.65 g/cm^3, and the wave impedance falling within 14,400–17,000 s/m * g/cm^3. Generally, the wave impedance of nonreservoir is greater than 17,000 s/m * g/cm^3, while that of mudstone is significantly less than 14,400 s/m * g/cm^3 (the sample wells are Well Gaoshi 16, Gaoshi 21, Hetian 1, Moxi 23, Moxi 41, Gaoshi 112, Gaoshi 113 and Tongtan 1). Petrophysical chart analysis shows that (Fig. 64): wave impedance inversion can be used to characterize the reservoirs.

5.2.2 Reservoir Predicted by Waveform Indication Inversion

The velocity will be significantly lower when encountering gas-bearing reefs, forming a large wave impedance difference with the surrounding rocks. The petrophysical chart analysis shows an obvious impedance difference between reef reservoir, nonreservoir and mudstone of Changxing Formation, which is the basis of wave impedance inversion. In this study, the inversion method of seismic waveform indication is adopted, and 3D seismic data and logging curves was used to carry out the

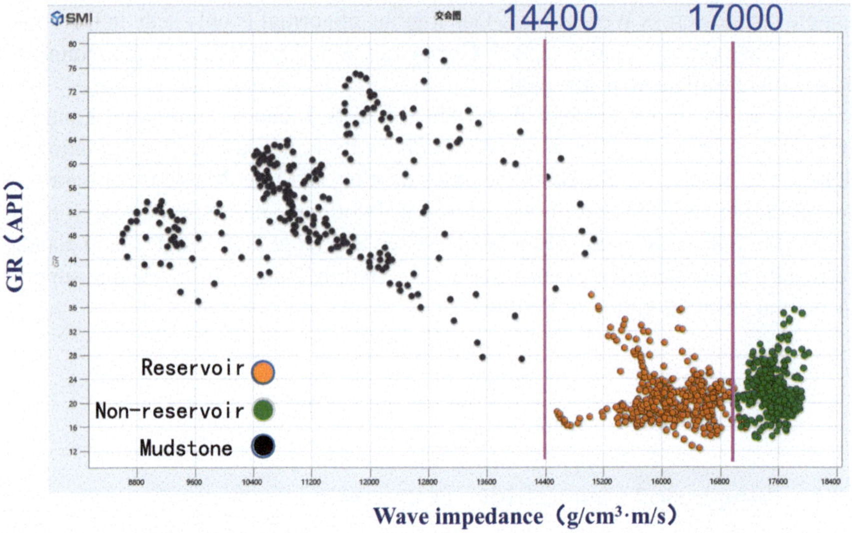

Fig. 64 Cross plot of wave impedance curve and GR curve of Changxing Formation

inversion. The reliability of the inversion results needs to be comprehensively considered in combination with other prediction methods as the carbonate reef beach reservoirs in Dengying Formation are developed with great lateral change. Figures 65 and 66 are the seismic and inversion profile across Well Gaoshi 112-Gaoshi 16-Tongtan 1-Gaoshi 113. It can be seen from the figure that the inversion results of the cross profile across a well are consistent near the well, and the reef boundary depicted by the wave impedance inversion results in the research area is basically consistent with the results in the seismic profile. Meanwhile, the wave impedance inversion results can clearly show the superposition relationship between reservoirs and interlayers in the longitudinal direction, so the wave impedance inversion results can better reflect the characteristics of reef beach reservoirs in Changxing Formation.

The reservoir thickness is extracted based on the wave impedance inversion results from the petrophysical chart. Through the comparison between the reservoir thickness interpreted by logging and that extracted from wave impedance inversion, it can be seen that the average thickness of reservoirs in Changxing Formation is 15.6 m, the average error of inversion prediction is 1.3 m, and 95.5% of the error is less than 10%, indicating that the seismic waveform inversion results are of high accuracy and can be used for the fine interpretation of reef beach reservoirs.

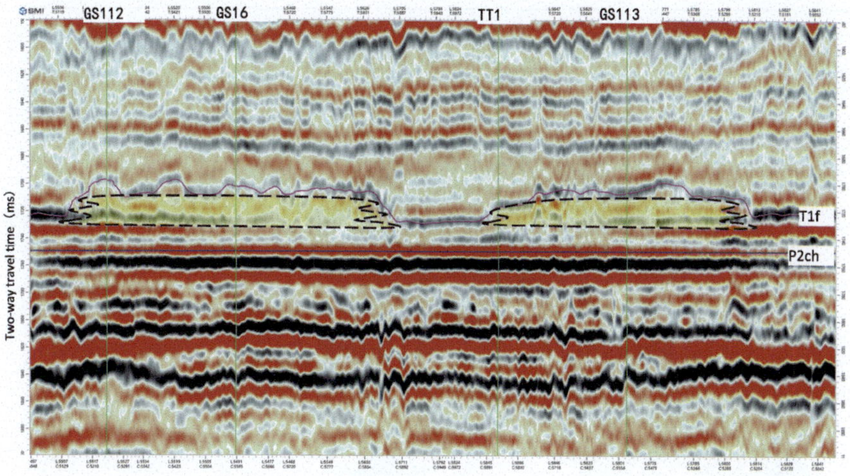

Fig. 65 Seismic profile of Well Gaoshi 112-Gaoshi 16-Tongtan 1-Gaoshi 113 (flattened P2ch)

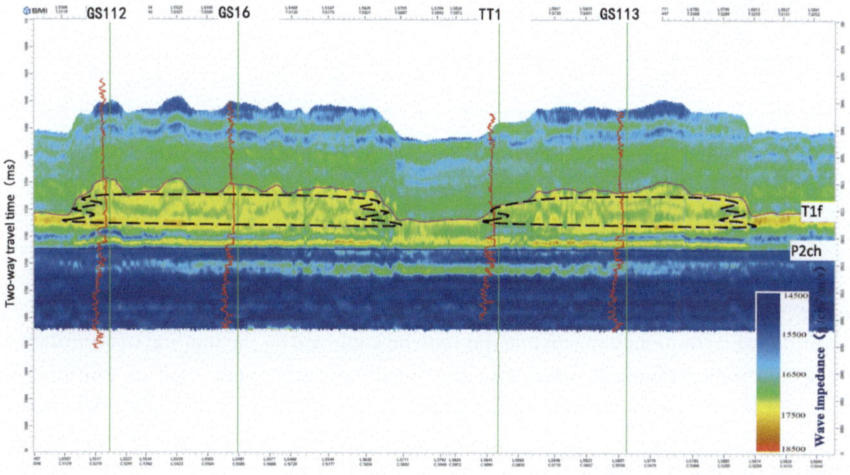

Fig. 66 Profile of wave impedance inversion in Well Gaoshi 112-Gaoshi 16-Tongtan 1-Gaoshi 113 (flattened P2ch)

Based on the inversion and prediction results from the petrophysical chart, the reservoirs of reef beach and bioclastic beach are interpreted in details. As a result, the bioclastic beach is developed with the area of 345 km^2, and the reef is developed with the area of 895 km^2, both of which are favorable exploration targets (Fig. 67).

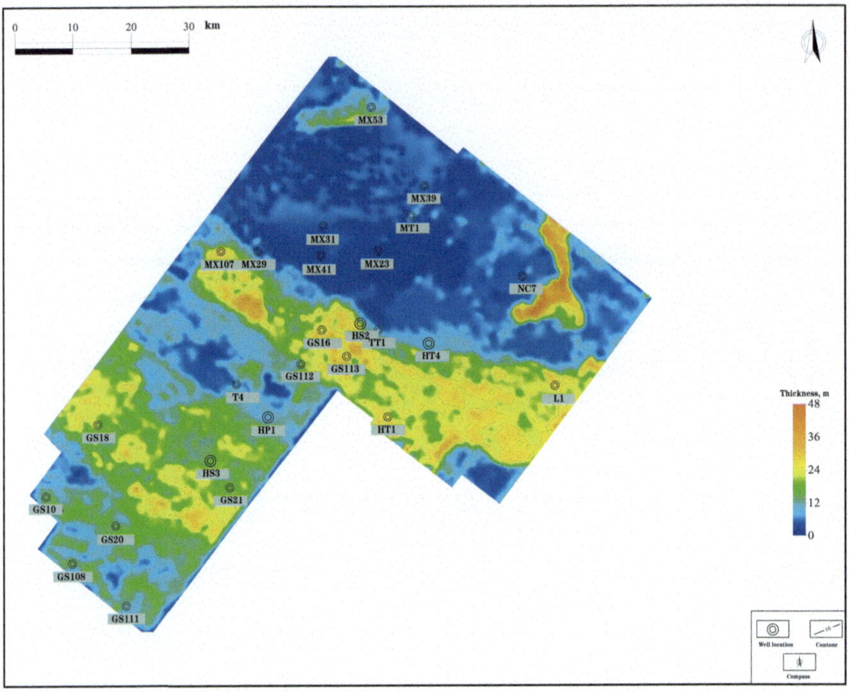

Fig. 67 Reservoir thickness of Changxing Formation in 3D area of Central Sichuan

5.3 Oil and Gas Detection

In the process of CM oil and gas detection, the characteristic informations such as amplitude and frequency of seismic data can be analyzed by the method of cumulative energy normal probability analysis. The attribute parameters such as cumulative energy of low-frequency and high-frequency bands and its ratio to the global energy can be calculated based on low-frequency and high-frequency bands determined by sensitive frequency points.

The analysis results of drilling frequency amplitude spectrum show that (Fig. 68), the daily gas production of Well Wangjia 1 is 5×10^4 m^3, showing typical bimodal characteristics. The relative amplitude peak in low-frequency band is 0.6–0.8, which is distributed in low-frequency band within 15 Hz, and the drop from the sub-peak in the high-frequency band is very small. The high-frequency band is distributed at 20–40 Hz. The daily gas production of Well Lai 1 is 1.1×10^4 m^3, showing similar characteristics to Well Wangjia 1. The Changxing Formation of Well Gaoshi 113 is tested as water layer, and of Well Moxi 7 is tested as dry layer with no prominent

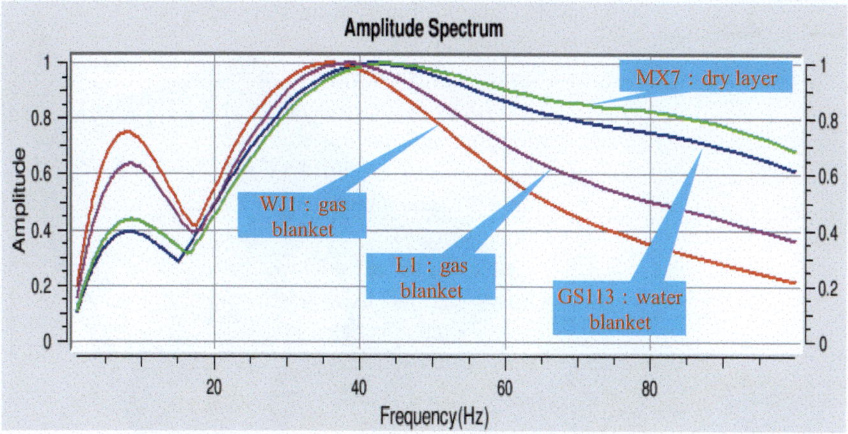

Fig. 68 Analysis of frequency amplitude spectrum in Changxing Formation

bimodal characteristics. The energy of the low-frequency band is weak, and the high-frequency band is mainly distributed at 40–80 Hz, which is 20–40 Hz higher than that of Well Wangjia 1 and Lai 1.

The gas-bearing test results of 42 wells in Changxing Formation show that 38 wells are in conformity and 4 wells are not, with a coincidence rate of 90% (Fig. 69).

6 Reservoir Characteristics and Distribution Prediction of the Triassic Leikoupo Formation

6.1 Genetic Types and Characteristics of Reservoirs

6.1.1 Petrological Characteristics of Reservoir

Based on the analysis of drilling outcrop data in Well Mo 27, Mo 29 and Sichuan Basin, the reservoir characteristics of Leikoupo Formation are mainly as follows: (1) sandy dolomite with dissolution pores; (2) dolomite with dissolution pores and caves related to weathering crust karstification.

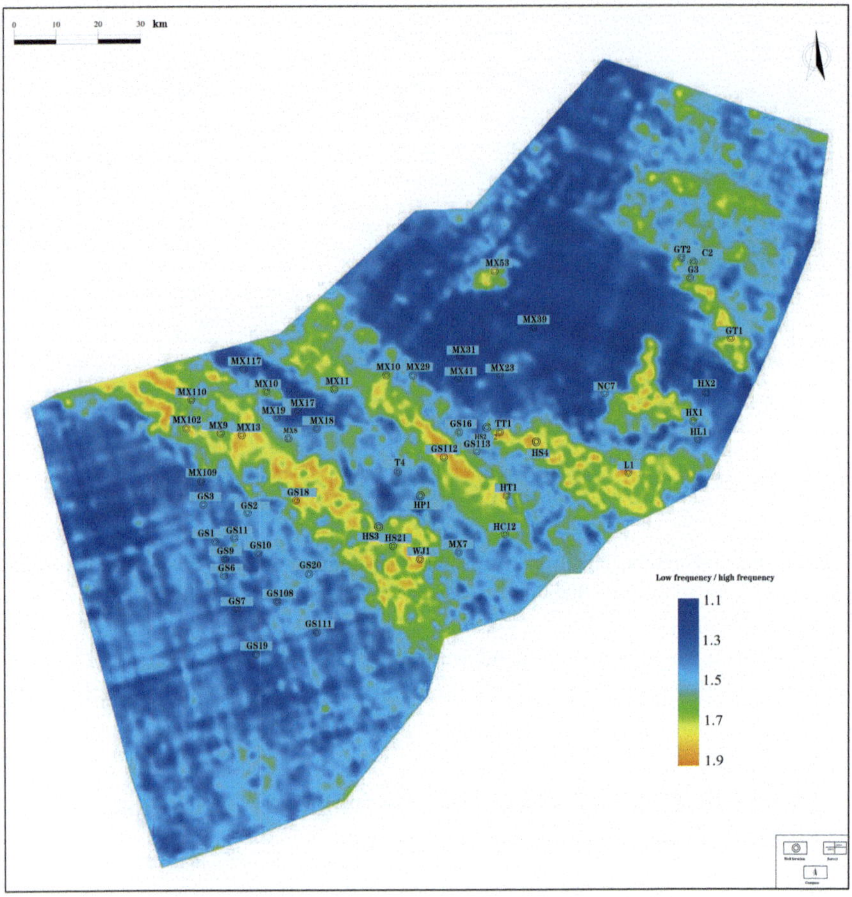

Fig. 69 Oil and gas detection of Changxing Formation in Central Sichuan

Doloarenite with Needle-Like Dissolution Pores

Doloarenites are deposited as high-energy intra-platform beaches with many needle-like dissolution pores (Fig. 70). Well Tongtan 1 is interpreted by well logging as gas layers of 45.5 m/7, poor gas layers of 29.5 m/5, and gas-bearing layers of 14.0 m/5 (Fig. 71). The gas reservoirs are mostly in Lei1 and Lei3 Member, which are the most favorable layers for doloarenite development according to sedimentological interpretation. In addition, the favorable reservoir of Lei3 Member is about 60 m away from the top unconformity, and the weathering crust karst may have little impact on it. Therefore, it can be considered that the main reservoir in the reservoir area is the doloarenite with dissolution pores, which is mainly developed in Lei1 and Lei3 Member.

Fig. 70 Petrological characteristics of reservoirs in Leikoupo Formation. **a** Doloarenite with needle-like dissolution pores, Well Mo 29, Lei1 Member, 2–38/58. **b** Doloarenite with needle-like dissolution pores, profile in Hanzeng, Jiangyou, Lei3 Member

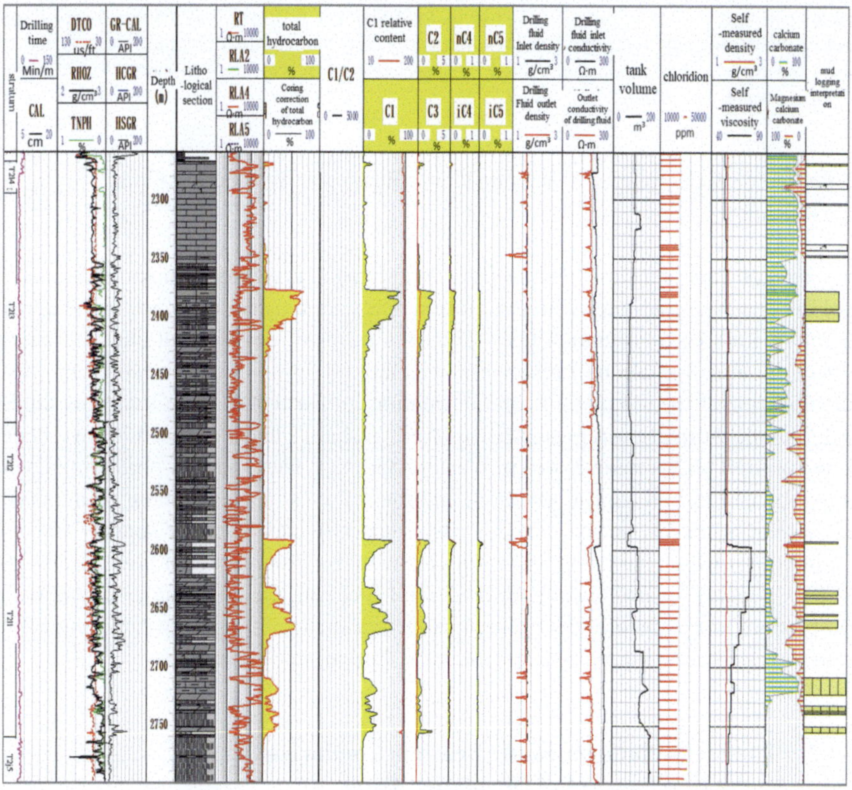

Fig. 71 Log interpretation of Leikoupo Formation in Well Tongtan 1

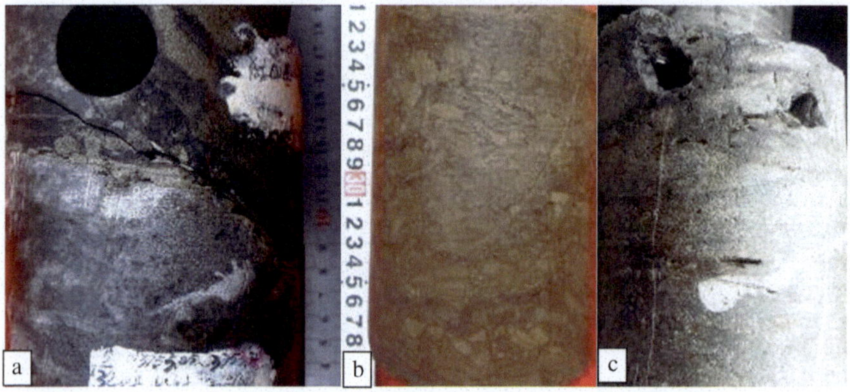

Fig. 72 Evidence of weathered crust karst. **a** Brownish-gray fine-grained dolomite, with seepage muddy siltstone, Well Longgang 19, 3753.15 m, Lei4 Member. **b** Brown-gray dissolved deposition breccia, Well Longgang 160, 3705.5 m, Lei4 Member. **c** Gray fine-grained dolomite, with dissolution pores and irregular dissolution caves, filled with seepage sand and calcite. Well Yashen 1, 5722.3 m, top of Leikoupo Formation

Dolomite with Dissolution Pores and Caves Related to Weathering Crust Karstification

Regional coring data show that seepage muddy silts, dissolved deposition breccias and fine-grained dolomites with dissolution pores are mostly developed in the upper or top of Leikoupo Formation (Fig. 72). This series of petrological characteristics reflect the transformation of reservoirs in Leikoupo Formation by weathering crust karstification at the top of Leikoupo Formation. In particular, large-scale gas fields in Leikoupo Formation represented by Well Yashen 1 and Pengshen 1 have been found in SINOPEC research area in Western Sichuan. The main gas-producing layer is the top of the Leikoupo Formation, and the reservoir space is mainly composed of dissolution pores and caves. This discovery provides valuable support for the exploration of Sichuan basin and even Hechuan-Tongnan research area. Although there is a lack of petrological data in Central Sichuan, the dolomite reservoir with dissolution pores shouldn't be ignored as the regional karst of the weathering crust at the top Leikoupo Formation can affect the formation of reservoirs.

6.1.2 The Intra-Platform Sandy Beach is Conducive to the Development of Dissolution Pores

High-energy facies belt is not only the material basis but also the soul of high-quality carbonate reservoirs. The above Dengying Formation, Longwangmiao Formation, Xixiangchi Formation, Qixia Formation and Changxing Formation are all controlled by high-energy facies belt, and it is also the case for Leikoupo Formation. According to the well logging of Well Tongtan 1, gas layers of 45.5 m/7, poor gas layers of

29.5 m/5, and gas bearing layers of 14.0 m/5 can be interpreted. The gas reservoirs are mostly in Lei1 and Lei3 Member, and the sedimentological interpretation shows that these two members are favorable layers for the development of doloarenite. Therefore, it can be considered that the main reservoir in the research area is reservoir of doloarenite with dissolution pores, which is mainly developed in Lei1 and Lei3 Member, and is mainly controlled by the high-energy facies of the intra-platform beach.

6.1.3 Pore-Cave-Fracture Reservoir Can Be Formed by Weathering Crust Karstification

Regional coring data shows that seepage muddy silts, dissolved deposition breccias and fine-grained dolomites with dissolution pores are mostly developed in the upper or top of Leikoupo Formation. This series of petrological characteristics reflect the transformation of reservoirs in Leikoupo Formation by weathering crust karstification at the top of Leikoupo Formation. Generally speaking, weathering crust karstification can actively transform carbonate reservoir forming reservoir with dissolution pore-cave-fracture. The above-mentioned Top Dengying Formation and Top Maokou Formation in Sinian are affected by this action. A regional unconformity surface is developed at the top of the Leikoupo Formation, and the weathering crust karstification can be superimposed and transformed. Although there is a lack of petrological data in the work area, this kind of reservoir can not be ignored. Despite the lack of petrological data, the reservoir of dolomite with dissolution pore-cave-fracture cannot be ignored in the research area.

6.1.4 Reservoir Correlation and Distribution

The reservoir of Lei1 Member is mainly controlled by high-energy facies belt, and the reservoir in Top Leikoupo Formation is mainly controlled by weathering crust karstification. Through the reservoir correlation profiles across Well Gaoshi 21—Hetan 1—Lai 1—Hualai 1—Huaxi 1—Guangtan 1 and Well Moxi 29—Gaoshi 16—Gaoshi 113—Hetan 1, it can be seen that Lei1 Member is developed as a thin reservoir of intra-platform beach with certain horizontal continuity, small single-layer thickness and small cumulative thickness. The Top Leikoupo Formation is mainly developed as a thin layered weathering crust reservoir with a thickness of s 20–50 m, and the weathering crust reservoir is poor.

6.2 Reservoir Prediction

6.2.1 Petrophysical Characteristics of Reservoir

The overall velocity of dolomite in Leikoupo Formation is higher than that of shale at the bottom of Xujiahe Formation, while the velocity of gypsum in Leikoupo Formation is slightly lower than that of limestone, resulting in little difference in wave impedance between limestone and gypsum. Therefore, it is difficult to identify the karst reservoir by wave impedance, while the density can better identify the karst reservoir. As a result, the porosity curve (sample wells: Well Gaoshi 16, Gaoshi 21, Hetian 1, Moxi 23, Nanchong 7, Gaoshi 112, Gaoshi 113 and Tongtan 1) is obtained by the establishment of the linear relationship between density curve and porosity. The threshold value of reservoir porosity is 4.5%, therefore, the karst weathering crust reservoir at the top of Leikoupo Formation can be identified by porosity curve (Fig. 73).

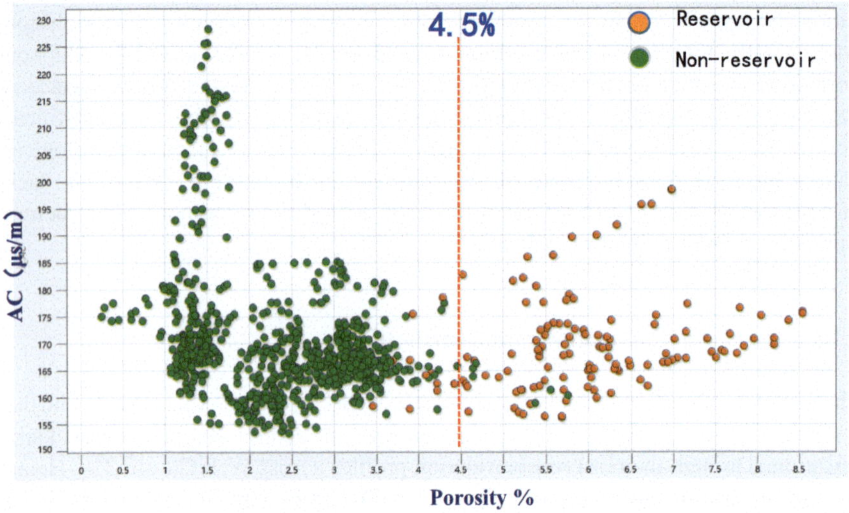

Fig. 73 Cross plot of AC and porosity of Lei3^3 Submember and Lei4 Member in Leikoupo Formation

6.2.2 Reservoir Predicted by Waveform Indication Inversion

The pores and fracture spaces are developed in the karst residual mound. The petrophysical chart analysis shows obvious differences in the porosity between karst residual mound reservoir, nonreservoir and mudstone, which is the basis for reservoir prediction through porosity inversion. In this study, the inversion method of seismic waveform indication is adopted, and the inversion is carried out by 3D seismic data and logging curves. The reliability of the inversion results needs to be comprehensively considered in combination with other prediction methods as the carbonate karst residual mound reservoirs are developed with great lateral change. Figure 74 is the porosity inversion profile across Well Moxi 107-Gaoshi 16-Gaoshi 113. It can be seen from the figure that the superposition relationship between reservoirs and interlayers in the longitudinal direction can be clearly seen from the inversion results, so the wave impedance inversion results can better reflect the characteristics of karst residual mound reservoirs.

By further extracting the porosity plan of two 3D areas in Lei3^3 Submember and Lei4 Member (Fig. 75), it can be seen that the areas with high porosity are mainly distributed in the well area of Well Gaoshi 112-Gaoshi 16, Well Tongtan 1, Well Moxi 41-Moxi 39, and Well Moxi 23-Nanchong 7. Compared with the porosities reflected by logging interpretation of 13 wells, the results show that the inversion coincidence rate of Lei3^3 Submember and Lei4 Member is 91.6%.

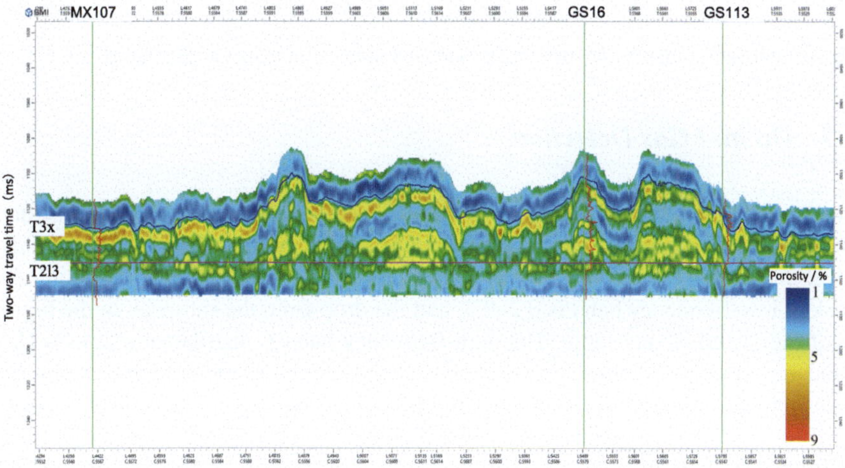

Fig. 74 Porosity inversion results across Well Moxi 107-Gaoshi 16-Gaoshi 113

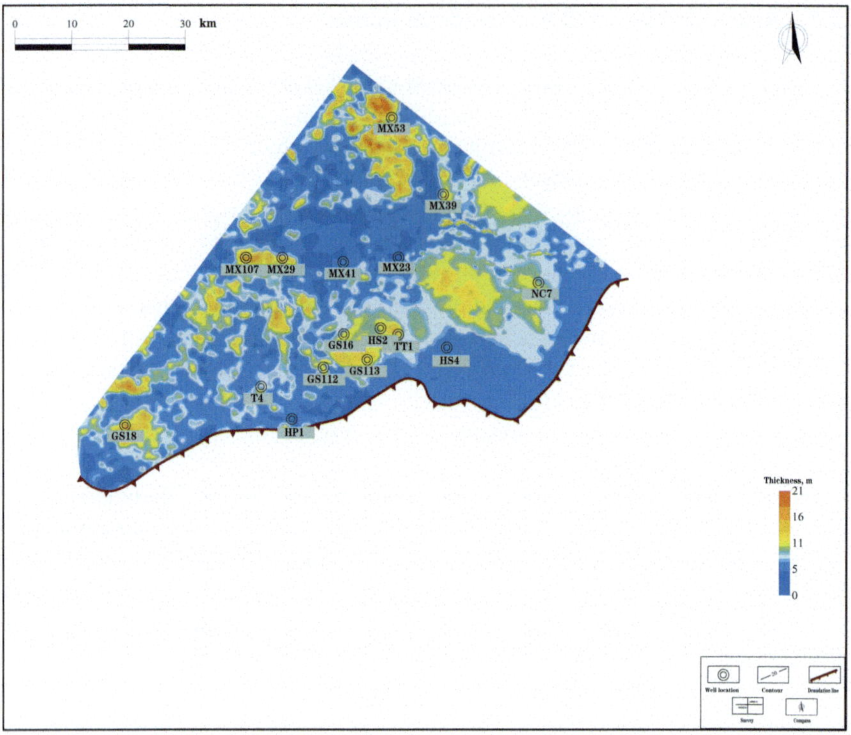

Fig. 75 Porosity plan of Lei3^3 Submember and Lei4 Member in Leikoupo Formation

6.3 *Oil and Gas Detection*

The wells drilled to Lei3^3 Submember and Lei4 Member in the research area are all interpreted as water layers. The analysis results of drilling frequency amplitude spectrum show a typical single-peak characteristic (Fig. 76). The relative amplitude peak in low-frequency band is 0.2–0.3, and the drop from the sub-peak in the high-frequency band is very large. The high-frequency band is distributed at 20–70 Hz, indicating poor gas content which is consistent with the test results. It can also be seen from the low/high frequency energy ratio attribute profile across Well Gaoshi 16 (Fig. 77) that the value is mainly distributed between 1.5 and 2, indicating poor gas constant.

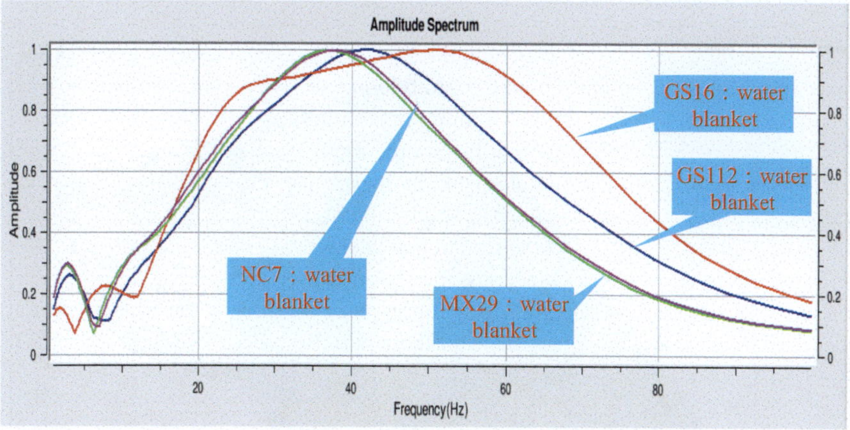

Fig. 76 Analysis of frequency amplitude spectrum in Leikoupo Formation

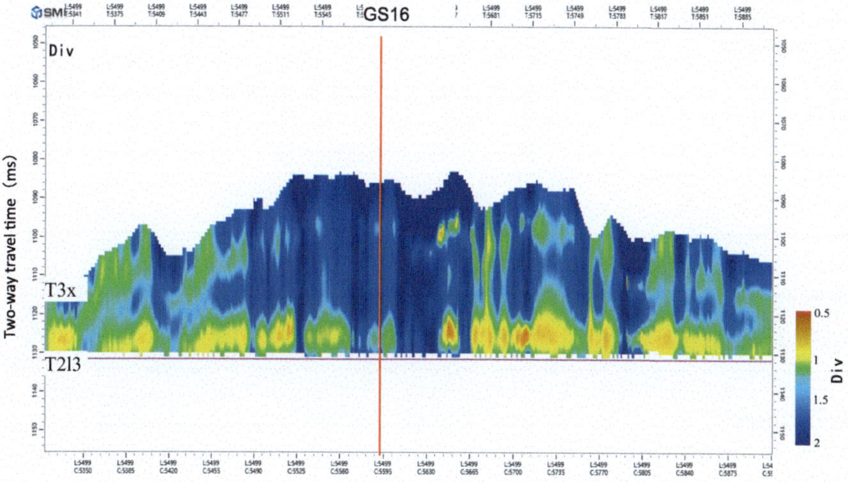

Fig. 77 Low/high frequency energy ratio attribute profile across Well Gaoshi 16

It can be seen from the oil and gas detection distribution of Lei3^3 Submember and Lei4 Member in Leikoupo Formation (Fig. 78) that the overall gas content of Lei3^3 Submember and Lei4 Member is poor. By comparing test results and logging interpretation results of 17 wells, 13 wells are consistent, and the coincidence rate is 82.4%.

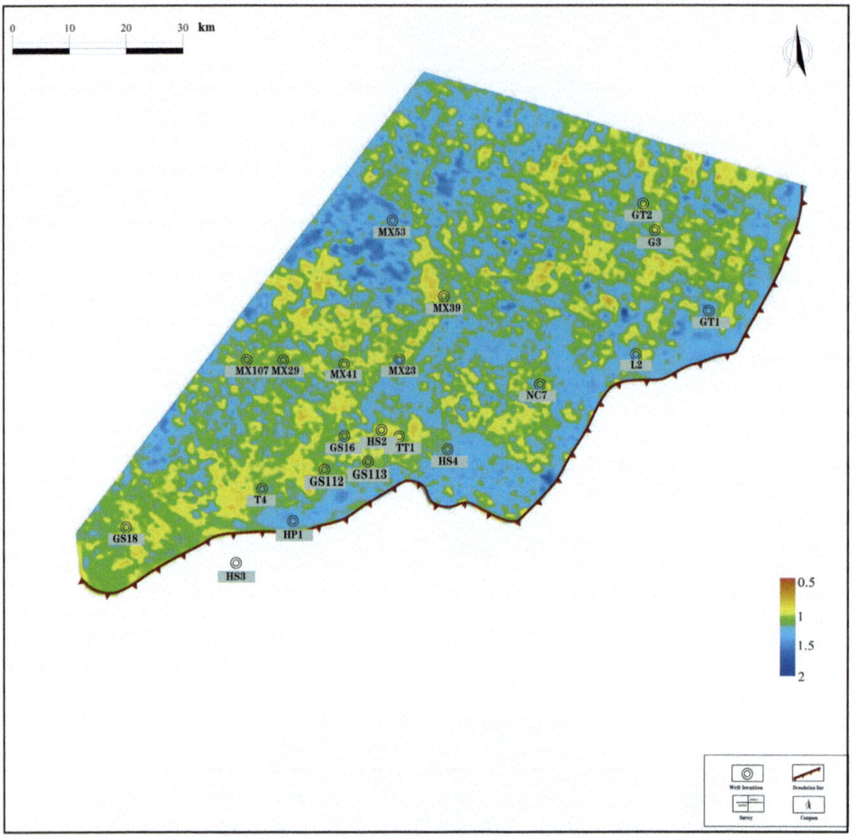

Fig. 78 Oil and gas detection of Lei3^3 Submember and Lei4 Member in Leikoupo Formation

References

Choquette, P.W. and James, N.P., 1990. Limestones, the burial diagenetic environment [J]. In: I.A. Mcllreath and D.W. Morrow (Editors), Diagenesis. Geoscience Canada, Reprint Series, 4: 75–111.

Davies G R, Smith L B. 2006. Structurally controlled hydrothermal dolomite reservoir facies: An overview [J]. AAPG Bulletin, 90(11): 1641–1690.

Tan X Ch, Xiao D, Chen J Sh, et al. 2015. New advance and enlightenment of eogenetic karst ification [J]. Journal of Palaeogeography, 17(4): 441–456.

Chapter 5
Evaluation of Natural Gas Accumulation Conditions

1 Characteristics of Accumulation Elements

1.1 Source Rock

Many sets of high-quality source rocks are developed in Central Sichuan, especially multiple sets of organic-rich muddy layers and shale layers formed during Sinian, Early Cambrian, Early Silurian, Late Permian and Late Triassic, which turn out to be important source rocks in the research area. Some limestone and coal-measure layers in the research area can also be good source rocks. The source rocks in the research area are mainly composed of organic-rich mudstones, shales and limestones (Table 1), including four sets of source rocks, such as muddy shale of Deng3 Member in Upper Sinian, black shale of Qiongzhusi Formation in Cambrian, Longmaxi Formation of Lower Silurian and Longtan Formation of Upper Permian.

1.2 Reservoir

In the Central Sichuan region, many sets of reservoirs were developed. Carbonate reservoirs in this area mainly include Dengying Formation of Sinian, Longwangmiao Formation of Lower Cambrian, Xixiangchi Formation of Upper Cambrian, Tongzi Formation and Honghuayuan Formation of Lower Ordovician, Shiniulan Formation of Lower Silurian, Qixia Formation and Maokou Formation of Middle Permian, Changxing Formation of Upper Permian, Jialingjiang Formation of Lower Triassic and Leikoupo Formation of Middle Triassic.

Q. Li and T. Wang, *Paleogeography and Gas Exploration of Marine Multiple Formations in Central Sichuan Basin*, https://doi.org/10.1007/978-981-99-2528-5_5

Table 1 Characteristics of source rocks in research area (some data are quoted from Wei et al. 2017; Huang et al. 2019)

Source rock	Lithology	Sedimentary environment	Thickness	TOC	Ro/Rb (%)	Types of organic matter	Maturity
Deng3 Member	Black shale	Marine deep-water shelf	15–74	0.04–4.73	3.16–3.21	I	Overmature
Qiongzhusi Formation	Dark grey and grey-black shale	Marine deep-water shelf	108–505	0.46–4.9	3.0–4.0 (Ro) 2.95–4.05 (Rb)	I	Overmature
Wufeng Formation	Black mudstone and shale	Marine deep-water shelf	4.2–9	0.85–3.16	2.16–3.38	I	Overmature
Longmaxi Formation	Black mudstone and shale	Marine deep-water shelf	0–80.7	0.50–8.75	2.0–3.6 (Ro) 3.14–3.38 (Rb)	I	Overmature
Liangshan Formation	Black mudstone, shale and coal bed	Tidal flat-shore	4–14	1.0–4.4	> 2.0	III	Overmature
Qixia Formation	Muddy limestone, bioclastic limestone	Carbonate platform	90–305	0.4–2.0	< 2.0	$I–II_1$	Overmature
Longtan Formation	Grey-black mudstone, black carbonaceous mudstone	Tidal flat, shore, carbonate platform	111–176	1.30–14.36	1.5–2.15	II	Overmature
Xujiahe Formation	Grey-black mudstone and shale mixed with coal bed	Fluvial facies	50–100	1.08–3.56	1.25–1.80	$II_2–III$	Mature-high mature
Lower Jurassic	Dark grey mudstone, black shale	Lacustrine facies	20–30	0.1–1.9	0.8–1.2	III	Mature

Dengying Formation of Upper Sinian is mainly composed of algal dolomites, doloarenites, silty-fine grained dolomites, micritic dolomites and other carbonate rocks, with dissolution fractures, caves, caved algal dolomites and doloarenites serving as the main reservoirs of this Formation. The reservoir pores of Dengying Formation are diverse, including primary pores such as intercrystal pores, intergranular pores and mold pores, as well as secondary dissolution pores such as intercrystal dissolution pores, intergranular dissolution pores, intragranular dissolution pores and fractures, and the secondary dissolution pores, caves and fractures are the main reservoir space.

Dolomite reservoirs in the Longwangmiao and Xixiangchi Formations are among the Cambrian reservoirs. The lower part of Longwangmiao Formation is mainly gray and dark-gray limestone, and the upper part is dolomite mixed with muddy dolomite, with a thickness of 87 ~ 127 m. The northwest margin of the research area is the denudation area of Longwangmiao Formation, and the thickness of dolomite gradually increases southeastwards. The Longwangmiao Formation in Southeast Sichuan is mainly developed with main reservoir pores such as intercrystal pores, intercrystal dissolution pores, intergranular pores and intergranular dissolution pores, followed by such secondary reservoir spaces as structural fractures, dissolution fractures and pressure-solution seams. Among them, doloarenite is the most important reservoir of this Formation.

Xixiangchi Formation is mainly composed of gray and grayish-brown thick micritic dolomites, microcrystal dolomites and silty dolomites. Some areas are developed with a small amount of siltstones and dolomitic siltstones, and the bottom is mainly composed of fine-grained quartz sandstones mixed with dolomitic mudstones. The thickness of Xixiangchi Formation is 8 ~ 256 m, and this Formation is gradually thickened from northwest to southeast under the influence of differential uplift and denudation.

The Ordovician is mainly composed of carbonate reservoirs with tight lithology. The reservoirs in Ordovician are developed with low porosity and low permeability and an average porosity of less than 1%. The Ordovician's main reservoirs in the research region are established in the carbonate layers of the Nanjinguan and Baota Formations. Nanjinguan Formation is also known as Tongzi Formation. The reservoirs of Nanjinguan Formation are mainly composed of microcrystal limestones, micritic-silty dolomites and a small amount of sandy limestone. The reservoirs of Baota Formation are mainly composed of dark-gray nodular bioclastic limestones, bioclastic micritic limestones and gray-white mud-bearing micritic limestones.

The Permian reservoirs are mainly composed of carbonate reservoirs of Qixia-Maokou Formation of Middle Permian. Mao1 Member is mainly composed of black-gray medium and thick micritic limestone, bioclastic limestone and marl, locally intercalated with shale and muddy dolomite. It is in conformable contact with the underlying Qixia Formation, and the mud content is high as a whole. Mao2 Member is mainly composed of gray and gray-white thick limestone, the lower part is mainly composed of gray bioclastic micritic limestone and marl, the middle and upper part is mainly composed of thick bioclastic limestone and micritic limestone, mixed with a small amount of chert strips and mud strips, and the mud content decreases gradually. Mao3 Member is characterized by light-gray, gray-white massive bioclastic limestone and sparry granular limestone, and is locally composed of dolomite, algal limestone and nodular limestone. It is developed in sedimentary environment of high water energy, and the strata of Mao3 Member are relatively thin. In some areas, the Mao3 Member is denuded. Mao4 Member is mainly black-gray and dark-gray micrite limestone and bioclastic limestone, mixed with algal limestone and shale. The top of Mao4 Member is widely denuded, generally containing pyrite with massive and dense granular aggregates. Mao4 Member is developed with local dolomitization and great change in thickness. The strata of Mao4 Member in the research area are greatly denuded and are only remained in some areas. The denudation thickness decreases gradually towards northwest and southeast. The karst weathering zone at the top of the Maokou Formation is a high-quality oil and gas reservoir in the research area, and the main reservoir spaces are karst fractures and caves. Fractured reservoirs are developed in some areas with strong heterogeneity in distribution.

Triassic carbonate reservoirs include carbonate reservoirs in Jialingjiang and Leikoupo Formation. Among them, the Jialingjiang and Leikoupo Formation reservoirs are mainly composed of granular limestone and granular dolomite, and the reservoir spaces are mainly secondary pores, including internal dissolution pores and mold pores, intergranular dissolution pores, intercrystal dissolution pores, gypsum dissolution pores and karst pores. At the same time, affected by tectonic uplift, karst pores, caves and structural fractures are widely developed, which are also the important reservoir spaces.

1.3 Caprock

The caprocks in Central Sichuan mainly include gypsum-salt rocks, silty mudstones, mudstones and shales, and the gypsum-salt rocks are developed with the best sealing performance. Gypsum-salt layers are mainly developed in Triassic, including the gypsum-salt layers of Jia2, Jia4 and Jia5 Member of Jialingjiang Formation. They are characterized by thin single-layered thickness, cyclic development, and interbedded with mudstones and dolomites. In addition, thin gypsum-salt layers are still developed in Leikoupo Formation as local sealing layers.

Muddy shale and silty mudstone are also developed with good sealing performance and they can be used as caprocks in the research area with the large distribution range, great thickness variation, and the function of source rock. The silty mudstones are developed with sealing indexes slightly lower than mudstone, and can be considered as regional good caprocks. The muddy caprocks are widely distributed in the plane, and are developed in all strata. The muddy, shale and carbonate layers in Ordovician are relatively tight and can be considered as regional caprocks. The Wufeng Formation of Upper Ordovician and the Longmaxi Formation of Lower Silurian can also be considered as regional caprocks. The Hanjiadian Formation of Middle and Lower Silurian is mainly composed of gray-green tight mudstone, which is also an important set of hydrocarbon sealing layer. Regardless of the thin thickness, Longtan Formation's shale in Permian can still be used as a set of regional caprock with good sealing and stable distribution. Mud, shale and muddy dolomites of Feixianguan Formation, muddy, shale and coal layers of Xujiahe Formation in Triassic, and multiple sets of mudstone and silty mudstone in Jurassic can be used as regional caprocks.

2 Source Rock Burial and Hydrocarbon Generation Evolution

The burial evolution of source rocks and the thermal history of basins are important aspects of oil and gas resource exploration and evaluation. The thermal history determines the hydrocarbon generation and evolution process of source rocks and has an important impact on the processes of oil and gas generation, preservation and liquid oil cracking into gas. In this research, the burial thermal evolutions of four sets of source rocks in 8 wells of Central Sichuan are analyzed, laying a foundation for regional oil and gas accumulation analysis.

Multi-stage tectonic movements have been developed in Central Sichuan, and the numerical values calculated according to the denudation amount are given to 8 simulated wells (Table 2). By consulting relevant literatures, it can be seen that when restoring the paleogeothermal temperature of Lower Paleozoic in Sichuan Basin, the newer the stratigraphic age, the higher the paleogeothermal temperature (Zhang et al. 2007). According to previous studies and the characteristics of paleo-uplifts in Central Sichuan, there was a value of high heat flow in the Permian. The maximum value of heat flow was 70 ~ 110 mW/m^2, and then gradually decreased to present 55 ~ 65 mW/m^2. As a result, the regional heat flow evolution history is obtained (Table 3). Boundary condition paleo-water depth and sedimentary water interface temperature (SWIT) are necessary parameters in PetroMod simulation, and play an auxiliary role in the restoration of sedimentary history. The determination of the two is difficult as there is no relevant literature. In this research, the relevant parameters of paleo-water depth are given according to the sedimentary facies, and the surface temperature is set as 20 °C (Table 4).

Since the Sinian, the source rocks in Central Sichuan have experienced a long process of sedimentary evolution, and the evolution of organic matters is closely related to the tectonic sedimentary evolution. The thermal evolution of organic matters is indirectly controlled by the sedimentary burial process of source rocks controlled by the tectonic evolution of paleo-uplifts. Taking Well Hetan 1 as an example, it can be seen from the map of sedimentary burial history (Fig. 1) that the source rocks deposited before the Longmaxi Formation of Silurian had experienced the process of burial-uplift (Caledonian)-reburial-reuplift (Himalayan movement), resulting in four main stages of hydrocarbon generation (Fig. 2). ① Initial hydrocarbon generation stage, corresponding to the sedimentary burial period from Sinian

Table 2 Statistics of basin simulated denudation of 8 wells in Central Sichuan

Tectonic period	Start and ending time (Ma)	GS16	GS21	GT2	HT1	L1	MX17	NC7	TT1
Himalayan	100–0	1600	1500	1900	1900	2200	1200	2150	1900
Late Indosinian	208–201	185	185	90	90	160	160	105	130
Early Indosinian	241–237	230	322	190	190	310	210	242	230
Dongwu II	253.9–251.9	132	105	25	25	65	37	25	60
Dongwu I	266.7–259.1	10	55	138	138	55	95	171	75
Hercynian	307–299	100	200	100	100	100	100	200	100
Caledonian	470–358.9	900	750	850	850	650	890	820	920
Yeli	494–485.4	411	174.4	472.9	472.9		213	178.4	280
Yunnan-Guizhou	509–500	290	303	138	138		226		205
Tongwan	535–526.5	168	508.76	161	161		170		

Table 3 Statistics of paleo-heat flow in Central Sichuan

Time (Ma)	Heat flow (mW/m^2)	Time (Ma)	Heat flow (mW/m^2)
510	50	210	70
480	51	180	66
360	53	150	65
300	58	120	64
270	77	0	63
240	75		

Table 4 Statistics of boundary conditions simulated by software in Central Sichuan

Time (Ma)	Water depth (m)	Time (Ma)	Surface temperature (°C)
65	0	65	20
140	0	140	20
210	20	210	20
235	50	235	20
259	0	259	20
295	80	295	20
580	100	580	20

to Silurian. During this period, the strata were buried continuously with denudation by many movements. ② Stagnant stage of hydrocarbon generation, which is shown as shallower burial of source rocks and even the denudation of source rocks in

Longmaxi Formation caused by the Caledonian tectonic uplift. ③ Secondary hydrocarbon generation stage, corresponding to the period that Permian-Cretaceous sedimentation was locally received after the Yunnan movement. Although multi-period movements caused strata denudation, the denudation thickness was not huge, as a result, the hydrocarbon generation evolution of source rocks continued. ④ Stage of slow hydrocarbon generation. During this period, the hydrocarbon generation of source rocks temperature slowly increases with time due to the uplift of strata in the Yanshan-Himalayan period and paleogeothermal temperature decline. The source rocks of the Permian and later period did not experience Caledonian movement, as a result, the sedimentary burial process of these source rocks only included the later burial-uplift stage. Therefore, the evolutions of source rocks only include the later stages of secondary hydrocarbon generation and slow hydrocarbon generation.

The thermal evolution process of source rocks, which is controlled by paleo-uplift evolution and morphological changes, has both similarity and significant differences in time and place. The buried depths, paleogeothermal temperatures and thermal evolution processes of source rocks in different parts of the research area are different in time and space due to the fold deformation and axis migration of Leshan-Longnvsi paleo-uplift. During the geological period, the thermal evolution of organic matters of the source rocks in the low part of the structure is higher than that in the high part of the uplift area. The source rocks in the low part is deeply buried and the paleogeothermal temperature is relatively high. As a result, the source rocks first entered the stages of oil and gas generation, such as Well Hetan 1 and Well Guangtan 2. The Ro of the source rocks in Deng3 Member has reached 0.5% before the Caledonian movement

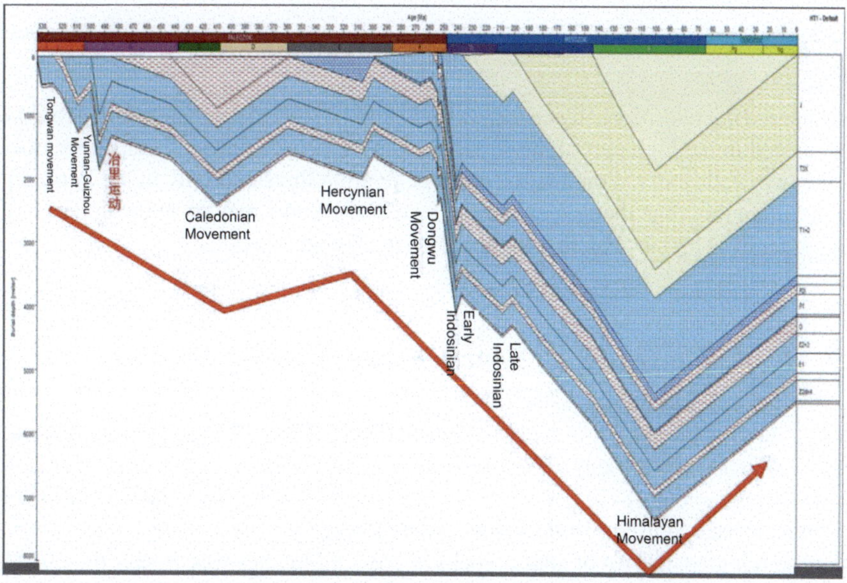

Fig. 1 Stratigraphic burial evolution of Well Hetan 1 in Central Sichuan

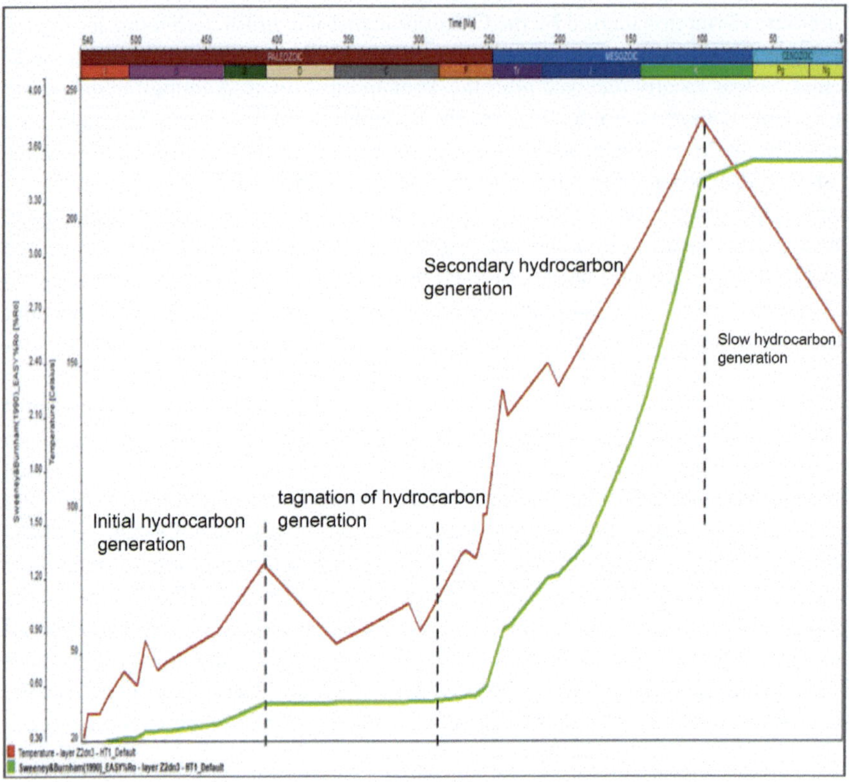

Fig. 2 Evolution of paleotemperature and Ro of source rocks in Deng3 Member of Well Hetan 1

and entered the hydrocarbon generation threshold. Well Gaoshi 16 and Well Moxi 14 are located in the high part of the structure, and the source rocks of Deng3 Member in the two wells entered the hydrocarbon generation threshold only in the Permian. At the same time, the thermal evolution degree of organic matters in the underlying source rocks is higher than that in the overlying source rocks.

2.1 Evolution History of Source Rocks in Deng3 Member

Before the Caledonian movement, the burial depth of Deng3 Member was shallow, and the source rocks in Deng3 Member were generally immature. Before the Yunnan-Guizhou movement, Ro of source rocks was less than 0.4%. With the deposition of Ordovician and Silurian, in the east of Well Guangtan 2-Nanchong 7-Hetan 1 in the research area, the Ro of source rocks in Deng3 Member was greater than 0.5% (Fig. 3), as a result, the source rocks started to mature and entered the hydrocarbon generation threshold, and the hydrocarbon generation processes are mainly oil generation. With the sedimentary burial of Permian strata, the Ro of source rocks in the research area before the period of Dongwu I was greater than 0.5%, and the source rocks entered the hydrocarbon generation threshold. At the end of Permian, in the southeast of the research area where the source rocks are deeply buried, the Ro of the source rocks was greater than 0.7%, indicating that the source rocks entered the mature stage (Fig. 4). With the deposition and burial of thicker Triassic strata, in the mature stage, marked by the Ro of 0.7–1.0%, the source rocks of Deng3 Member in Early Indosinian mainly generated liquid oil. In the Late Indosinian, the rapid hydrocarbon generation evolution of source rocks was developed by the thicker deposition of Xujiahe Formation. The Ro of source rocks in the southeast of Well Guangtan 2-Gaoshi 16 was greater than 1.3, indicating that the source rocks entered the stage of wet gas, and source rocks in other areas mainly generate oil. During the period of Yanshan-Himalayan, the source rocks in research area entered the foreland tectonic stage. After the deposition of the Cretaceous, the deposit and burial of a number of strata led to the rapid hydrocarbon generation and evolution of source rocks, especially in the southeast. The Ro of source rocks in the west of Well Guangtan 2-Nanchong 7 was greater than 4.0, indicating that the source rocks entered the over-mature stage. The Ro of source rocks in the east was between 2.8 and 4.0%, which indicated that the source rocks were in the dry gas stage.

2.2 *Evolution History of Source Rocks in the Qiongzhusi Formation*

Before the Caledonian movement, the Qiongzhusi Formation was buried shallow, and the source rocks were generally immature. The Ro of source rocks was less than 0.4% before the Yunnan-Guizhou movement. With the deposition of Ordovician and Silurian, the Ro of source rocks in the Qiongzhusi Formation before Caledonian movement was still less than 0.5%, as a result, the source rocks were immature (Figs. 5 and 6). With the sedimentary burial of Permian strata, in the east of Well Guangtan 2-Nanchong 7-Hetan 1-Gaoshi 21, the Ro was greater than 0.5%, indicating that the source rocks entered the hydrocarbon generation threshold, entered the stage of early mature, and mainly generated oil. However, the source rocks in the west were still immature. At the end of Permian, in the research area, except the source rocks around Well Moxi 17 with shallow burial depth and immature state, most source rocks were developed with Ro more than 0.5% and a mature state. With the deposition and burial of thicker Triassic strata, in the Early Indosinian, the Ro of source rocks in Qiongzhusi Formation of Well Moxi 17 and Well Nvji in the northwestern research area was less than 0.7, indicating that the source rocks were in the stage of early mature. And the source rocks in other areas were in the stage of mature with the Ro of 0.7 ~ 1.0%, and mainly generated liquid oil. In the Late Indosinian, the thicker deposition of Xujiahe Formation led to the rapid hydrocarbon generation evolution

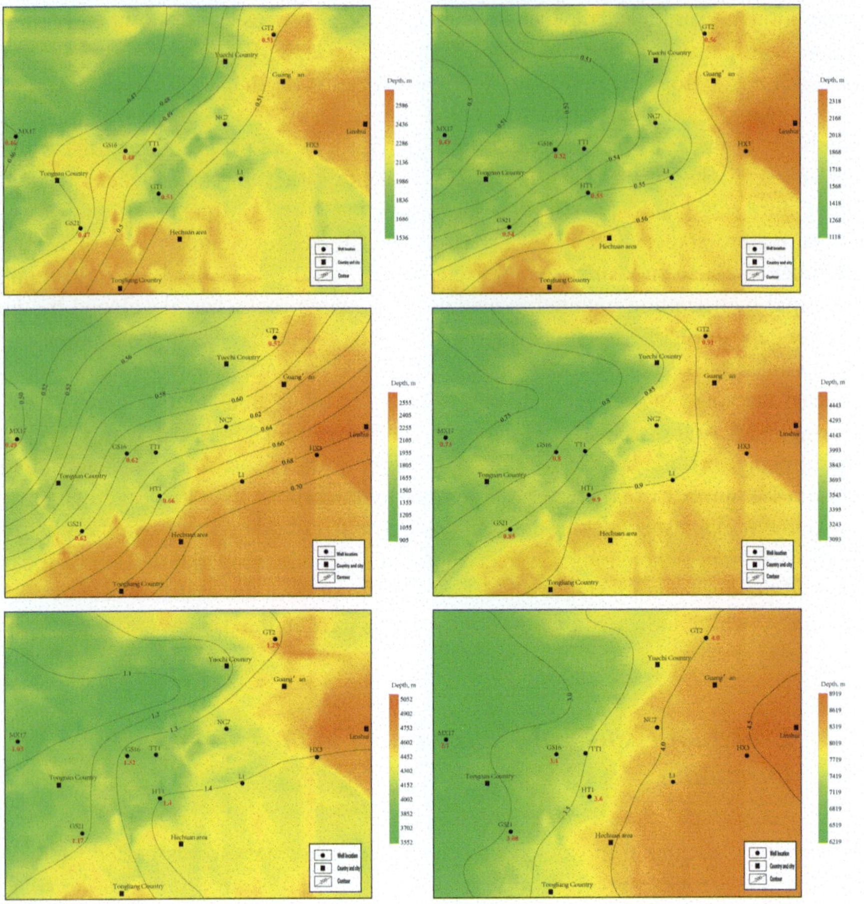

Fig. 3 Ro contours of source rocks in key period before denudation in Deng3 Member of Central Sichuan (before Caledonian, before Dongwu I episode, before Dongwu II episode, before Early Indosinian, before Late Indosinian, before Yanshan-Himalayan)

of source rocks. The source rocks in the east of Well Tongtan 1-Gaoshi 16-Gaoshi 21 were developed with the Ro more than 1.0, indicating the late mature stage. In other areas, the source rocks were developed with the Ro of 0.8 ~ 1.0%, indicating the mature stage, during which the source rocks mainly generated oil. During the period of Yanshan-Himalayan, research area entered the foreland tectonic stage. After the deposition of Cretaceous, the deposition and burial of a large number of strata led to the rapid hydrocarbon generation and evolution of source rocks, especially in the eastern research area. On the whole, the Ro of source rocks in Qiongzhusi Formation was 2.5 ~ 4.0%, indicating the dry gas stage. In the eastern part, the Ro of the Huayingshan-Linshui area was greater than 4.0, indicating that the source rocks entered the over-mature stage.

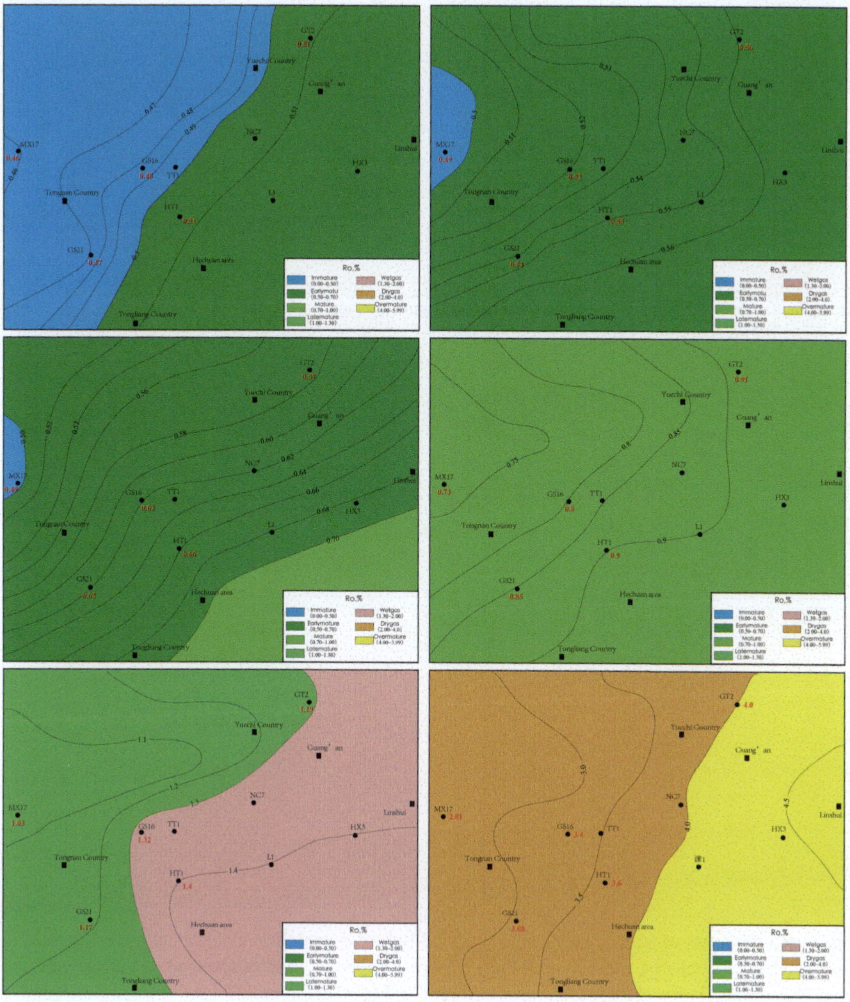

Fig. 4 Hydrocarbon generation evolution of source rocks before denudation in Deng3 Member of Central Sichuan (before Caledonian, before Dongwu I episode, before Dongwu II episode, before Early Indosinian, before Late Indosinian, before Yanshan-Himalayan)

2.3 Evolution History of Source Rocks in the Longmaxi Formation

The Longmaxi Formation in the northwest of the research area was totally denuded by the Caledonian movement, and the residual strata were evolved gradually with the deposition and burial of Permian strata. However, the Ro of source rocks in the research area before Dongwu II was less than 0.4%, indicating that the source rocks were immature (Figs. 7 and 8). With the deposition and burial of thick Triassic strata,

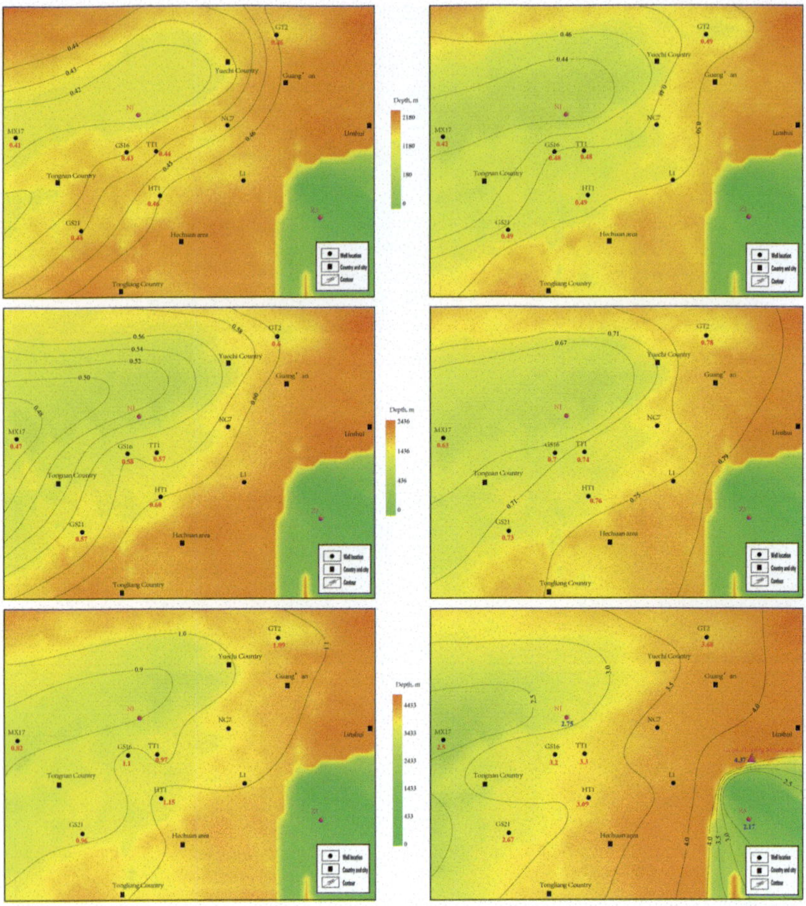

Fig. 5 Ro contours of source rocks in key period before denudation of Qiongzhusi Formation in Central Sichuan (before Caledonian, before Dongwu I episode, before Dongwu II episode, before Early Indosinian, before Late Indosinian, before Yanshan-Himalayan)

the Ro of source rocks in Longmaxi Formation of Early Indosinian was 0.5 ~ 0.6, indicating that the source rocks were in the early mature stage and mainly generated liquid oil. In the Late Indosinian, the thicker deposition of Xujiahe Formation led to the rapid hydrocarbon generation evolution of source rocks. The Ro of source rocks was greater than 0.7, indicating that the source rocks were in the mature stage and mainly generated oil. The thermal evolution degree of source rocks in the northern part was slightly higher than that in the southern part. During the period of Yanshan-Himalayan, the source rocks in the research area entered the foreland tectonic stage. After the deposition of the Cretaceous, the sedimentary burial of a large number of strata led to the rapid hydrocarbon generation and evolution of the source rocks in Longmaxi Formation, especially in the northern part. Bounded by Well Zuo 3-Gaoshi

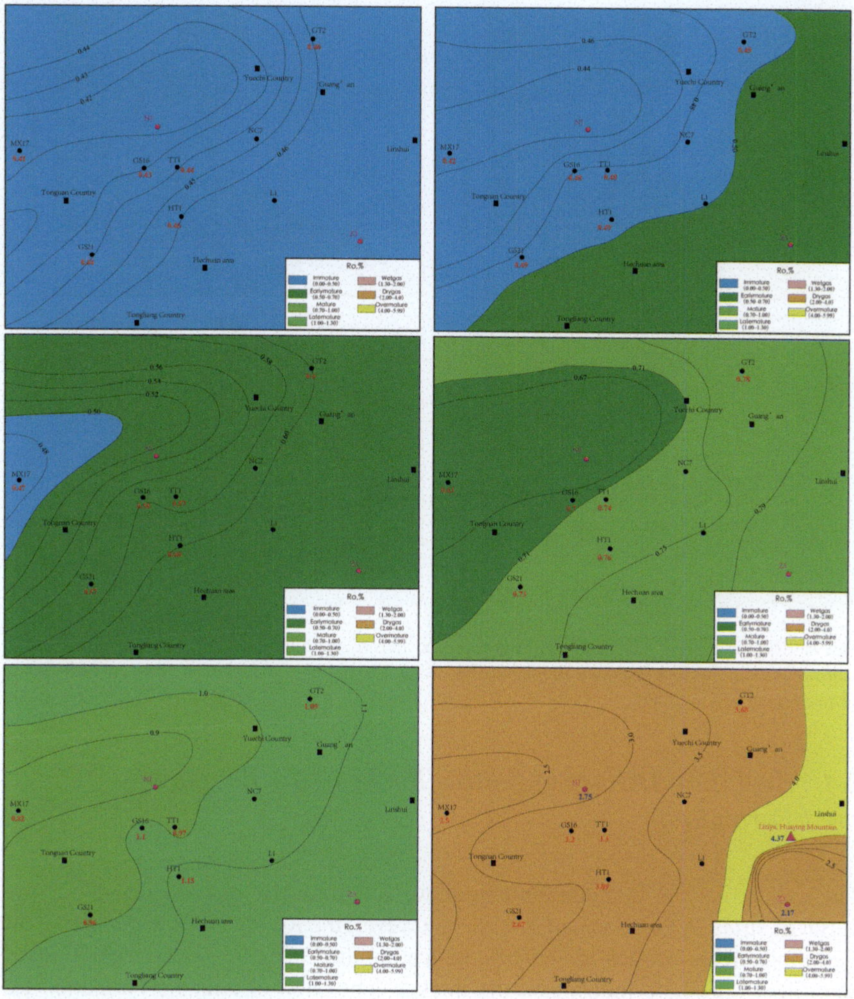

Fig. 6 Hydrocarbon generation evolution of source rocks in Qiongzhusi Formation of Central Sichuan (before Caledonian, before Dongwu I episode, before Dongwu II episode, before Early Indosinian, before Late Indosinian, before Yanshan-Himalayan)

21, the Ro of the source rocks in the north was 2.0 ~ 4.0%, indicating the dry gas stage, and the Ro in the south was less than 2.0, indicating the wet gas stage.

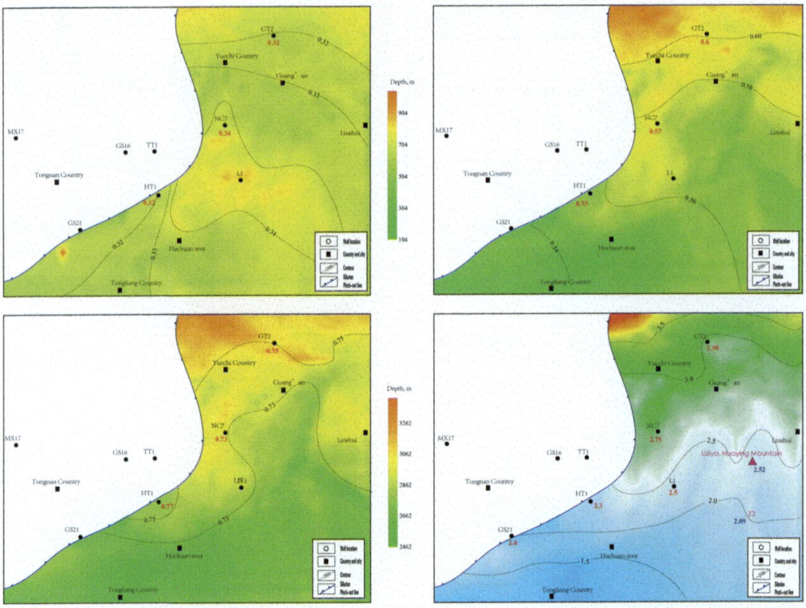

Fig. 7 Ro contours of source rocks in key period before denudation of Longmaxi Formation in Central Sichuan (before Dongwu II episode, before the Early Indosinian, before the Late Indosinian, before Yanshan-Himalayan)

2.4 Evolution History of Source Rocks in the Longtan Formation

The source rocks in Longtan Formation evolved gradually with the deposition and burial of strata in Permian and Triassic. However, the Ro of source rocks before the Early Indosinian was less than 0.5%, indicating the immature source rocks (Figs. 9 and 10). In the Late Indosinian, the thicker deposition of Xujiahe Formation led to the rapid hydrocarbon generation evolution of source rocks. The Ro of source rocks was greater than 0.6, indicating that the source rocks were in the early mature stage and mainly generated oil. The thermal evolution degree of source rocks in the northern part was slightly higher than that in the southern part. During the period of Yanshan-Himalayan, the source rocks in the research area entered the foreland tectonic stage. After the deposition of Cretaceous, the sedimentary burial of a large number of strata led to the rapid hydrocarbon generation and evolution of the source rocks in Longtan Formation, especially in the northern part. The Ro of the source rocks in the northern part was 2.0 ~ 3.0%, indicating the dry gas stage, and the Ro in the south was less than 2.0, indicating the wet gas stage.

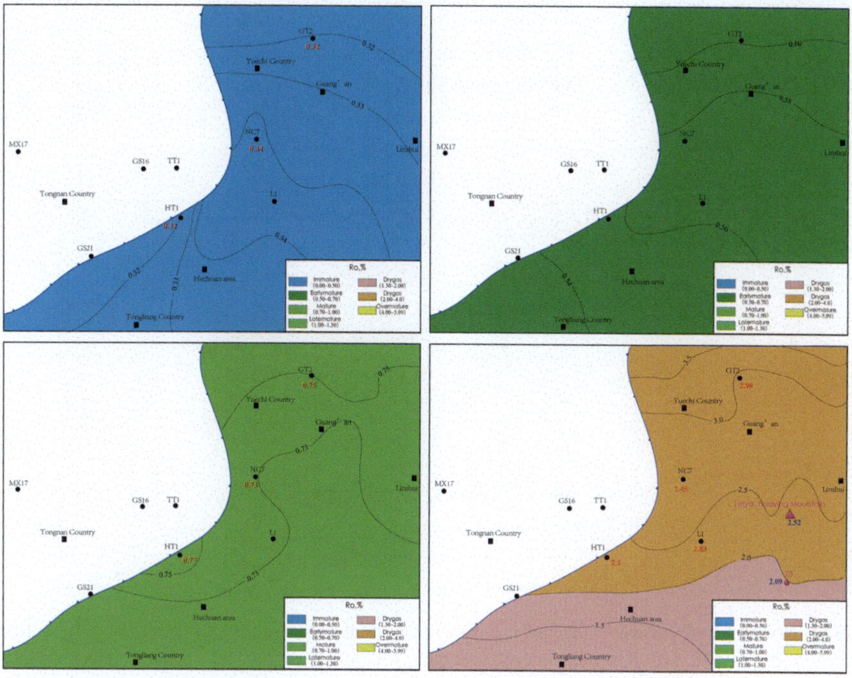

Fig. 8 Hydrocarbon generation evolution of Longmaxi Formation source rocks in Central Sichuan (before Dongwu II episode, before the Early Indosinian, before the Late Indosinian, before Yanshan-Himalayan)

2.5 Hydrocarbon Generation and Evolution of Regional Source Rocks

According to the number and hydrocarbon generation evolution of the main source rocks in the research area, the hydrocarbon generation evolution of regional source rocks is drawn (Fig. 11 shows the nearly NS profile).

The main source rocks in the research area are Deng3 Member, Qiongzhusi and Longtan Formation since the Longmaxi Formation was denuded in the Caledonian. The source rocks in Deng3 Member and Qiongzhusi Formation generally began to mature and generated liquid hydrocarbons in the Dongwu period, and the Ro of source rocks was greater than 2.0% in the Yanshanian, indicating that the source rocks enter the dry gas stage. The mature period of source rocks in the Longtan Formation was a little later, and the source rocks began to mature in the Early Indosinian to generate liquid hydrocarbons and entered the dry gas stage in the Yanshanian. Figure 12 is a nearly NS evolution profile. It can be seen from the map that the source rocks in the Deng3 Member entered the early maturity stage and began to generate liquid hydrocarbons with the deposition of Ordovician and Silurian in Caledonian. After the Caledonian movement, the Longmaxi Formation was not completely denuded.

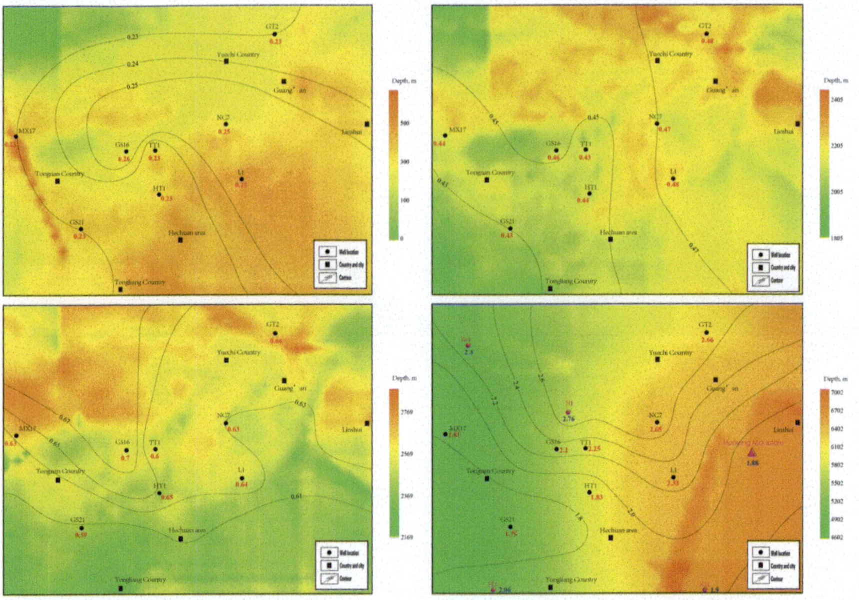

Fig. 9 Ro contours of source rocks in key period before denudation of Longtan Formation in Central Sichuan (before Dongwu II episode, before the Early Indosinian, before the Late Indosinian, before the Yanshan-Himalayan)

The source rocks and the strata of Qiongzhusi Formation didn't begin to mature and generate liquid hydrocarbons until they were in the Dongwu period, while the source rocks in Longtan Formation didn't enter the early mature stage until it was in the Early Indosinian. Generally, four sets of source rocks were rapidly buried and evolved in Indosinian, and the Deng3 Member, Qiongzhusi Formation and Longmaxi Formation then entered the dry gas stage successively in Yanshanian with Ro greater than 2.0%. However, the source rocks in Longtan Formation were developed with slightly lower maturity and the Ro of less than 2.0%, indicating the wet gas stage.

3 Paleogeotemperature and Hydrocarbon Phases of Reservoirs

In this research, the phases of hydrocarbon fluid in the six sets of karst reservoirs, such as Deng4 Member, Longwangmiao Formation, Xixiangchi Formation, Maokou Formation, Changxing Formation and Leikoupo Formation in Central Sichuan are predicted by calculation and estimation of reservoir temperatures in each key period. By sorting out the previous researches of paleotemperature in this area (Tables 5 and 6), combined with the calculation results of denudation, the paleotemperature at the top of the reservoir in the geological history process is obtained by multiplying the

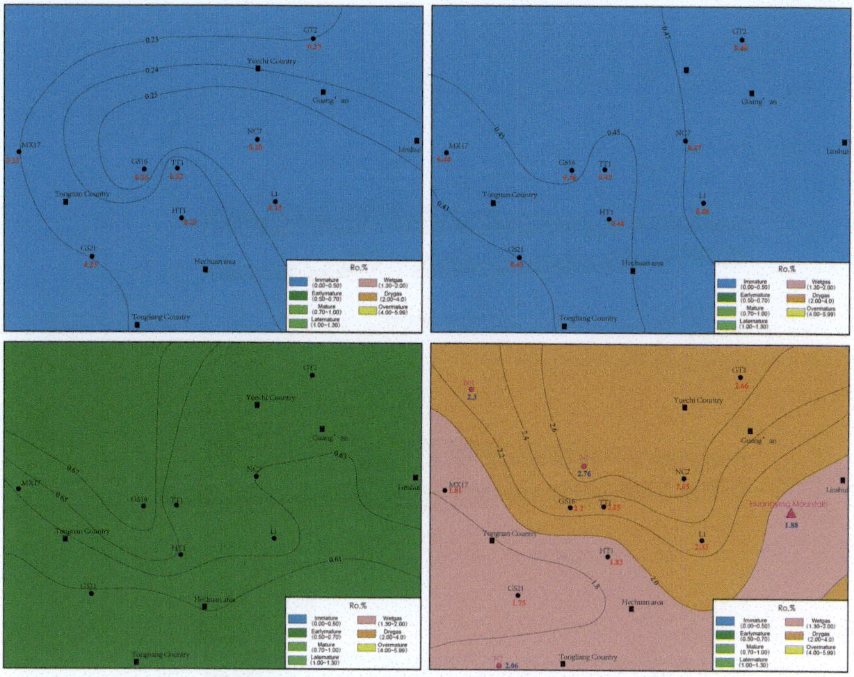

Fig. 10 Hydrocarbon generation evolution of source rocks in Longtan Formation in Central Sichuan (before Dongwu II episode, before the Early Indosinian, before the Late Indosinian, and before the Yanshan-Himalayan)

depth by the paleogeothermal gradient, and the result matches the results calculated based on heatflow in single well simulation.

Oil cracked gas results from cracking of crude oil preserved under prolonged dry conditions, under the continuous increase of thermal stress due to the increase of buried depth, which is a complex process of chemical reaction. The thermodynamic characteristics of oil cracking reaction are largely determined by the thermal stability of hydrocarbons as they are the main components of crude oil. Most scholars mainly focus on cracking characteristics of underground hydrocarbons and have made a large amount of experimental researches on the factors that affect the cracking threshold temperature, such as crude oil properties, pressure, formation water and surrounding rock minerals with quite a lot of research findings being achieved.

Crude oil cracking experiment generally requires a temperature of above 300 °C, much higher than that required for the underground crude oil cracking (generally 120 ~ 250 °C). By combining the dynamics simulation of crude oil cracking and thermodynamics simulation of petroleum inclusion, it can be seen that the reservoirs with formation temperature of 160 ~ 190 °C are ordinary pressure or low pressure reservoirs, and the reservoirs with formation temperature over 190 °C are overpressure or ultrahigh pressure reservoirs (Ping et al. 2014). Researches show that

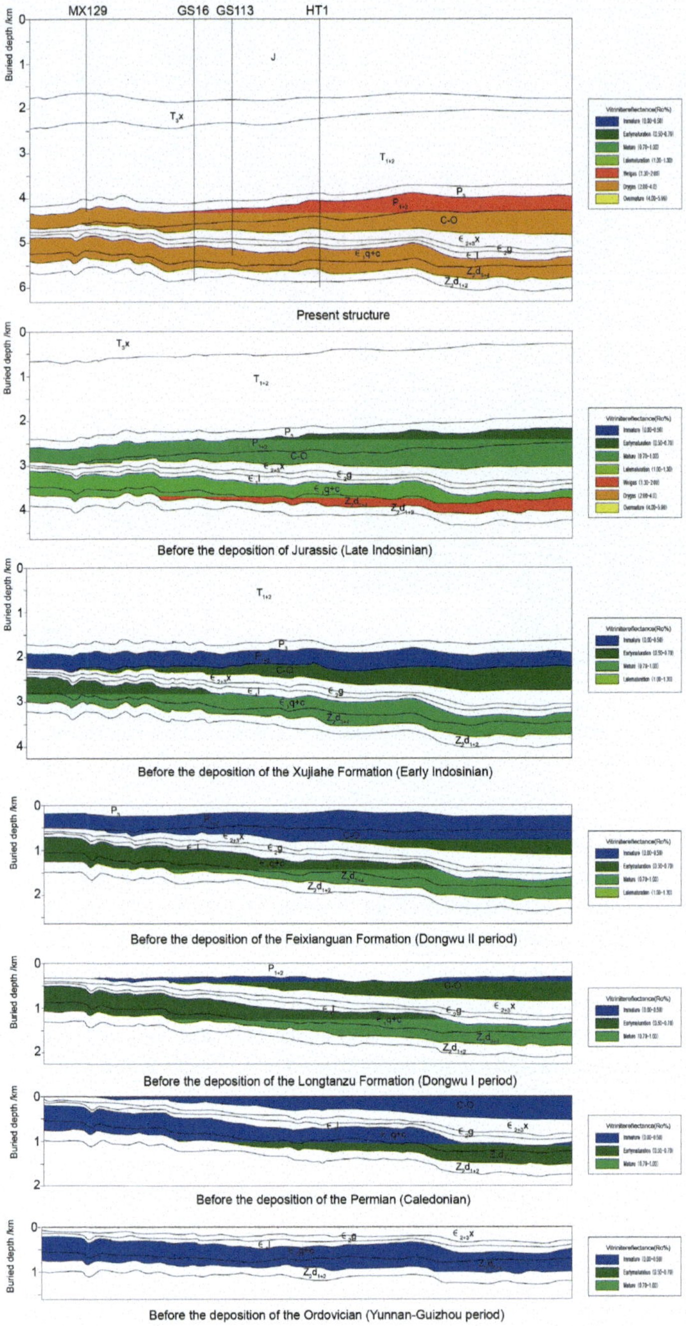

Fig. 11 SE profile of hydrocarbon generation and evolution of source rocks in Central Sichuan

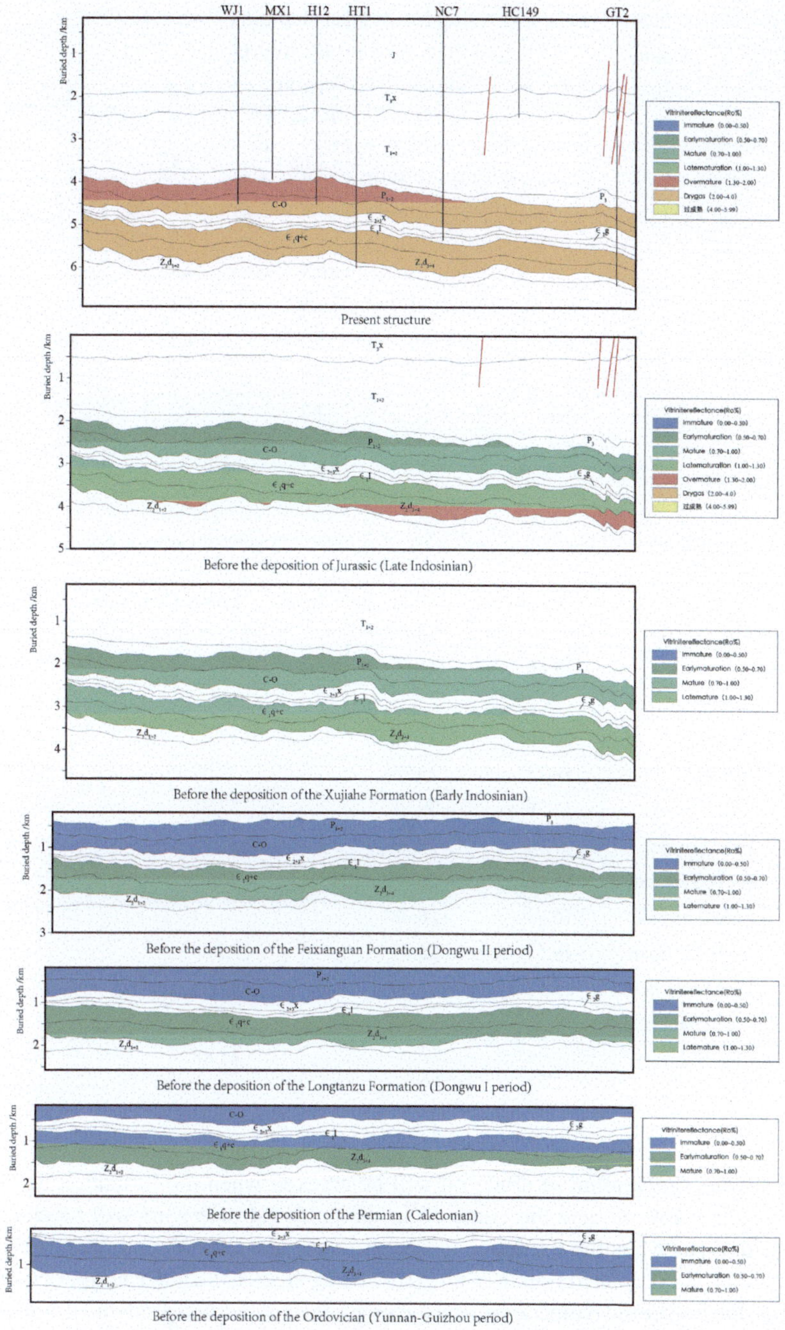

Fig. 12 NE profile of hydrocarbon generation and evolution of source rocks in Central Sichuan

Table 5 Paleogeothermal gradient in central part of Sichuan Basin (Wang et al. 1998)

Time	Statum					
	Z-C	P	T1 + 2	T3-J3	K-E	N-Q
J2 + 3				2.25	2.2	2.14
J1				2.65	2.55	2.5
T3				2.6	2.53	2.5
T2			2.6	2.5	2.4	2.38
T1j			2.5	2.4	2.35	2.3
T1f			3.9	3.5	3.3	3.2
P2		6.5	5.5	5.3	4.2	3.6
P1		6	5.2	4.8	4.1	3.35
C	5.1	5.1	4.4	4.2	4	3.35
S	5.2	5.2	4.5	4.1	3.6	3.35

Table 6 Paleogeothermal gradient of each layer of Anpingdian-Gaoshiti structural belt in Central Sichuan (Xu 2009)

Time		Ordovician	Middle and Upper Cambrian	Lower Cambrian	Sinian
0		28.86	31.23	32.29	29.33
52	Paleogene				38.26
140	Cretaceous	21.36	31.13	35	34.08
200	Jurassic	18.15	29.75	38.08	28.37
212	Late Triassic	17.79	29.21	37.81	27.44
220	Late Triassic	17.94	29.23	37.75	27.68
248	Early Triassic	16.78	26.38	36.1	25.44
260	Late Permian	16.9	26.48	36.14	25.65
285	Early Permian	18.51	25.77	35.68	25.68
450	Late Ordovician	18.58	26.09	35.74	25.2
500	Late Cambrian		25.39	35.25	24.31
550	Late Sinian			41.88	25.44
570	Late Sinian				25.38

generally a large amount of crude oil will be cracked when the formation temperature reaches 160 °C, and the crude oil in the paleo-oil reservoirs will be cracked into natural gas when the formation temperature exceeds 200 °C, and the process of oil cracking is completed (Duan et al. 2013; He et al. 2016). Under normal circumstances, the higher the temperature and the larger the geothermal gradient, the greater the cracking rate and the crude oil conversion rate. However, due to the constraints of geologic conditions, crude oil properties and experimental differences, the temperatures to start crude oil cracking are largely different (Fig. 13). Currently, there is no

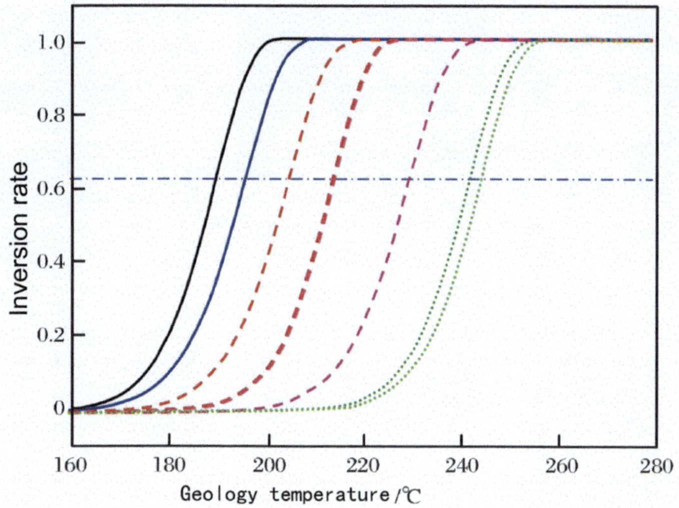

Fig. 13 Comparison of cracking conversion rates of different crude oils (Li et al. 2013)

definite conclusion on the threshold temperature of crude oil cracking. It is generally believed that oil cracking occurs when the formation temperature is more than 150 °C. Experiments show that in some areas oil cracked gas doesn't begin to form until the formation temperature is higher than 190 °C (Li et al. 2013). The crude oil of Well Hade 11 in Tarim Basin strarts cracking on large scale at the temperature of 190 °C and ends at 230 °C. However, the reservoir temperature of marine cracked gas in the platform-basin area of Tarim Basin is greater than 210 °C (Zhu et al. 2012). Therefore, in this research, the starting and ending temperature of crude oil cracking in the reservoirs is set as 190 °C and 250 °C, respectively.

3.1 *Paleogeotemperature and Hydrocarbon Phases in Top of Deng4 Member*

The carbonate rocks of Deng4 Member entered the stage of burial diagenetic evolution after the formation of karst reservoirs by uplift and denudation in Tongwan period. The overall burial trend is that the burial depth and temperature in the eastern part are greater than those in the western part (Fig. 14). Before the Late Indosinian, the formation temperature was less than 190 °C, and the main fluid of hydrocarbons was liquid crude oil. In the Late Indosinian, the southeast of the research area was deeply buried. As a result, the reservoir temperature exceeded 190 °C, and the liquid hydrocarbon in the reservoirs began to crack into natural gas. In the Early and Middle Cretaceous, the reservoir temperature was generally greater than 220 °C. In the west area of Well Guangtan 2-Nanchong 7-Hechuan area, the temperature was less than

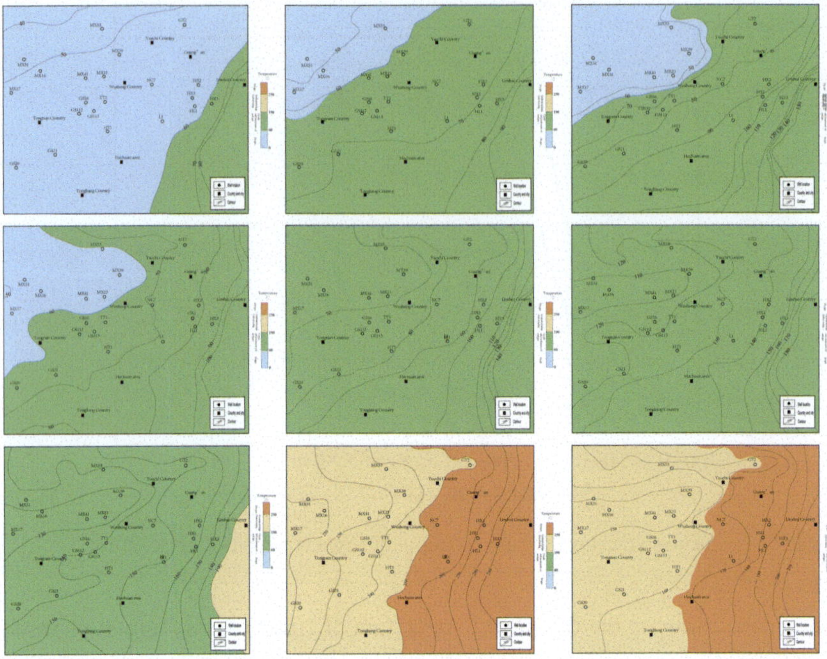

Fig. 14 Temperature and hydrocarbon phase evolution of reservoirs in top Deng4 Member in Central Sichuan (before Yunnan-Guizhou movement, before Yeli movement, before Caledonian movement, before Dongwu I episode, before Dongwu II episode, before Early Indosinian, before Late Indosinian, before Early-Middle Cretaceous, present)

250 °C, and the oil cracked condensate oil and gas were formed. While in the east area of the line, the temperature was greater than 250 °C, and the reservoir fluid was dry gas. At present, with a large number of strata being denuded, the formation temperature drops rapidly. Taking the connecting line of Well Hualai 1 and Well Huaxi 2 as the boundary, the temperature in the west was less than 190 °C, and the temperature in the east was more than 190 °C. The reservoir fluid should be consistent with the fluid type in the previous stage.

3.2 *Paleogeotemperature and Hydrocarbon Phases in the Longwangmiao Formation*

Longwangmiao Formation is mainly composed of layered porous dolomite reservoirs. The reservoir pores mainly include intergranular pores, intercrystal pores and dissolution pores. The development of reservoirs is mainly controlled by sedimentary facies and karstification. The Longwangmiao Formation entered the stage of burial diagenetic evolution after the formation of karst reservoirs developed by uplift and denudation as a result of Yunnan-Guizhou movement. The overall burial trend is similar to that of Deng4 Member, and the burial depth and temperature in the east are greater than those in the west (Fig. 15). Before the Early and Middle Cretaceous, the temperature of the reservoir was less than 190 °C, and the main fluid type of hydrocarbons was liquid crude oil. In the Early and Middle Cretaceous, with the accumulation and burial of a large number of sediments in the foreland basin, the reservoir temperature rose sharply, generally more than 200 °C. Taking Well Guangtan 2-Lai 1 as the boundary, the reservoir temperature in the west was less than 250 °C, and the reservoir liquid hydrocarbon was cracked into natural gas. While the reservoir temperature in the east area of the line was greater than 250 °C, and the reservoir fluid was dry gas. At present, with a large number of denudation of the overlying strata, the reservoir temperature drops rapidly. Taking the connecting line of Well Huaxi 2 and Huaxi 3 as the boundary, the temperature in the west is less than 190 °C, and the temperature in the east is greater than 190 °C. The reservoir fluid should be consistent with the fluid type in the previous stage.

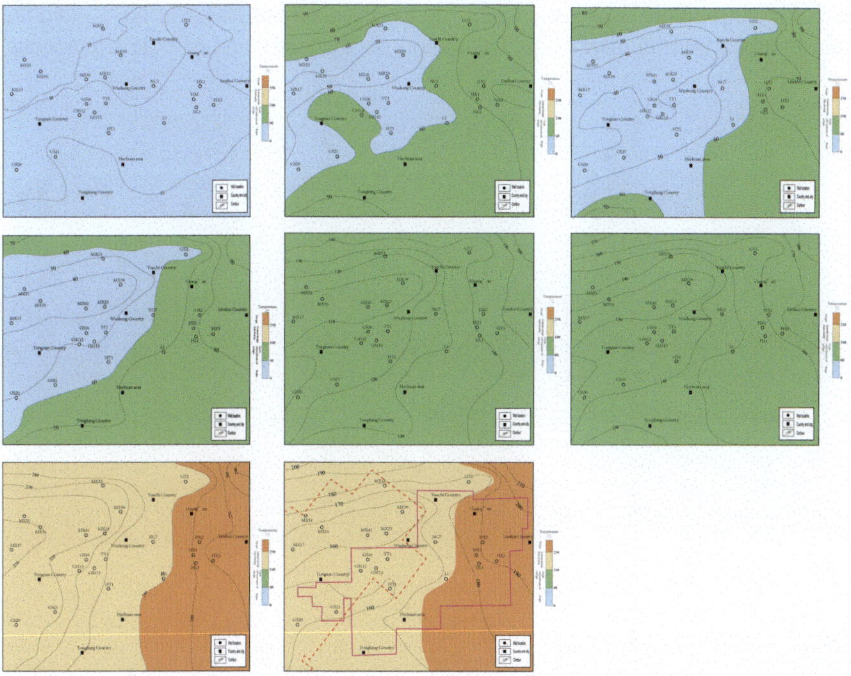

Fig. 15 Reservoir temperature and hydrocarbon phase evolution at the top of Longwangmiao Formation in Central Sichuan (before Yeli movement, before Caledonian movement, before Dongwu I episode, before Dongwu II episode, before Early Indosinian, before Late Indosinian, before Early-Middle Cretaceous, present)

3.3 Paleogeotemperature and Hydrocarbon Phases in the Xixiangchi Formation

The Xixiangchi Formation entered the stage of burial diagenetic evolution after the formation of karst reservoirs developed by uplift and denudation as a result of Yeli movements. The overall burial trend is that the burial depth and temperature in the east are greater than those in the west (Fig. 16). Before Caledonian, the reservoirs were bounded by Well Moxi 16-Well Moxi 53-Well Guangtan 2-Linshui County. The reservoirs in the northern part were developed with the buried temperature of greater than 60 °C, and mainly generated liquid hydrocarbon. The buried depth of the reservoirs before Dongwu I did not exceed that before the Caledonian movement, so the distribution pattern of reservoir temperature did not change. In Dongwu II, the reservoirs were bounded by Well Moxi 16-Well Moxi 53-Well Guangtan 2-Well Lai 1-Hechuan County. The burial temperature of the reservoirs in the eastern part was more than 60 °C, and the main hydrocarbons fluid was liquid hydrocarbon.

The temperature in the western reservoirs was less than 60 °C. However, the reservoir temperature during the Early and Late Indosinian was over 60 °C, and the main hydrocarbon fluid was liquid crude oil. In the Early and Middle Cretaceous,

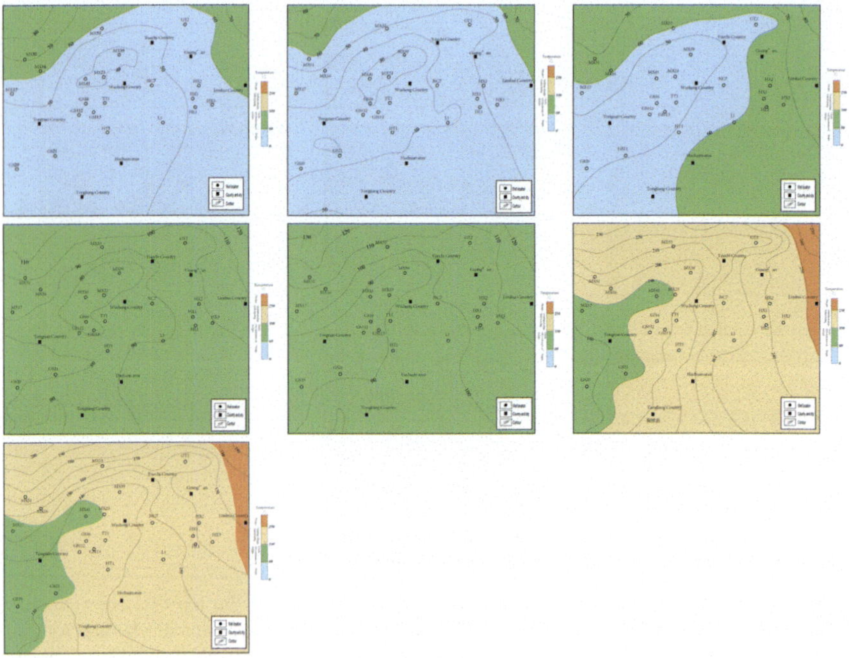

Fig. 16 Reservoir temperature and hydrocarbon phase evolution at the top of Xixiangchi Formation in Central Sichuan (before Caledonian movement, before Dongwu I, before Dongwu II, Indosinian, Early-Middle Cretaceous, present)

with the accumulation and burial of a large amount of sediments in the foreland basin, the reservoir temperature rose sharply. Bounded by Well Moxi 17-Moxi 41-Gaoshi 21, the reservoir temperature of the western part was lower than 190 °C, and the main fluid of hydrocarbons was liquid crude oil. The reservoir temperature in the eastern area was more than 190 °C, and the reservoir liquid hydrocarbon was cracked to natural gas. While the reservoir temperature in the northeast corner of the research area was greater than 250 °C, and the reservoir fluid was dry gas. At present, with a large number of denudation of the overlying strata, the reservoir temperature drops rapidly. The reservoir temperature is generally less than 190 °C, and it is greater than 190 °C only in some areas of the northeast and northwest corners. The reservoir fluid should be consistent with the fluid type of the previous stage.

3.4 *Paleogeotemperature and Hydrocarbon Phases in the Maokou Formation*

From the deposition of the Maokou Formation in Hechuan-Tongnan area to the second episode of Dongwu movement, the temperature of the top of carbonate reservoirs was less than 60 °C, and the mainly hydrocarbon was biogas (Fig. 17). After the Indosinian movement in two stages, the reservoir temperature has comprehensively exceeded 60 °C, and the main fluid of hydrocarbons was liquid crude oil. In the Early and Middle Cretaceous, with the accumulation and burial of a large amount of sediments in the foreland basin, the reservoir temperature rose sharply. Only the hydrocarbons in reservoirs of regional southwest corner with temperature less than 190 °C were liquid crude oil. In other areas, the reservoir temperature exceeded 190 °C, and the reservoir liquid hydrocarbons were cracked to natural gas. At present, with a large number of denudation of the overlying strata, the reservoir temperature drops rapidly. The reservoir temperature is generally less than 190 °C, showing a situation of high in the north and low in the south. The reservoir fluid should be consistent with the fluid type in the previous stage.

3.5 *Paleogeotemperature and Hydrocarbon Phases in the Changxing Formation*

Changxing Formation in Hechuan-Tongnan area was buried rapidly after deposition, and the temperature of the top of carbonate reservoir in Early Indosinian exceeded 60 °C (Fig. 18). The main fluid of hydrocarbon was liquid crude oil. In the Late Indosinian, with further burial, the reservoir temperature increased to 120 ~ 160 °C, and it was lower in the north and higher in the south, and the reservoir hydrocarbons were mainly liquid crude oil. During the Early and Middle Cretaceous, when a large amount of material was accumulated and buried in the foreland basin, the

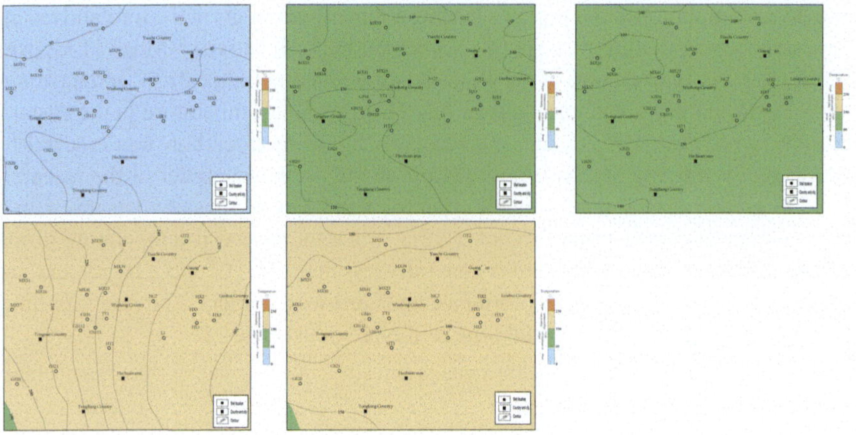

Fig. 17 Reservoir temperature and hydrocarbon phase evolution of top Maokou Formation in Hechuan-Tongnan area (before the Dongwu II, Early Indosinian, Late Indosinian, Early-Middle Cretaceous, present)

reservoir temperature increased dramatically. Only the hydrocarbons in reservoirs of regional southwest corner with temperature less than 190 °C were liquid crude oil. In other areas, the reservoir temperature exceeded 190 °C, and the reservoir liquid hydrocarbons were cracked to natural gas. Currently, with a large number of denudation of the overlying strata, the reservoir temperature drops rapidly. The reservoir temperature is high in the north and low in the south except for the reservoir in the northwest corner with the temperature generally less than 190 °C. The reservoir fluid should be consistent with the fluid type in the previous stage.

3.6 Paleogeotemperature and Hydrocarbon Phases in the Leikoupo Formation

Carbonate reservoirs in Hechuan-Tongnan were usually less than 60 °C during the Leikoupo Formation deposition period (Fig. 19), and the mainly hydrocarbon was biogas. In the Early and Middle Cretaceous, with the accumulation and burial of a large amount of sediments in the foreland basin, the reservoir temperature increased sharply and was more than 60 °C. The hydrocarbon was liquid crude oil. At present, with a large number of denudation of the overlying strata, the reservoir temperature drops rapidly, but is still greater than 60 °C, and the reservoir fluid is still liquid crude oil.

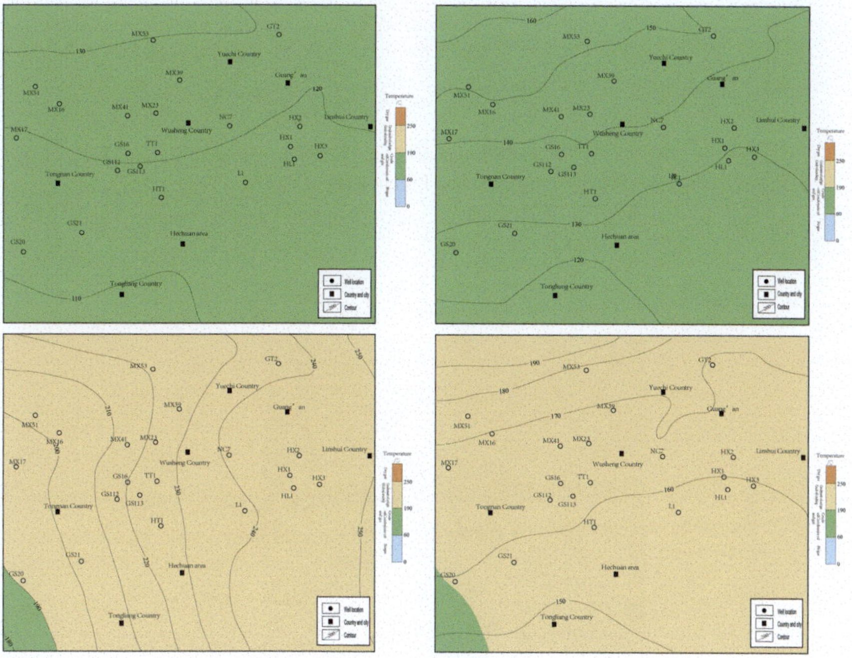

Fig. 18 Reservoir temperature and hydrocarbon phase evolution of top Changxing Formation in Hechuan-Tongnan area (Early Indosinian, Late Indosinian, Early-Middle Cretaceous, present)

4 Hydrocarbon Accumulation Process

4.1 History of Oil and Gas Filling

Inclusions could provide a series of original information about mineral formation, such as temperature, composition and salinity, as they can reflect the essential characteristics of diagenetic and metallogenic fluids. Organic inclusion refers to the hydrocarbons captured and encapsulated by minerals in each stage of maturity and evolution of organic matters, and in the process of oil and gas migration and accumulation. It records the history of hydrocarbon evolution and migration.

In Well Anping 1 (Xu 2009), the inclusions in the pore and fracture fillings of Sinian and Cambrian are mainly salt water and hydrocarbon-bearing brine. The homogenization temperature is 120–250 °C and is mostly concentrated in 190–220 and 240–250 °C. The approximate formation period ranges from Late Jurassic to Early Cretaceous. The homogenization temperature of the first-phase inclusions in some fractures is 120–140 °C, and the formation period is approximately Triassic.

Yuan et al. (2014) determined there are four stages of oil and gas accumulation in reservoirs of Longwangmiao Formation according to the filling sequence and types of minerals in pores, caves and fractures of the reservoirs, and the development

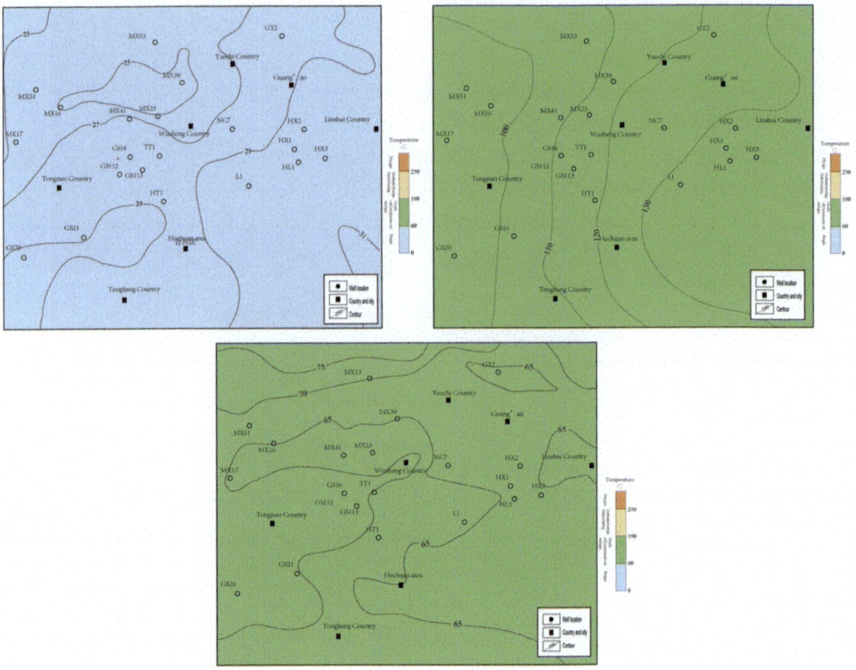

Fig. 19 Reservoir temperature and hydrocarbon phase evolution of top Leikoupo Formation in Hechuan-Tongnan area (Late Indosinian, Early-Middle Cretaceous, present)

characteristics of inclusions in different stages. The first time hydrocarbon filling occurred in the Middle-Late Triassic, and it was recorded by the homogenization temperature (110 ~ 133 °C) of the inclusions in the fine grain-microcrystal dolomites. The second time of hydrocarbon filling occurred in the Early-Middle Jurassic and was recorded by the homogenization temperature (143 ~ 167 °C) of the inclusions in the first generation of dolomites in pores and caves. The third time of hydrocarbon filling was a process that the paleo-oil reservoir was cracked into gas under high temperature and was recorded by the second generation bituminous minerals filling in the pores, caves and fractures of the reservoirs. The fourth time of hydrocarbon filling occurred in the Middle-Late Cretaceous and was recorded by the homogenization temperature (190 ~ 210 °C) of the fluid inclusions in the third generation quartz.

Zhang et al. (2017) studied two wells around the research area, Well Moxi 13 and Well Gaoshi 17. The inclusions mainly include asphaltic gas inclusions and gas inclusions captured in various burial depth, which also confirm that the current gas reservoir is developed from crude oil cracking. According to the minerals contained in inclusions and the contact relationship with asphalt, combined with the homogenization temperature of inclusions, the fluid filling in the reservoirs of Longwangmiao Formation is divided into four stages. ① Stage I: it is mainly found in Well Gaoshi 17, with little inclusions, no oil inclusions, and inclusions are sporadically distributed on the secondary sparry margin of early dolomites. A small amount of asphalt inclusions can be seen through Raman spectroscopy, and the homogenization temperature is 100 ~ 120 °C. It is speculated that the inclusions were captured in the early filling stage. ② Stage II: it is mainly found in Well Gaoshi 17 with few inclusions. The Stage II is speculated as the formation of migration channel and the gas reservoir is not retained. Inclusions are mainly developed in fracture-cavity dolomite cement or dolomite vein filled in structural fractures. There are few gaseous hydrocarbon inclusions and much less associated brine inclusions. It is difficult to measure temperature, and the homogenization temperature is 130 ~ 170 °C. It is speculated that the inclusions were captured in the middle filling stage. ③ Stage III: it is found in Well Moxi 13. At that time, the inclusions were mainly developed in the dolomite cement or in the contact part with asphalt representing the period of oil cracking. There are many gaseous hydrocarbon inclusions in good shape which are convenient for temperature measurement. The homogenization temperatures of associated brine inclusions are all 90 ~ 210 °C. It is speculated that the inclusions were captured in the filling stage of a large amount of oil cracked gas. ④ Stage IV: no gaseous hydrocarbon inclusions were found in this stage. The fluid inclusions in this stage are mainly gas–liquid H_2O inclusions, hydrocarbon-water inclusions and pure methane inclusions. The homogenization temperature of brine inclusions is 210 ~ 230 °C. In this stage, the three temperature levels represent three processes of stage filling, and the fourth stage was formed in the tectonic uplift stage.

Xu (2018) researched inclusion petrography and fluid properties during natural gas filling of gas reservoirs in the Cambrian Longwangmiao Formation drilled by Well Moxi 12, Moxi 13, Moxi 17 and Moxi 21 in the research area. It is found that there are three stages of fluid inclusions in the dolomite reservoir of Cambrian Longwangmiao Formation in Moxi structure. The fluid inclusions of Stage I were developed in microcrystal and fine-crystal dolomites, with GOI value of 80% and homogenization temperature of 110 ~ 133 °C. The fluid inclusions of Stage II were developed during the precipitation of coarse-crystal dolomites filled in fractures and caves. The inclusions were clustered or evenly distributed in dolomites filled in fractures and caves. The GOI value was about 60%, and the homogenization temperature of brine inclusions was 143 ~ 167 °C. The fluid inclusions of Stage III were developed during the late precipitation and filling period of quartz crystal in fractures and caves, and the inclusions were distributed in groups or evenly in the quartz minerals filled in the fractures and caves in middle and late stages. The GOI value was 1 ~ 2%. The homogenization temperature of brine inclusions in quartz minerals of Stage III is 170 ~ 195 °C.

Using Well Tongtan 1 as an example (Fig. 20), together with regional tectonic evolution and prior study findings, it is believed that the oil and gas were filled during the Middle-Late Triassic, Early-Middle Jurassic, and Early-Middle Cretaceous periods. The three periods of accumulation were formed in the structural period with continuous buried depth, and the Early-Middle Cretaceous is the key period of gas accumulation in the research area.

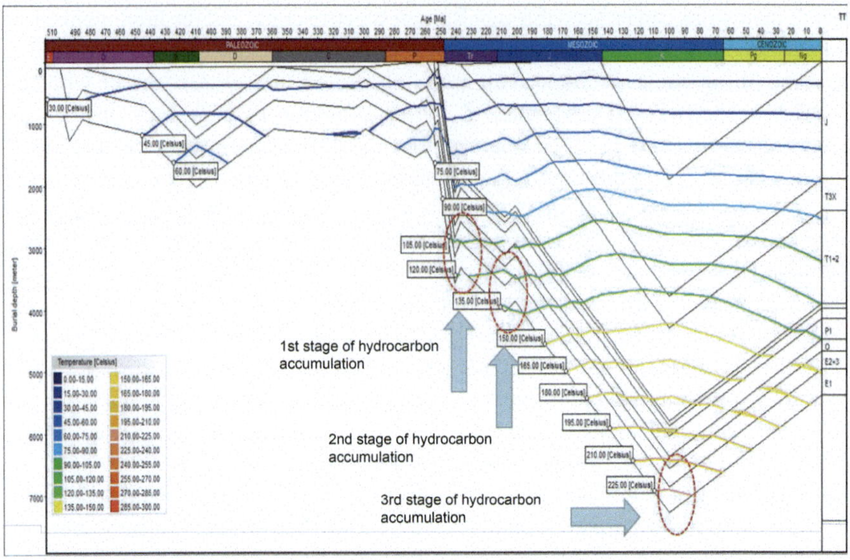

Fig. 20 Hydrocarbon accumulation periods of Well Tongtan 1 in Central Sichuan

4.2 *Oil Source Correlation*

4.2.1 Dengying Formation-Longwangmiao Formation

Genetic Types of Natural Gas

Ethane carbon isotope shows high inheritance from original parent materials, although it is also affected by the thermal evolution degree of source rocks, it is much less affected than methane carbon isotope. Therefore, ethane carbon isotope is the most commonly used effective index to distinguish coal gas and oil-type gas (Dai et al. 2005). By studying the composition of ethane carbon isotope corresponding to coal gas and oil-type gas in the same evolution stage, it can be found that ^{12}C is significantly more enriched in oil-type gas than coal gas (Liu et al. 2004). After studied the geochemical characteristics of Jurassic-Sinian natural gas in Sichuan Basin, Wang (1994) pointed out that $\delta^{13}C_2$ of coal gas is greater than $-29‰$, and $\delta^{13}C_3$ is greater than $-27‰$. According to the research of Gang et al. (1997), $\delta^{13}C_2$ is sensitive to the parent material types of natural gas, and it is more than $-29‰$ in humic natural gas, and less than $< -29‰$ in sapropel natural gas. Dai et al. (1992) comprehensively studied China's natural gas and pointed out that the $\delta^{13}C_2$ of oil-type gas is less than $-28.8‰$, and the $\delta^{13}C_2$ of coal gas is more than $-28.5‰$. By synthesizing the empirical statistical results of multiple scholars, this research selected $\delta^{13}C_2 = -29‰$ as the boundary between coal gas and oil-type gas. The $\delta^{13}C_2$ of coal gas is generally more than $-29‰$, and of oil-type gas is generally less than $-29‰$.

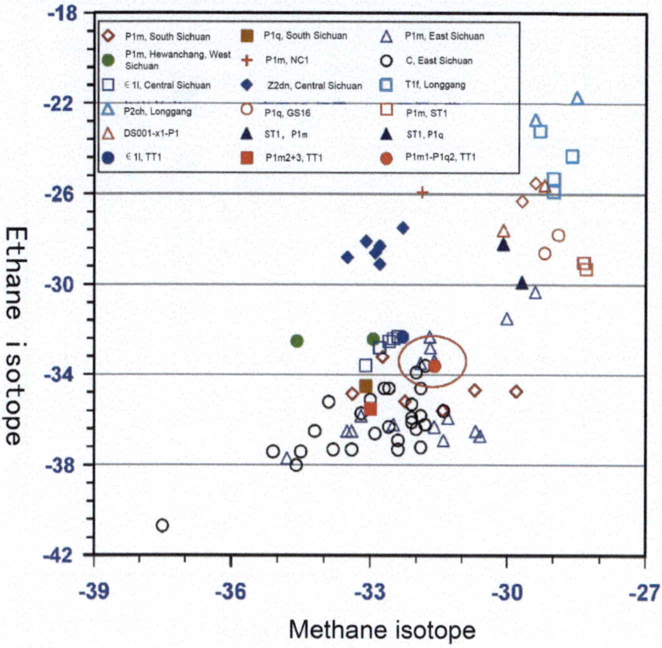

Fig. 21 Correlation of methane-ethane isotopes of natural gas in Sinian-Permian of Sichuan Basin

The ethane carbon isotope of the natural gas in the Cambrian Longwangmiao Formation of Central Sichuan is − 35 ~ − 31‰, and the ethane carbon isotope of the Longwangmiao Formation in Well Tongtan 1 is − 32.3‰ (Fig. 21), both of the two values are obviously less than − 29‰, indicating that the natural gas in Longwangmiao Formation is typical oil-type gas. The ethane carbon isotope value of natural gas in Dengying Formation (− 30 ~ − 27.5‰) is significantly heavier than that in Longwangmiao Formation. Although the main value is more than − 29‰, based on other geochemical indicators, the natural gas in Dengying Formation is oil-type gas.

Through correlation of natural gas $\delta^{13}C_1$—C_1/C_{2+3} (Fig. 22), it can be seen that the data of natural gas in Longwangmiao Formation of Central Sichuan are distributed in the area of type II kerogen, indicating that the natural gas should be sapropel gas, while the data of natural gas in Longwangmiao Formation of Well Tongtan 1 are distributed in the extension area of type II kerogen, indicating the oil-type gas. The data of natural gas in Dengying Formation of Central Sichuan are distributed in the extension area of type II kerogen, indicating the source of sapropel organic matter.

The discrimination diagram of $\ln(C_1/C_2)$ versus $\ln(C_2/C_3)$ of kerogen-degraded gas and oil cracked gas in different evolution stages of sapropel organic matters are established by Li et al. (2017), with the analysis of composition characteristics of original kerogen, crude oil and residual kerogen (residual sample without liquid hydrocarbon) by carrying out the experiments of high-temperature and high-pressure

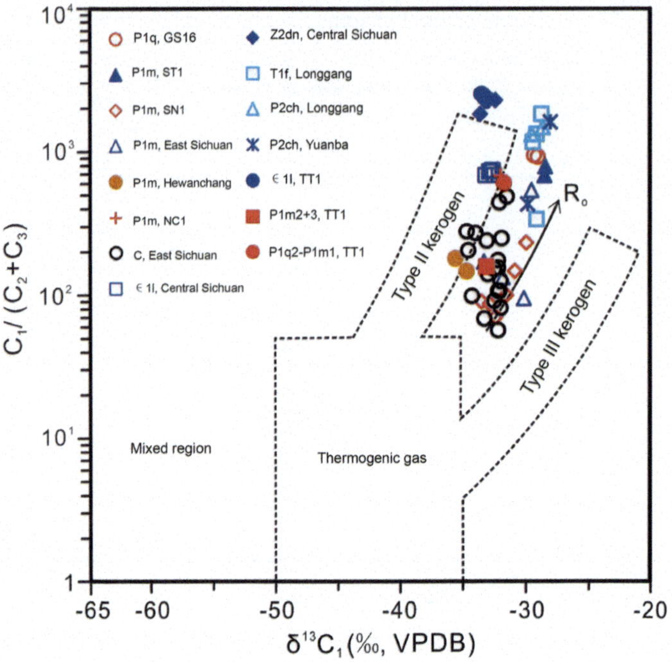

Fig. 5.22 Correlation of natural gas $\delta^{13}C_1$—C_1/C_{2+3} in Sinian-Permian of Sichuan Basin

gold tube and autoclave thermal simulation (Fig. 23). The $\ln(C_1/C_2)$ value of Sinian-Cambrian natural gas in Gaoshiti-Moxi area (Gao-Mo area for short) of Sichuan Basin is 6.35 ~ 7.85, and the $\ln(C_1/C_2)$ value is 3.11 ~ 4.69, which basically fall into the range that the Ro value of oil cracked gas is greater than 2.5%, indicating that the Sinian-Cambrian natural gas is mainly oil cracked gas. The content of propane in Longwangmiao Formation of Well Tongtan 1 is too low to detect, but the $\ln(C_1/C_2)$ is 7.84, which is very close to the value of Sinian-Cambrian natural gas in Gao-Mo area in Central Sichuan, indicating that the natural gas in Longwangmiao Formation is crude oil cracked gas.

Based on the thermal simulation experiment, Li et al. (2017) established a discrimination diagram for aggregated and dispersed liquid hydrocarbon cracking gas identified by light hydrocarbon. They believe that the value of methylcyclohexane/nC7 in gas cracked by highly-evolved aggregated liquid hydrocarbon is more than 1.0, and the value of ΣC6–7 cycloalkanes/(nC6 + nC7) is more than 1.0; while the two geochemical indexes of dispersed liquid hydrocarbon cracked gas are less than 1.0. Through the analysis of light hydrocarbon of natural gas in Sinian Dengying Formation, Cambrian Longwangmiao Formation of Gaoshiti-Moxi area and Cambrian of Weiyuan area in Sichuan Basin, it can be seen that the two parameters of ΣC6–7 naphthenic/(nC6 + nC7) and methylcyclohexane/nC7 are generally greater than 1, and the maximum value is 4.55. All the data fall into the area of gas cracked by

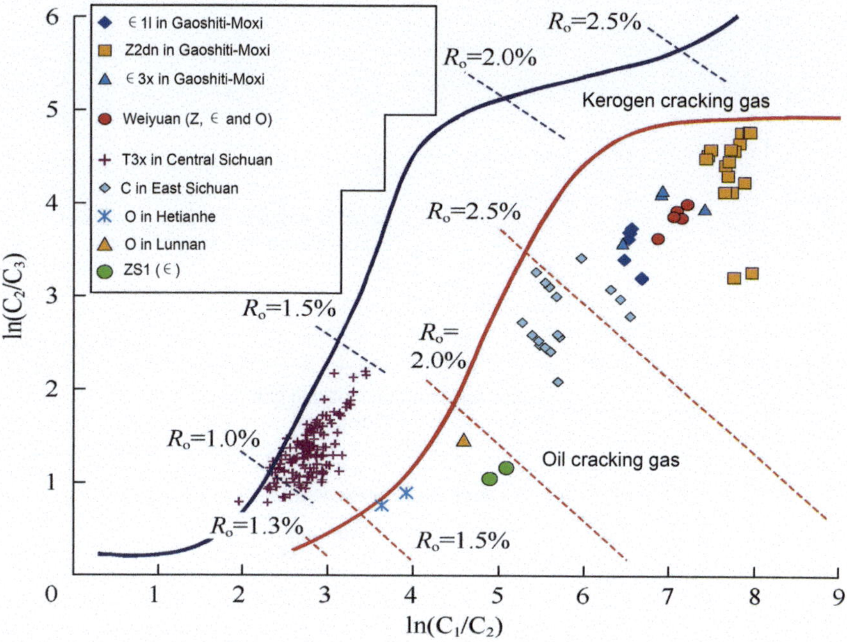

Fig. 23 Discrimination diagram of $\ln(C_1/C_2)$ versus $\ln(C_2/C_3)$ of kerogen-degraded gas and crude oil cracked gas in sapropel organic maters

highly evolved aggregated liquid hydrocarbon, indicating that the natural gas in this area is the gas cracked by aggregated liquid hydrocarbon (see Fig. 24).

Gas Source Analysis

The origin of natural gas in Anyue gas field is oil cracking (Du et al. 2015), but there are differences in the understanding of gas source. ① The natural gas all comes from the source rocks of Qiongzhusi Formation (Du et al. 2014). ② The natural gas of Dengying Formation comes from Qiongzhusi Formation and Deng3 Member, and the natural gas of Longwangmiao Formation comes from Qiongzhusi Formation (Zou et al. 2014). ③ The natural gas of Dengying Formation and Longwangmiao Formation is source-mixed gas from source rocks of Qiongzhusi Formation and Dengying Formation (Xu et al. 2014). The natural gas in high-mature and over-mature areas is single-component gas, and there is no geochemical index to directly compare natural gas and source rocks. As a result, it is difficult to compare gas and sources. In order to accurately and scientifically judge the gas source, multi-parameter method is adopted to combine geology and geochemistry. This comprehensive analysis shows that the natural gas in Anyue gas field mainly comes from the source rocks of Lower Cambrian

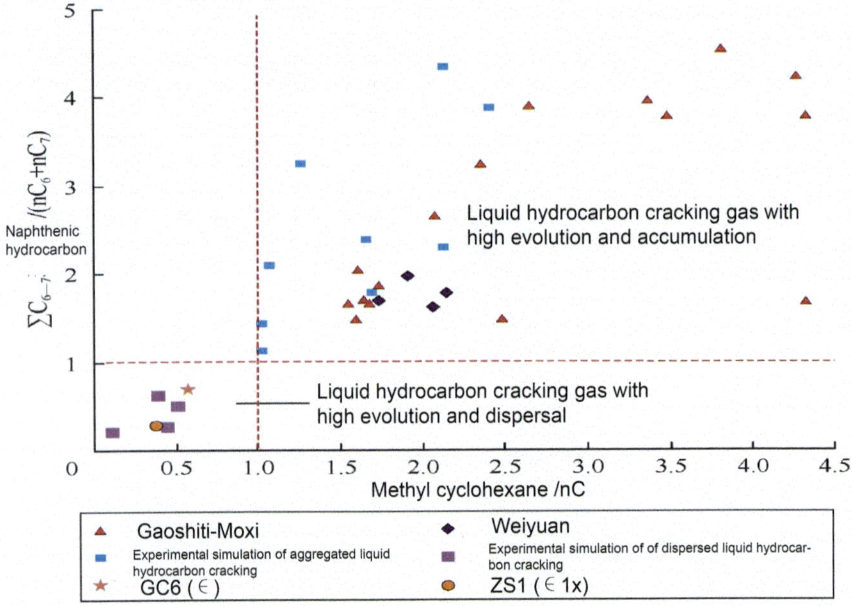

Fig. 24 Discrimination diagram of gas cracked by highly evolved aggregated and dispersed liquid hydrocarbon

Qiongzhusi Formation and Deng3 Member is a contributing source (Fig. 25). There are the following four aspects of evidence.

Carbon Isotope Evidence of Alkane Gas

The carbon isotope of ethane in the natural gas of the Cambrian Longwangmiao Formation is – 35 ~ – 31‰ that is close to the carbon isotope of kerogen in the Qiongzhusi Formation ($\delta^{13}C_{ker}$, – 36.4 ~ – 30‰), indicating that the natural gas mainly comes from the mudstone of Qiongzhusi Formation. Compared with Longwangmiao Formation, the ethane carbon isotope of natural gas in Dengying Formation (– 30 ~ – 27.5‰) is significantly heavier than that in Longwangmiao Formation, indicating that natural gas in Dengying Formation may come from different gas sources.

Carbon Isotope of Asphalt

Reservoir asphalt of Longwangmiao Formation (– 35.4 ~ – 33.1‰) is obviously related to $\delta^{13}C_{ker}$ in Qiongzhusi Formation (Fig. 25). From the comparison between $\delta^{13}C$ value of reservoir asphalt in Dengying Formation (– 36.8 ~ – 34.5‰), $\delta^{13}C_{ker}$ value in Qiongzhusi Formation (– 36.4 ~ – 30‰) and $\delta^{13}C_{ker}$ value in Deng3 Member (– 34.5 ~ – 29.0‰), it can be seen that the result conforms to the inheritance

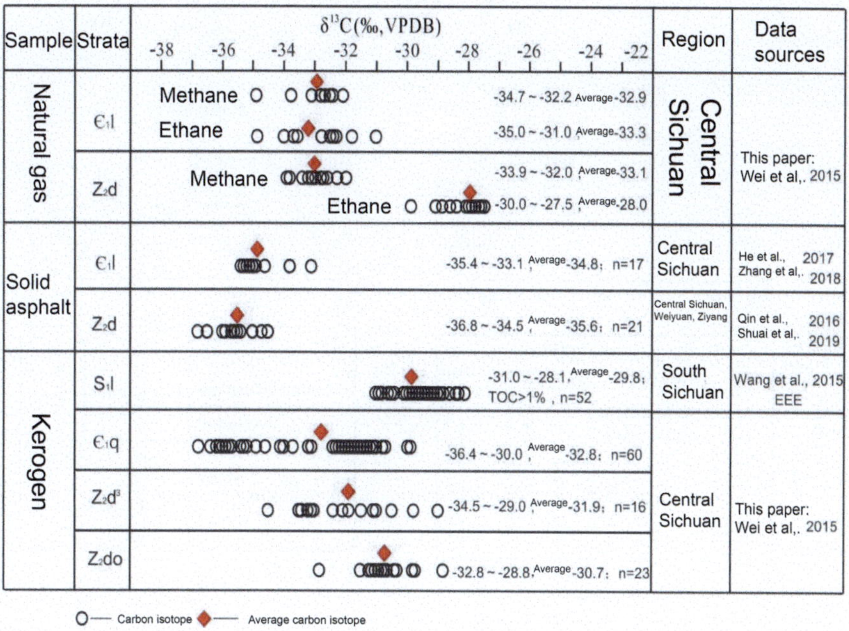

Fig. 25 Carbon isotope comparison of Sinian-Triassic source rocks and solid asphalt in Sichuan Basin

relationship between asphalt and parent kerogen, indicating that the natural gas of Dengying Formation may come from the above two sets of source rocks.

Hydrogen Isotope Evidence of Alkane Gas

The hydrogen isotope of natural gas is affected not only by maturity but also by the salinity of water during deposition. Generally, gammacerane with high abundance is regarded as an important indicator of high salinity of sedimentary water (Peters et al. 1993). The value of gammacerane/C_{30} hopane of the source rocks in the Qiongzhusi Formation is higher than that of the source rocks in the Deng3 Member (Chen et al. 2017), indicating that the water salinity of the former is higher than that of the latter. Most of the methane carbon isotopes of the Longwangmiao Formation and Dengying Formation are distributed between − 34 ~ − 32‰, indicating that there is little difference in maturity between the two formations. The hydrogen isotope (− 138 ~ − 132‰) of methane in Longwangmiao Formation is significantly heavier than that in Deng4 Member (− 147 ~ − 135‰), and that in Deng2 Member is the lightest (− 150 ~ − 141‰), indicating that the natural gas of Longwangmiao Formation should come from Qiongzhusi Formation with high salinity, the natural gas of Dengying Formation should be source-mixed gas, and the contribution of source rocks in Qiongzhusi Formation to the natural gas in Deng4 Member is higher than that of source rocks in Deng3 Member.

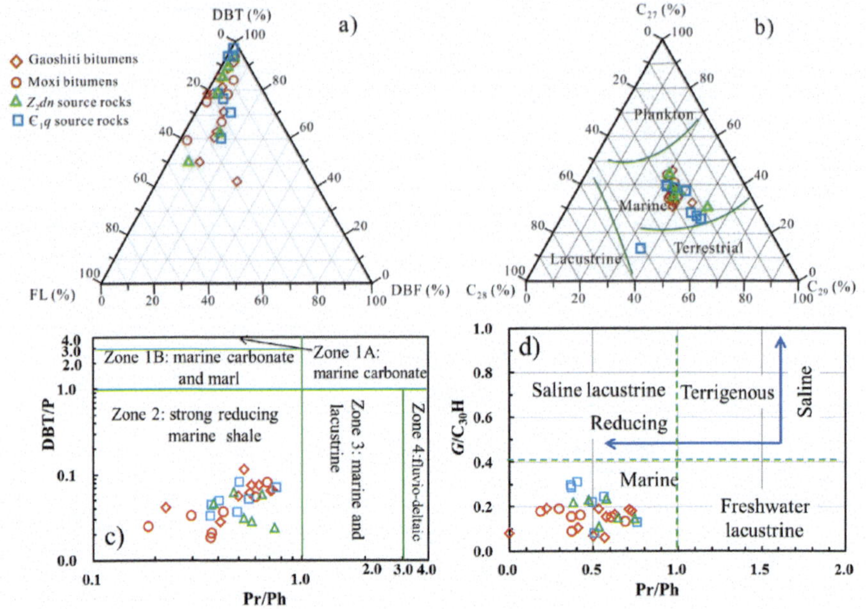

Fig. 26 Compositions of biomarkers of asphalt in Dengying Formation and source rocks in Qiongzhusi Formation and Dengying Formation, Sichuan Basin (**a** ternary of furan-dibenzothiophene-dibenzofuran; **b** ternary of C_{27}–C_{28}–C_{29} regular sterane; **c** correlogram of Pr/Ph vs. DBT/P; **d** Pr/Ph vs. G/C_{30}H) (according to Chen et al. 2017)

The source rocks in the Qiongzhusi Formation are much better than the source rocks in the Dengying Formation in terms of organic matter abundance, thickness and hydrocarbon generation intensity. Therefore, it is judged that the Sinian natural gas in Central Sichuan mainly comes from the Qiongzhusi Formation, and a small amount from the Deng3 Member and the natural gas of the Longwangmiao Formation comes from the Qiongzhusi Formation.

Biomarker Composition

By analyzing the relative composition of biomarkers of saturated hydrocarbon extracted from asphalt, aromatics, pristane and phytane of Dengying Formation, it can be seen that the asphalt of Dengying Formation comes from marine mudstone and the value is obviously close to the characteristic compound ratio of source rocks in Cambrian Qiongzhusi Formation and Dengying Formation (Fig. 26), indicating that the asphalt of Dengying Formation comes from mudstone source rocks of Qiongzhusi Formation and Dengying Formation.

4.2.2 Maokou Formation of Lower Permian

Genetic Types of Natural Gas

The carbon isotope composition of natural gas in Maokou Formation of different areas is quite different. The methane carbon isotope in the Maokou Formation of Southern Sichuan is − 33.4 ~ − 29.8‰, and the average value is − 31.9‰. The carbon isotope of ethane is − 35.6 ~ − 33.2‰, and the average value is − 34.7‰. The carbon isotope range of propane is − 32.5 ~ − 30.2‰, and the average value is − 31.5‰. The methane carbon isotope in Maokou Formation of Eastern Sichuan is − 34.8 ~ − 29.4‰, and the average value is − 31.8‰. The carbon isotope of ethane is − 37.7 ~ − 30.3‰, and the average value is − 34.9‰. The carbon isotope of propane is − 35.9 ~ − 26.3‰, and the average value is − 31.9‰. The methane carbon isotope in Maokou Formation of Western Sichuan is − 39.5 ~ − 30.4‰, and the average value is − 34.8‰. The carbon isotope of ethane is − 35.4 ~ 24.1‰, and the average value is − 30.8‰. The carbon isotope of methane in Qixia-Maokou Formation of Central Sichuan is − 35.5 ~ − 28.9‰, and the average value is − 32.6‰. The carbon isotope of ethane is − 35.4 ~ − 25.9‰, and the average value is − 30.4‰.

According to the carbon isotopic composition of ethane in Maokou Formation, the carbon isotopes of ethane in Southern Sichuan, Eastern Sichuan and Western Sichuan are basically less than − 28‰, indicating that the natural gases are mainly oil-type gas. The carbon isotope values of ethane in Shuangyushi area of Northwest Sichuan and Daxingchang of Southwest Sichuan are relatively heavy. The carbon isotopic composition of ethane in Daxingchang is similar to that of the natural gas in Changxing-Feixianguan Formation of Longgang area, which is characterized as coal gas.

The carbon isotope values of ethane of natural gas in Mao2-Mao3 Member and Qi2-Mao1 Member of Well Tongtan 1 are − 35.5‰ and − 33.6‰ respectively, indicating typical oil-type gas.

From the discrimination diagram of natural gas generation of $\delta^{13}C_1$—C_1/C_{2+3}, it can be seen that the natural gas in Maokou Formation of Well Tongtan 1 is distributed close to type II kerogen, which is another evidence of oil-type gas. The propane content in Mao2-Mao3 Member and Qixia Formation-Mao1 Member of Well Tongtan 1 is too low to detect, and the value of $\ln(C_2/C_3)$ should also be very low. And the value of $\ln(C_1/C_2)$ is 5.18 and 6.41 respectively. Input the data into Fig. 3, it can be seen that the data should be distributed within the distribution of crude oil cracked gas.

Gas Source Analysis

The distribution ranges of carbon isotopes of methane and ethane of gas in Western Sichuan are the widest, indicating that the sources of natural gas in different structures are quite different. The distribution range in Eastern Sichuan is the second, and that in Southern Sichuan is the narrowest, indicating that the sources of natural gas in Southern Sichuan are generally the same, while there are differences between different structures in Eastern Sichuan, but it is not as obvious as that in Northwest Sichuan.

The methane carbon isotope of Carboniferous natural gas is − 37.5 ~ − 31.3‰, and the average value is − 32.8‰. The carbon isotope of ethane is − 40.7 ~ − 33.6‰, whereas its average value is − 36.2‰. The carbon isotope of propane is − 36.9 ~ − 27.1‰, and the average value is − 33.3‰. Through the analysis of reservoir asphalt, biomarker composition of source rocks and carbon isotopic composition of alkane gas, it is found that the Carboniferous natural gas in Eastern Sichuan comes from the source rocks of Silurian Longmaxi Formation, and this understanding has been acknowledged among researchers (Hu et al. 1997; Dai et al. 2010). Comparing the carbon isotopes of methane and ethane of the natural gas in Maokou Formation and Carboniferous Huanglong Formation of Eastern Sichuan (Fig. 4.99), it is found that they have relatively similar composition. However, the carbon isotope of Carboniferous natural gas is generally lighter than that of Maokou Formation, indicating that the natural gas of Maokou Formation in Eastern Sichuan mainly comes from the source rocks of Silurian Longmaxi Formation, and also comes from a small amount of source rocks in other layers.

The kerogen carbon isotope of Silurian source rocks in the Upper Yangtze region is − 30.5 ~ − 27.5‰, while the kerogen carbon isotope of source rocks in Lower Permian Qixia-Maokou Formation is − 30 ~ − 26‰ (Liang et al. 2009), that is, the latter is significantly heavier than the former (Fig. 27). According to the isotopic inheritance effect in the process of organic matter generation, the natural gas generated by the source rocks with heavier kerogen carbon isotopic composition should also have heavier carbon isotopic composition under the same or similar maturity. Therefore, it is believed that natural gas of Maokou Formation in Eastern Sichuan partially comes from the muddy carbonate source rocks of Lower Permian.

The carbon isotopic composition of alkane gas in Qixia-Maokou Formation of Southern Sichuan is very close to that of Carboniferous in Eastern Sichuan, and that of Southern Sichuan happens to be the hydrocarbon generation center of Silurian Longmaxi Formation, indicating that most of the natural gas comes from Silurian Longmaxi Formation.

The ethane carbon isotope of natural gas from Mao2-Mao3 Member of Well Tongtan 1 is very similar to that of Carboniferous natural gas, and has genetic link with kerogen carbon isotope of source rocks in Longmaxi Formation, indicating that the natural gas in Mao2-Mao3 Member comes from source rocks of Longmaxi Formation. The ethane carbon isotope of natural gas from Qixia Formation-Mao1 Member of Well Tongtan 1 is slightly heavier than that of natural gas from Mao2-Mao3 Member, and is lighter than that of natural gas from Maokou Formation of Dongyuezhai gas field, which only comes from source rocks of Maokou Formation, indicating that the natural gas from Qixia Formation-Mao1 Member of Well Tongtan 1 should be a mixture of natural gas from source rocks of Longmaxi Formation and Mao1 Member, and is mainly contributed by source rocks of Mao1 Member.

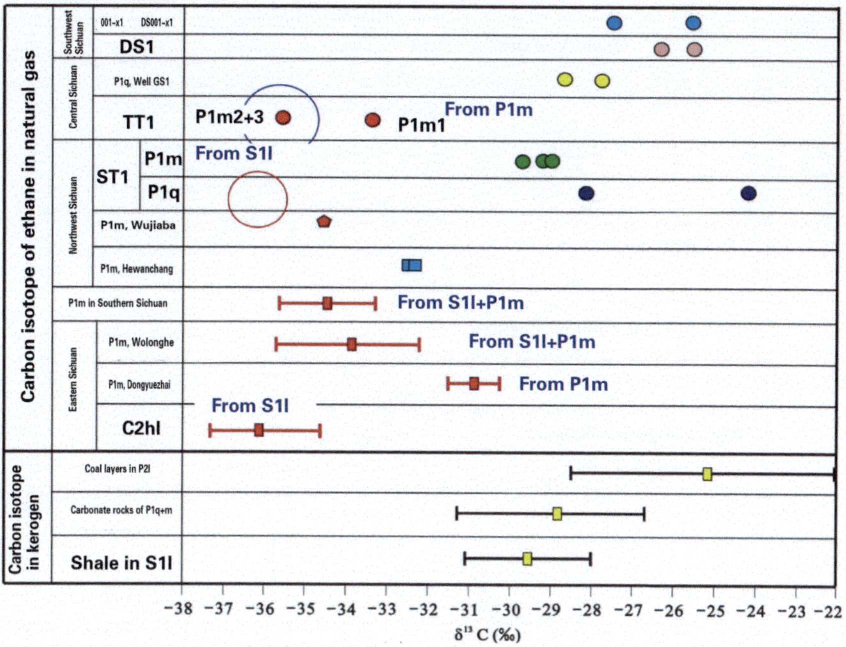

Fig. 27 Carbon isotopic composition of ethane of natural gas and of kerogen in source rocks of Qixia-Maokou Formation in different areas of Sichuan Basin

4.2.3 Changxing Formation

The source of gas reservoir in Changxing Formation of Central Sichuan has always been controversial. The natural gas of drilled wells such as Well Moxi, Lai 1 and Wangjia 1 is typical oil-type gas (Fig. 28). The isotopic values of the natural gas in the three wells are reversed, which are very similar to the characteristics of the natural gas from Longmaxi Formation (Fig. 29). The values are related to the kerogen carbon isotope of the source rocks from Longmaxi Formation, indicating that the source rocks of Longmaxi Formation to some extent contribute the natural gas in Changxing Formation. As a result, it is considered that the potential source rocks of gas reservoir in Changxing Formation of Hechuan-Tongnan are source rocks from Lower Silurian Longmaxi Formation and Permian Longtan Formation.

4.2.4 Leikoupo Formation

Since the breakthrough made in Zhongba Gas Field, probing the source of gas reservoir of Leikoupo Formation in Sichuan Basin has been paid much attention, but it is still controversial. The natural gas in Moxi gas reservoir is oil-type gas, which has the characteristics of high drying coefficient, high ethane carbon isotope value

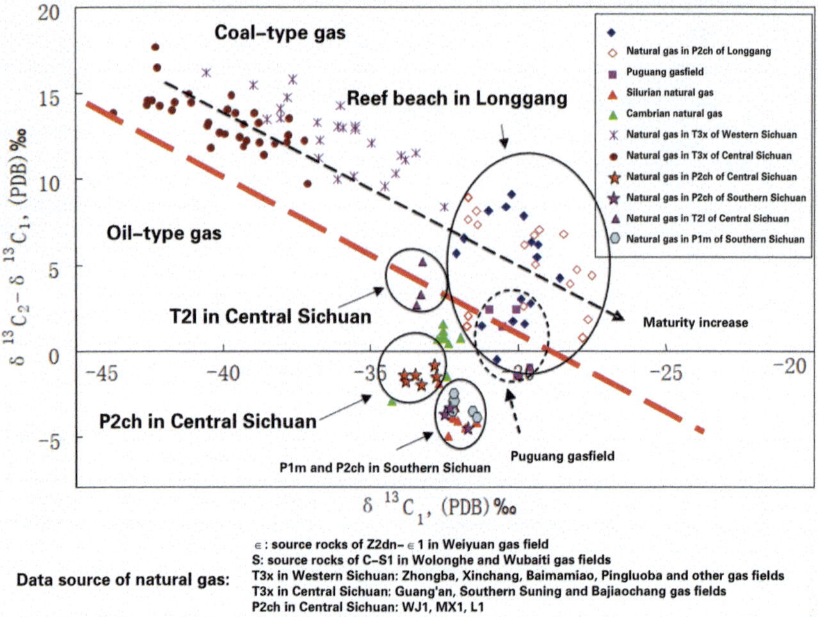

Fig. 28 Correlation of natural gas $\delta^{13}C_2$-$\delta^{13}C_1$—$\delta^{13}C_1$ in Sichuan Basin

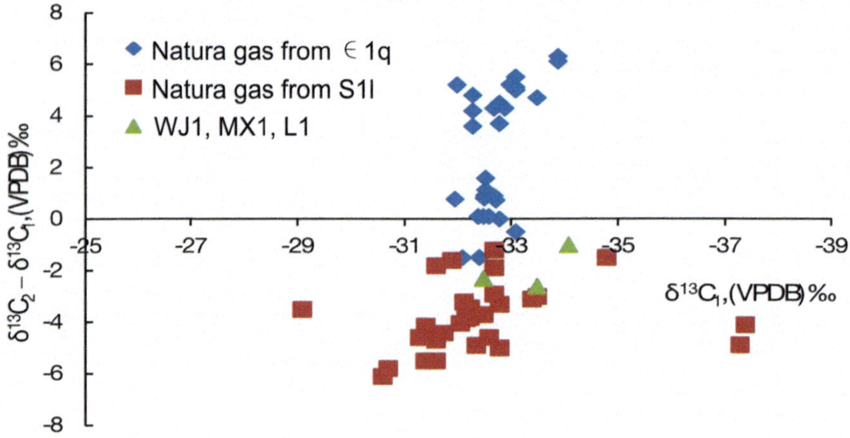

Fig. 29 Comparison between some natural gas of Changxing Formation in Central Sichuan and natural gas from source rocks of Qiongzhusi and Longmaxi Formation

(with the characteristics of coal gas), high hydrogen sulfide content and TSR reaction (Wang et al. 1997; Sun 2014). However, the reservoir is not in direct contact with the source rocks and lacks obvious large faults. Therefore, although it is considered that the source rocks should come from the deep Permian, the corresponding migration channels are lacked. Sun and other researchers believe that the natural gas in the Moxi gas reservoir is migrated from Longnvsi area through faults and then moved to the Moxi area. For oil-type gas in Moxi gas reservoir migrated from deep strata to Leikoupo Formation, it have to pass through two sets of carbonate strata of Lower Triassic Feixianguan Formation and Jialingjiang Formation. At the same time, with the development of gypsum salt rocks of the Jialingjiang Formation, only through faults can large-scale oil and gas migration occur. The potential source rocks of Leikoupo Formation gas reservoir in Hechuan-Tongnan are source rocks from underlying Silurian Longmaxi Formation and Permian Longtan Formation.

4.3 Hydrocarbon Supply of Source Rocks and Accumulation Model

The hydrocarbon reservoirs in Central Sichuan should be gas reservoirs, and the natural gas comes from source rocks and crude oil cracking. The crude oil cracked gas is mainly the natural gas and asphalt formed by the cracking of the original reservoir in high temperature. The source rocks in different intervals provide gas for reservoirs in different intervals (Fig. 30). The potential source rocks that may provide natural gas for gas reservoirs in Deng4 Member are three sets of source rocks of overlying Qiongzhusi Formation, underlying Deng3 Formation and underlying Doushantuo Formation. There is a ternary hydrocarbon supply model of overlying inverted irrigation and underlying dual-source supply, and the hydrocarbon sources are relatively rich. The potential source rocks of gas reservoirs in Longwangmiao Formation and Xixiangchi Formation are the underlying Qiongzhusi Formation, which may develop as lower generation and upper storage accumulation models. The potential source rocks of gas reservoir in Maokou Formation are Maokou Formation itself and the source rocks in underlying Qixia Formation, Longmaxi Formation and Qiongzhusi Formation, which may develop as the ternary hydrocarbon accumulation model of lower generation and upper storage. The potential source rocks of gas reservoirs in Changxing Formation are the source rocks in underlying Longtan Formation and Silurian Longmaxi Formation, which may develop as the accumulation model of lower generation and upper storage. The potential source rocks of gas reservoirs in Leikoupo Formation are source rocks in Lower Silurian Longmaxi Formation and Permian Longtan Formation, which may develop as the accumulation model of lower generation and upper storage. There is stratigraphic pinch out of Silurian Longmaxi Formation in the research area, and there may also be the accumulation model of side-generation and side-storage in gas reservoirs of Xixiangchi Formation and Maokou Formation at the high structural position to the west of pinch out line.

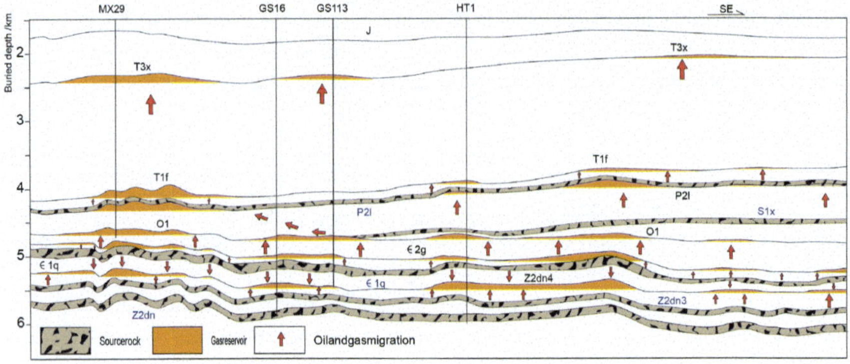

Fig. 30 Hydrocarbon supply and accumulation model of source rocks in Central Sichuan

4.4 Regional Hydrocarbon Accumulation Process

The hydrocarbon accumulation process of carbonate strata in Central Sichuan can be summarized as follows: the formation of early paleo-oil reservoir; the destruction of paleo-oil reservoir; the re-formation of paleo-oil reservoir; the transformation from paleo-oil reservoir to overpressure paleo-gas reservoir through crude oil cracking; present atmospheric residual gas reservoir. After the sedimentation of Sinian, Dengying Formation was obviously uplifted, exposed and denuded by Tongwan movement, especially in the core of paleo-uplift in Central Sichuan. Due to the exposure of the surface, the dissolution of atmospheric fresh water is developed in Dengying Formation, forming a porous reservoir with good physical properties. Therefore, the good reservoir conditions of Sinian Dengying Formation in Central Sichuan are inseparable from the pre-existing characteristics of high porosity and density caused by the dissolution of atmospheric fresh water in the sedimentary period of Paleo-Sinian. During the sedimentary period of Sinian, the overpressure strata have not yet formed due to the immature organic matters of the source rocks. The black argillaceous shale of Qiongzhusi Formation deposited later is an important source rock for oil and gas in Sinian and upper reservoirs in the research area. After the sedimentation of Cambrian, the underlying source rocks of Qiongzhusi Formation are not yet mature. At the end of Silurian sedimentation, the source rocks of Deng3 Member of Sinian and Qiongzhusi Formation of Cambrian in Moxi and Gaoshiti structural belts in the high part of Northwest Sichuan were not mature, while the Ro value of source rocks in other deep buried areas was 0.5%, indicating the early maturity stage. The oil and gas can be filled in the traps of Deng4 Member of Sinian, and overlying Longwangmiao and Xixiangchi Formation of Cambrian system along the faults. It is the formation stage of paleo-oil reservoir (Fig. 31). During the Caledonian movement, a large number of strata were uplifted and denuded. At this time, the hydrocarbon generation evolution of the source rocks stopped, and the early paleo-oil reservoir was damaged by adjustment, resulting in different degrees of biodegradation. Generally, asphalt originated from oxidative degradation was contained in the reservoir. The paleo-oil reservoir was adjusted and damaged again in Hercynian, and the buried depth of the strata was generally less than that before the Caledonian movement, as a result, the source rocks were in the stage of hydrocarbon generation stagnation. With the deposition of Permian and Triassic strata, the Ro values of all Sinian and Cambrian source rocks in the research area increased sharply. In addition, the source rocks in Silurian Longmaxi Formation and Permian Longtan Formation began to mature successively in the Triassic. This stage is the main period of crude oil filling in the research area. The Ro of Sinian Deng3 Member in the southeast of the research area is greater than 1.3%, which may lead to a small amount of wet gas filling to the reservoir. After entering the Jurassic, with a large number of stratigraphic deposition in the foreland tectonic stage, the source rocks quickly continued to generate hydrocarbons. Generally, before the uplift and denudation of the Early and Middle Cretaceous strata, except that the Longtan Formation, which was located in the high part of the northwest structure and was shallow, was in the wet gas stage, the Ro of all four sets of source rocks in other areas exceeded 2.0%, indicating the dry gas stage. This period is also an important period of crude oil cracking. In the period of gas generation, the liquid hydrocarbon in the previous reservoirs was cracked under high temperature, and the gas invasion was caused by strong natural gas filling cracked by source rocks. As a result, the deasphalting was formed with a small amount of asphalt precipitation. This period is the main period of natural gas accumulation and filling in the research area. The crude oil was cracked into natural gas and asphalt, and the crude oil in the paleo-oil reservoir was basically cracked into natural gas, which made the paleo-oil reservoir gradually disappear and transform into paleo-gas reservoir. Due to the volume expansion effect, the pressure was formed in the process of crude oil cracking into natural gas, which made the paleo-gas reservoir into overpressure reservoir. In the later stage of Himalayan movement, due to a large number of uplift and denudation of the strata, the evolution of source rocks stopped. The sharp decline of buried depth leads to the decrease of reservoir temperature and the release of formation pressure, as a result, the previous water-solution gas and oil-soluble gas may be exsolved, further adjusted and re-accumulated with the previous natural gas, forming the present oil and gas reservoirs.

Carbonate reservoirs in Central Sichuan should be dominated by gas reservoirs, containing a small amount of condensate reservoirs, with many vertical production layers and wide plane distribution. The main production layers are Qixia Formation, Maokou Formation and Changxing Formation of Permian system, Feixianguan Formation, Jialingjiang Formation and Leikoupo Formation of Triassic, Xixiangchi and Longwangmiao Formation of Cambrian and Deng4 Member of Sinian. Based on the structural evolution process and hydrocarbon generation evolution of source rocks, combined with multi-stage fluid filling and accumulation events, the accumulation models of the two profiles are analyzed.

After the deposition of Ordovician and Silurian strata in the NW–SE profile (Fig. 32), the Sinian-Cambrian source rocks were deeply buried in the southeast area, reaching the hydrocarbon generation threshold and entering the early maturity stage. The generated liquid hydrocarbons accumulated in Deng4 Member and the favorable traps of overlying Longwangmiao and Xixiangchi Formation along

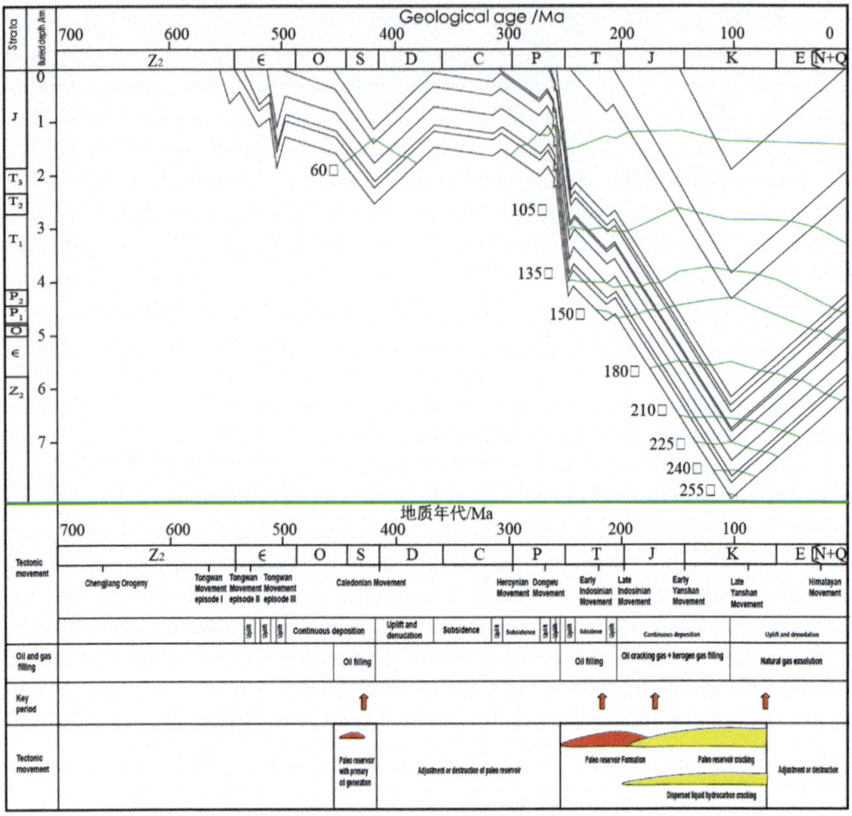

Fig. 31 Hydrocarbon accumulation events of Well Guangtan 2 in Central Sichuan

the faults and karst unconformities. During the period of Caledonian movement, the strata in the northwest area were uplifted and denuded, and the Xixiangchi Formation was exposed to the surface. The paleo-oil reservoirs were damaged, and the paleo-oil reservoirs of Deng4 Member, Longwangmiao and Xixiangchi Formation were adjusted and re-accumulated. At that time, the paleo-oil reservoirs were distributed in the favorable traps of various parts of the slope. With the burial of Permian strata, the Sinian-Cambrian source rocks reach the burial depth before Caledonian movement and continue to be buried. The maturity continues to increase and the source rocks enter the mature stage. The generated liquid hydrocarbons are mainly concentrated in the southeast and northwest areas, and there are few favorable traps in the middle part. During the whole Dongwu and Indosinian, the Sinian-Cambrian source rocks manly generated oil. In the early Indosinian, source rocks in Longmaxi and Longtan Formation were developed with the Ro reaching 0.5, and began to supply hydrocarbon for overlaying traps.

The paleo-oil reservoirs were finalized in the Early Indosinian, and were distributed in the favorable traps of each layer in the slope. In the Late Indosinian,

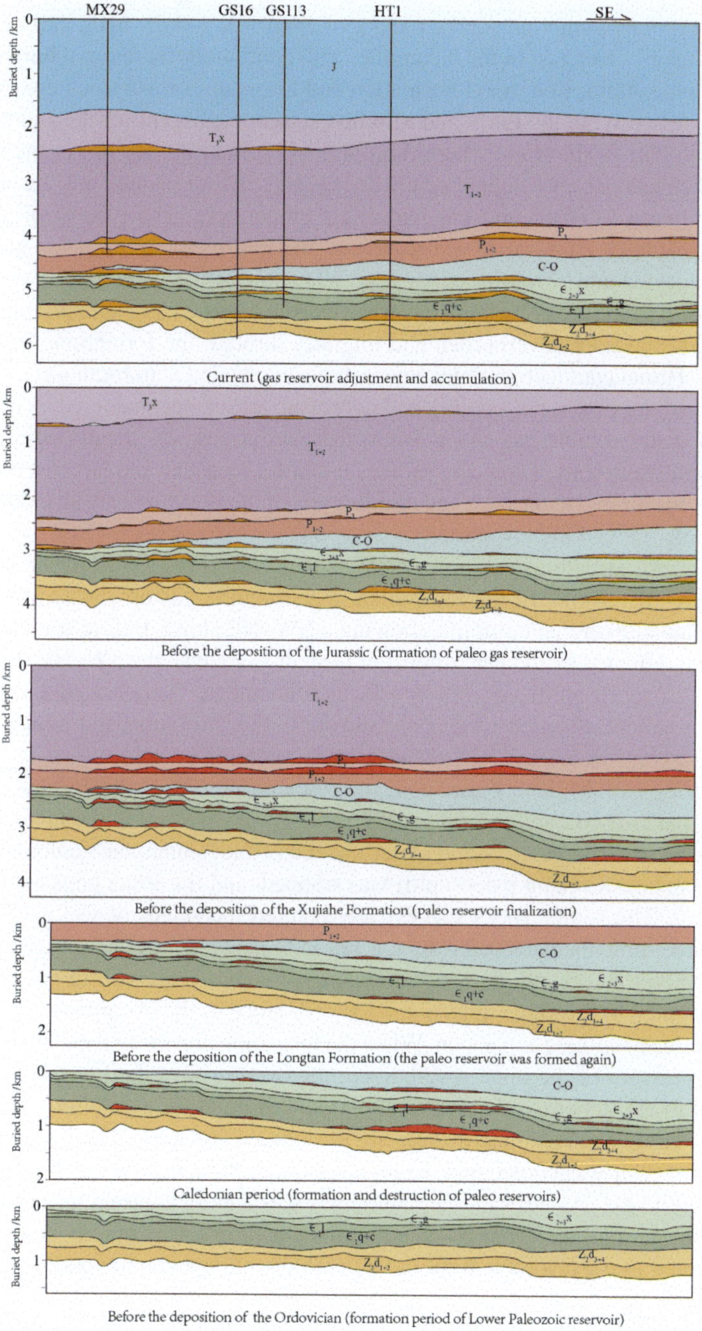

Fig. 32 Accumulation process and model of NW–SE oil and gas reservoirs in Central Sichuan

the Ro of Sinian-Cambrian source rocks exceeded 1.3, and the hydrocarbon generated was mainly wet gas. In the meantime, with continuous increase of buried depth of overlaying strata, Longmaxi Formation and Longtan Formation entered wet gas stage successively. In the Early and Middle Cretaceous, except for the source rocks in the high part of the slope where the Longtan Formation was still in the wet gas stage, other source rocks all entered the dry gas stage. At the same time, the paleo-oil reservoir began to crack into gas with high temperature. The type of oil and gas reservoir changed from oil to gas reservoir, and the overpressure paleo-gas reservoir was formed due to volume expansion. When the formation pressure exceeds the breakthrough pressure of the upper overburden, the overpressure fluid is episodically expulsed to the upper overburden and migrates through the formation. During the period of Himalayan tectonic movement, both Sinian and Cambrian gas reservoirs experienced the process of cooling and depressurization, accompanied by the exsolution of water-soluble gas. As a result, the natural gas was locally adjusted and formed into reservoirs. From the profile, it can be seen that the gas reservoirs are mainly concentrated in the high and middle parts of the slope, and are also distributed in favorable traps of the low part.

The overall structural position of SW–NE profile (Fig. 33) is lower than that of the NW–SE profile. In the Indosinian, the structure was obviously affected by the Luzhou-Kaijiang paleo-uplift, showing a SW–NE slope. Before the Indosinian movement, affected by the paleo-uplift of Leshan-Longnvsi, the SW–NE strata only showed as overall uplift and fall and local adjustment. Therefore, the accumulation environment changed little, only shown as the migration and adjustment of paleo-reservoirs in a small range and a large range of layered reservoirs. Before the deposition of Xujiahe Formation, the research area was affected by Luzhou-Kaijiang paleo-uplift, and the paleo-oil reservoir was finalized in the favorable trap of northeast inclined gentle slope. After the end of Indosinian, the geological influence of Luzhou-Kaijiang paleo-uplift was stopped, and the strata presented overall subsidence. The source rocks successively entered the stages of wet and dry gas. At the same time, the paleo-oil reservoir began to generate gas under high temperatures. The type of oil and gas reservoir changed from oil reservoir to gas reservoir, and distributed in favorable traps of the SW–NE slope as the overpressure paleo-gas reservoir due to volume expansion. When the formation pressure exceeded the breakthrough pressure of the upper overburden, the overpressure fluid was episodically expulsed to the upper overburden and migrated through the formation. During the period of Himalayan tectonic movement, the strata were uplifted and denuded, and the gas reservoir was cooled and depressurized, accompanied by the desolvation of water-soluble gas. As a result, the natural gas was locally adjusted to form a reservoir. From the profile, it can be seen that the gas reservoir is mainly stratified and concentrated in the favorable traps in the middle section and other parts.

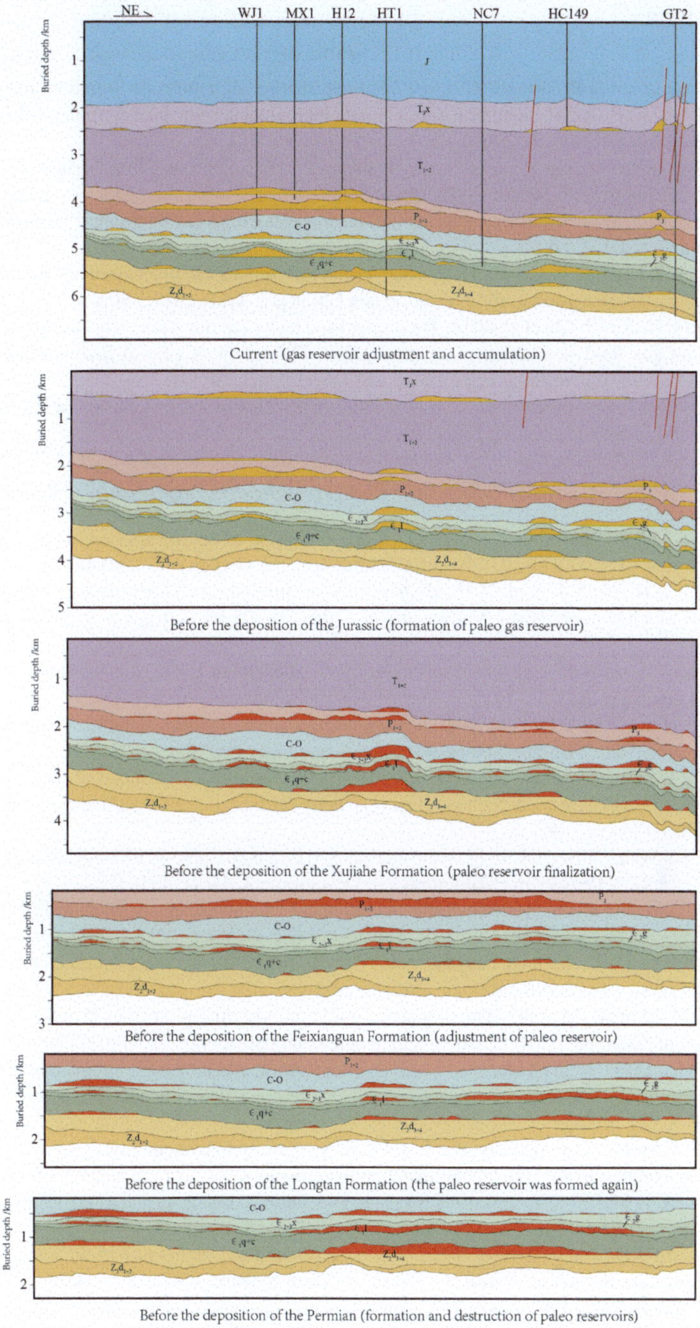

Fig. 33 Accumulation process and model of SW–NE oil and gas reservoirs in Central Sichuan

References

Chen Z H, Simoneit B R, Wang T G, et al. 2017. Biomarker signatures of Sinian bitumens in the Moxi-Gaoshiti Bulge of Sichuan Basin, China: Geological significance for paleo-oil reservoirs [J]. Precambrian Research, 296: 1–19.

Dai J X. 1992. Identification of various types of alkane gas [J]. Science in China Seris B, 2: 185–193.

Dai J X, Li J, Ding W W, et al. 2005. Geochemical characters of the giant gas accumulations with over one hundred billion cubic meters reserves in China [J]. Petroleum Exploration and Development, 45(3): 16–23.

Dai J X, Ni Y Y, Huang Sh P. 2010. Discussion on the carbon isotopic reversal of alkane gases from the Huanglong Formation in the Sichuan Basin China [J]. Acta Petrolei Sinica, 31(5): 710–717.

Du J H, Zou C N, Xu Ch Ch, et al. 2014. Theoretical and technical innovations in strategic discovery of a giant gas field in Cambrian Longwangmiao Formation of central Sichuan paleo-uplift, Sichuan Basin [J]. Petroleum Exploration and Development, 41(3): 268–277.

Du J H et al. 2015. Geological theory and exploration practice of ancient carbonate gas field [M]. Petroleum Industry Publishing House, Beijing.

Duan J B, Li P P, Chen D, et al. 2013. Formation and evolution of the reef flat facies lithologic gas reservoir of Changxing Formation in Yuanba Gas Field, Sichuan Basin [J]. Lithologic Reservoirs, 25(3): 43–91.

Gang W Zh, Gao G, Hao Sh Sh, et al. 1997. Carbon isotope of ethane applied in the analyses of genetic types of natural Gas [J]. Experimental Petroleum Geology, 19(2): 164–167.

He Zh L, Jin X H, Wo Y J, et al. 2016. Hydrocarbon accumulation characteristics and exploration domains of ultra-deep marine carbonates in China [J]. China Petroleum Exploration, 21(1): 3–14.

Hu G C. 1997. A breakthrough of oil-gas exploration of Sichuan Basin- a case study [J]. Marine Oil-Gas Geology, 2(3): 52–53.

Huang H Y, He D F, Li Y Q, et al. 2019. Determination and formation of the Luzhou paleo-uplift in the southeastern Sichuan Basin [J]. Earth Science Frontiers, 26(1): 102–120.

Li L, Tan X C, Zeng W, et al. 2013. Development and reservoir significance of mud mounds in Sinian Dengying Formation, Sichuan Basin [J]. Petroleum Exploration and Development, 40(6): 666–673.

Li J, Li Zh Sh, Wang X B, et al. 2017. New indexes and charts for genesis identification of multiple natural gases [J]. Petroleum Exploration and Development, 44(4): 503–512.

Liang D G, Guo T L, Chen J P, et al. 2009. Geochemical Characteristics of Four Suits of Regional Marine Source Rocks, South China [J]. Marine Oil-Gas Geology, 2(3): 1–15.

Liu W H, Zhang D W, Wang X F, et al. 2004. Geochemistry study on Gas-source correlation of natural gas [J]. Acta Sedimentologica Sinica, 22(5): 27–32.

Peters KE, Moldowan JM. 1993. The biomarker guide: interpreting molecular fossils in petroleum and ancient Sediments [M]. Prentice-Hall, Englewood Cliffs.

Ping H W, Chen H H, Régis Thiéry, et al. 2014. Effects of oil cracking on homogenization temperature and trapping pressure of oil inclusion and its geological significance [J]. Journal of Earth Science, 39(5): 587–599.

Sun J T. 2014. Middle Triassic Leikoupo Formation source rocks characteristics and analysis of gas source in Sichuan Basin [D]. Chengdu: Chengdu University of Technology, 52–78.

Wang L S, Gou X M, Liu G Y, et al. 1997. Organic geochemical characteristics and origin of natural gas in Sichuan Basin [J]. Acta Sedimentologica Sinica, 15(2): 49–53.

Wang Y G, Yu X F, Yang Y, et al. 1998. Applications of fluid inclusions in the study of paleo geotemperature in Sichuan Basin [J]. Journal of Earth Science, 23(3): 285–288.

Wei G Q, Wang Z H, Li J, et al. 2017. Characteristics of source rocks, resource potential and exploration direction of Sinian and Cambrian in Sichuan Basin [J]. Natural Gas Geoscience, 28(1): 1–13.

Xu Y L. 2009. Study on Sinian Cambrian hydrocarbon accumulation conditions in Central Sichuan [D]. Chengdu: Chengdu University of Technology, 2009.

Xu Ch Ch, Shen P, Yang Y M, et al. 2014. Accumulation conditions and enrichment patterns of natural gas in the Lower Cambrian Longwangmiao Fm reservoirs of the Leshan-Longnvsi Paleohigh, Sichuan Basin. Natural Gas Industry 34(3): 1–7.

Xu F H, Yuan H F, Xu G Sh, et al. 2018. Fluid charging and hydrocarbon accumulation in the Cambrian Longwangmiao Formation of Moxi Structure, Sichuan Basin, SW China [J]. Petroleum Exploration and Development, 45(3): 426–435.

Yuan H F, Liu Y, Xu F H, et al. 2014. The fluid charge and hydrocarbon accumulation, Sinian reservoir, Anpingdian-Gaoshiti Structure, Central Sichuan Basin [J]. Acta Petrologica Sinica, 30(3): 727–736.

Wang Sh Q. 1994. Geochemical characteristics of Jurassic-Sinian gas in Sichuan Basin [J]. Natural Gas Industry 14(6): 1–5.

Zhang L, Wei G Q, Li X Z, et al. 2007. The thermal history of Sinian-Lower Paleozoic high/over mature source rock in Sichuan Basin [J]. Natural Gas Geoscience, 18(5): 726–731.

Zhang T, Ma X Zh, Zhao W W et al. 2017. Characteristics of paleo-oil reservoir of Longwangmiao Formation in Moxi- Gaoshiti Area [J]. Fault-block Oil & Gas Field, 24(4): 466–470.

Zhu G Y, Yang H J, Su J, et al. 2012. True exploration potential of marine oils in the Tarim Basin [J]. Acta Petrologica Sinica, 28(3): 1333–1347.

Zou C N, Du J H, Xu C C, et al. 2014. Formation, distribution, resource potential and discovery of the Sinian-Cambrian giant gas field, Sichuan Basin, SW China [J]. Petroleum Exploration and Development, 41(3): 278–293.

Chapter 6
Recent Key Exploration Fields and Zones

1 Exploration Fields and Zones of the Sinian Dengying Formation

1.1 Key Exploration Fields and Zones

The main reservoir accumulation types of Sinian Dengying Formation are karst reservoir with side-generation and side-storage, "sandwich reservoir", near-source dolomite reservoir with upper generation and lower storage, and karst reservoir with lower generation and upper storage. The major source rocks are developed in Cambrian Qiongzhusi Formation, and the potential source rocks are formed in Deng3 Member, Doushantuo Formation and pre-Sinian. The reservoirs are developed in Deng2 Member and Deng4 Member. Based on the current exploration results and the main understanding of the controlling factors of the gas reservoir, three favorable exploration zones of Deng2 Member in Central Sichuan intra-platform zone covering an area of 928 km^2 are evaluated. They are based on the mound-beach bodies, tectonic background and source rock conditions in Deng2 Member of Central Sichuan intra-platform zone, including two Class I favorable zones (Tongnan favorable zone and Northern Hechuan favorable zone, covering an area of 718 km^2), one Class II favorable area (Eastern Wusheng favorable zone, covering an area of 210 km^2) (Fig. 1). There are three favorable exploration zones in Deng4 Member, which collectively cover an area of 1105 km^2. These zones include one Class I favorable zone (Northern Hechuan favorable zone, covering an area of 765 km^2) and two Class II favorable zones (Southern Hechuan favorable zone and Eastern Wusheng favorable zone, covering an area of 340 km^2) (Fig. 2).

Q. Li and T. Wang, *Paleogeography and Gas Exploration of Marine Multiple Formations in Central Sichuan Basin*, https://doi.org/10.1007/978-981-99-2528-5_6

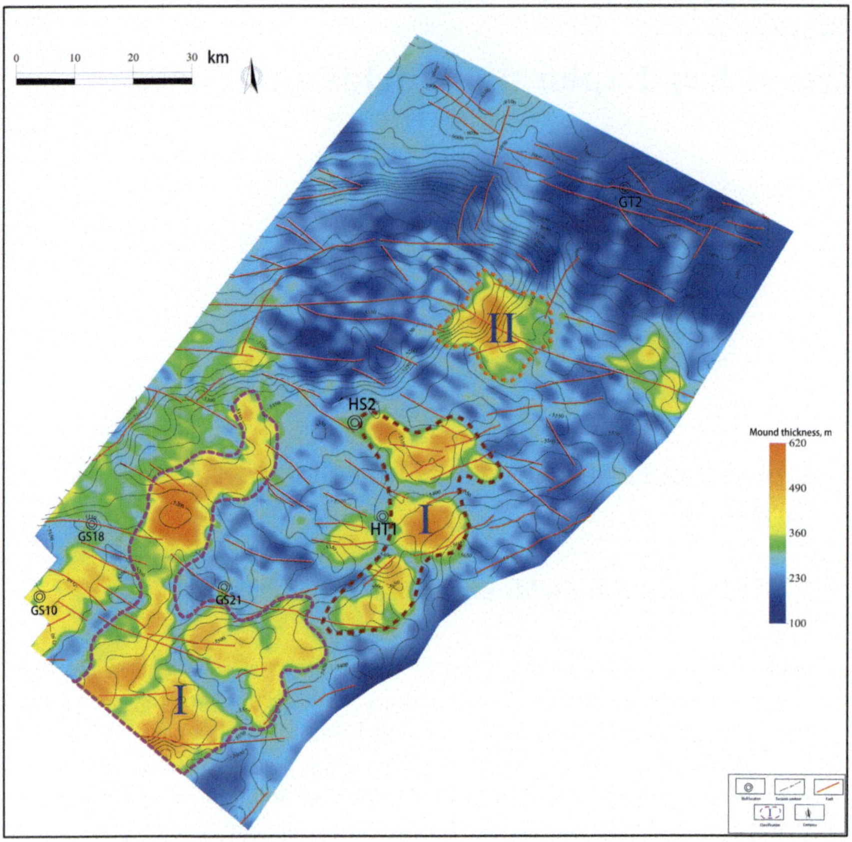

Fig. 1 Comprehensive evaluation map of favorable zones in Deng2 Member of intra-platform zone, Central Sichuan

1.2 Favorable Exploration Targets

For the mound-beach bodies in Dengying Formation, four favorable exploration targets are selected, such as Well Qiantan 1, Wutan 1, Longtan 1 and Antan 1, according to ① the thickness and area of mound-beach bodies are large and the shape of beach bodies are good; ② the tectonic background is located in the high part and periphery of continuous structure; ③ the mound-beach bodies are adjacent to the thick-layer area of source rocks in Deng3 Member, and potential source rocks are developed in Doushantuo Formation and pre-Sinian.

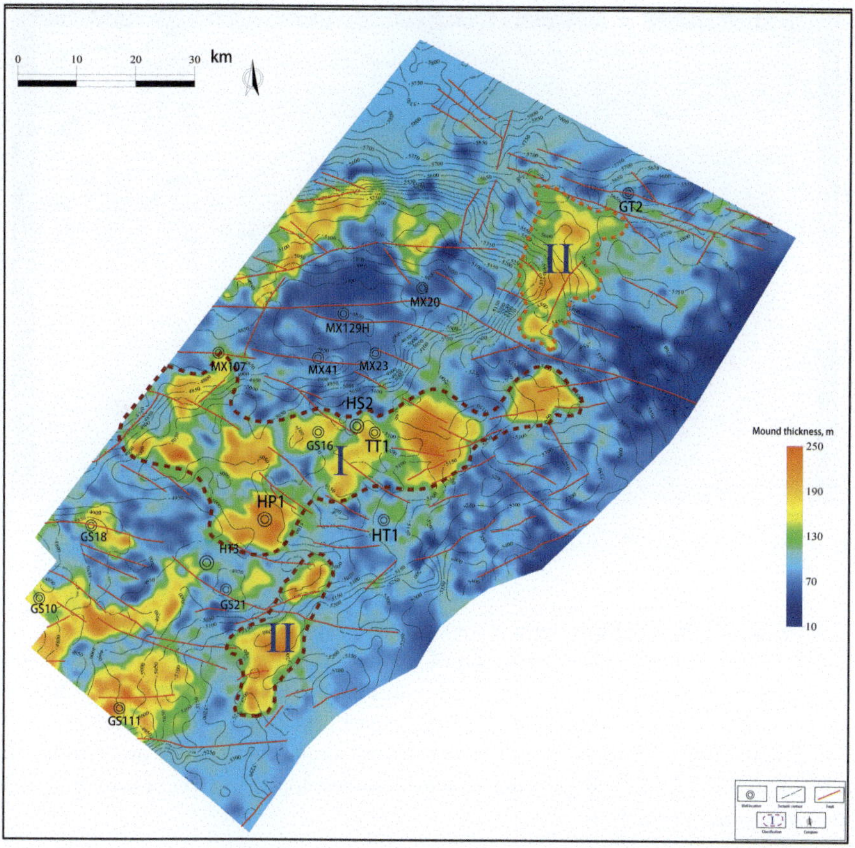

Fig. 2 Comprehensive evaluation map of favorable zones in Deng4 Member of intra-platform zone, Central Sichuan

Mound-beach bodies are developed in both Deng4 Member and Deng2 Member of Well Qiantan 1 (Fig. 3). The area of mound-beach bodies in Deng2 and Deng4 Member is about 163 km^2 and 175 km^2, respectively. Well Qiantan 1 is developed with tectonic background, and the Deng2 Member in the well is developed as an anticline trap with an area of 73.5 km^2, a minimum trap line of − 5370 m and a closure of 20 m (Fig. 4). The Deng4 Member in the well is also developed as an anticline trap with an area of 24.7 km^2, a minimum trap line of − 5100 m and a closure of 60 m. According to the reservoir thickness of seismic inversion, the reservoirs are well developed in Well Qiantan 1. The reservoirs in Deng2 Member of this well are developed with the thickness of about 70 m (Fig. 5), and the reservoirs in Deng4 Member of this well are developed with the thickness of 45 m and the porosity ranging between 4 and 8%. According to the seismic profile of hydrocarbon detection, there is a high possibility of gas content in Deng2 and Deng4 Member of Well Qiantan 1 (Fig. 6).

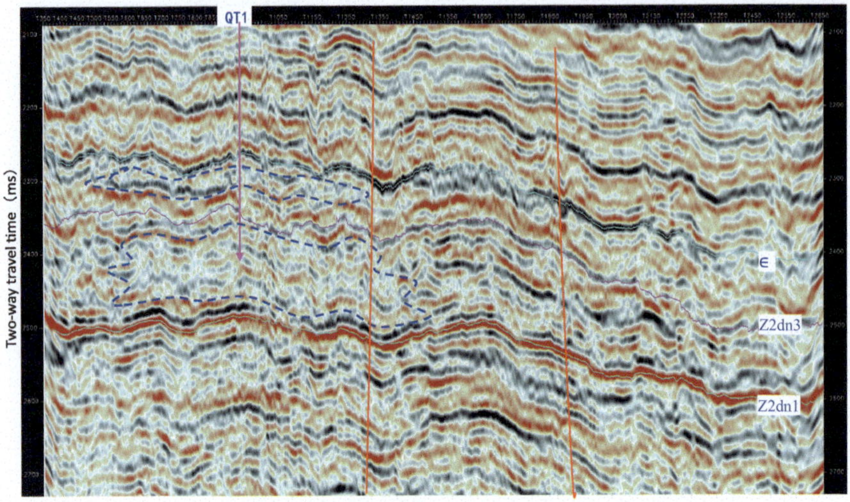

Fig. 3 Profile cross Well Qiantan 1 (L1170)

In Well Qiantan 1, three targets can also be explored: (1) fault-karst body of Maokou Formation, (2) granular beach of Qixia Formation, and (3) granular beach of Longwangmiao Formation (Fig. 7). It can be seen from Fig. 7 that the source rocks are developed in Silurian and Qiongzhusi Formation of Well Qiantan 1, oil source faults connecting Silurian are developed, fault-karst bodies are developed in Maokou Formation, and granular beaches are developed in Qixia Formation and Longwangmiao Formation.

2 Key Exploration Fields and Zones of Cambrian Strata

2.1 Key Exploration Areas and Zones

The reservoir accumulations in Longwangmiao and Xixiangchi Formation of Cambrian are developed with lower-generation and upper storage. The major source rocks are developed in Cambrian Qiongzhusi Formation, and the potential source rocks are developed in Deng3 Member, Doushantuo Formation and pre-Sinian. The reservoirs are developed in Longwangmiao Formation and Xixiangchi Formation. Based on the current exploration results and the main understanding of the controlling factors of the gas reservoir, focusing on the key factors controlling the gas reservoir, two favorable exploration zones of Longwangmiao Formation in Central

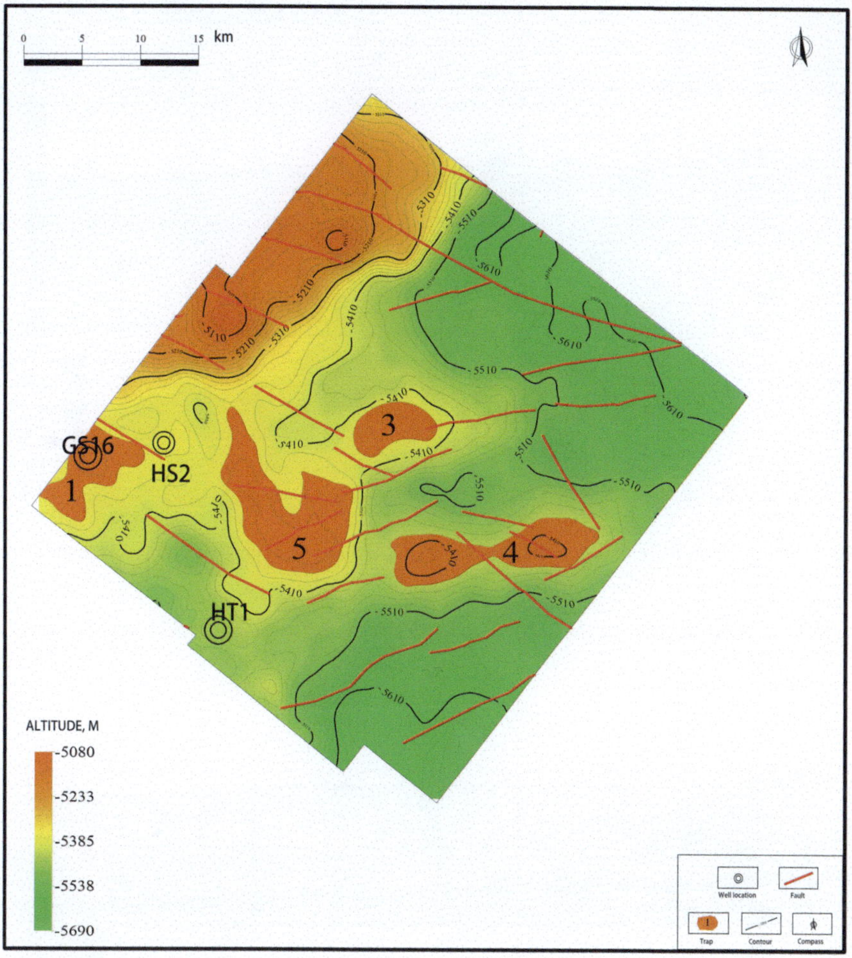

Fig. 4 Top structure map of Deng2 Member in Well Hechuan 125

Sichuan are evaluated covering an area of 1630 km^2. Based on the mound-beach bodies, tectonic background and source rock conditions, including one Class I favorable zone, covering an area of 480 km^2, and one Class II favorable area, covering an area of 1150 km^2 (Fig. 8). There are three favorable exploration zones in Xixiangchi Formation, covering an area of 1162 km^2, including two Class I favorable zone covering an area of 574 km^2, and one Class II favorable zone covering an area of 588 km^2 (Fig. 9).

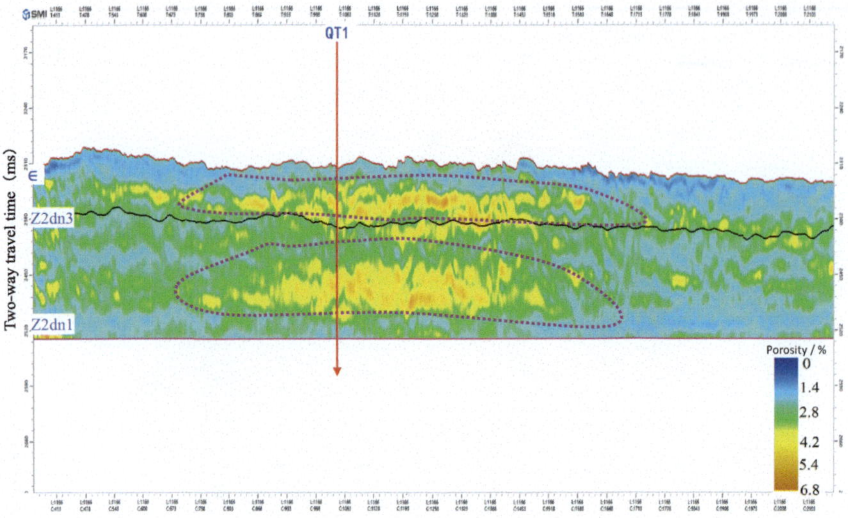

Fig. 5 Inversion profile cross Well Qiantan 1 (L1170, Bottom Dengying Formation flattened)

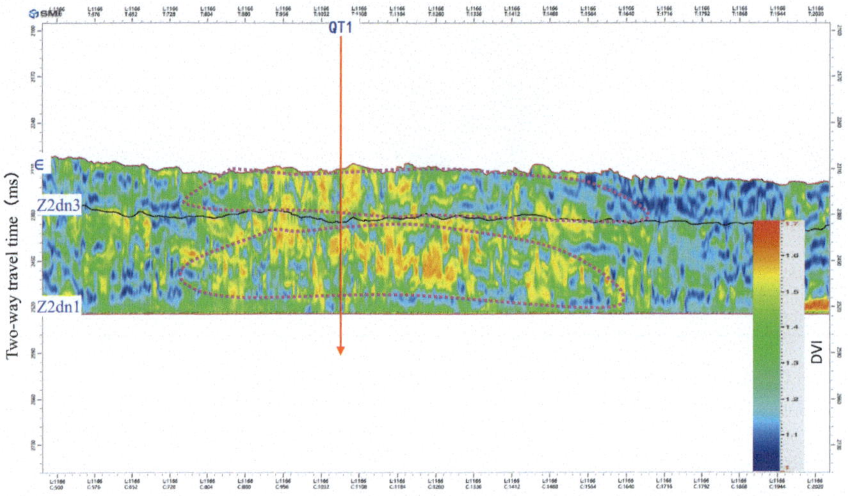

Fig. 6 Hydrocarbon detection profile cross Well Qiantan 1 (T1170, Bottom Dengying Formation flattened)

2.2 *Favorable Exploration Targets*

For the mound-beach bodies in Longwangmiao Formation, three favorable exploration targets are selected, such as Well Tongtan 2, Tongtan 3 and Tongtan 5 (Fig. 8),

Fig. 7 3D seismic profile cross Well Hechuan 125–Hechuan 125 (L1158)

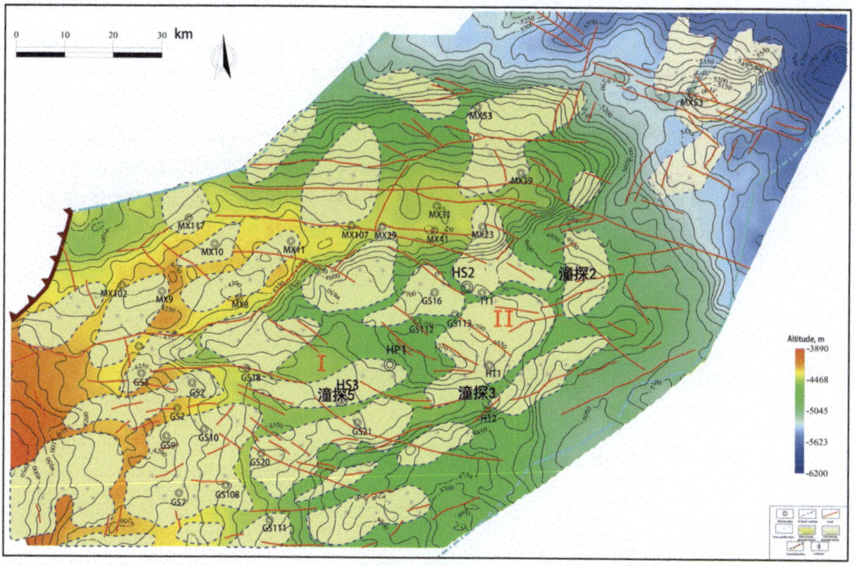

Fig. 8 Comprehensive evaluation map of favorable zones of Longwangmiao Formation in Central Sichuan

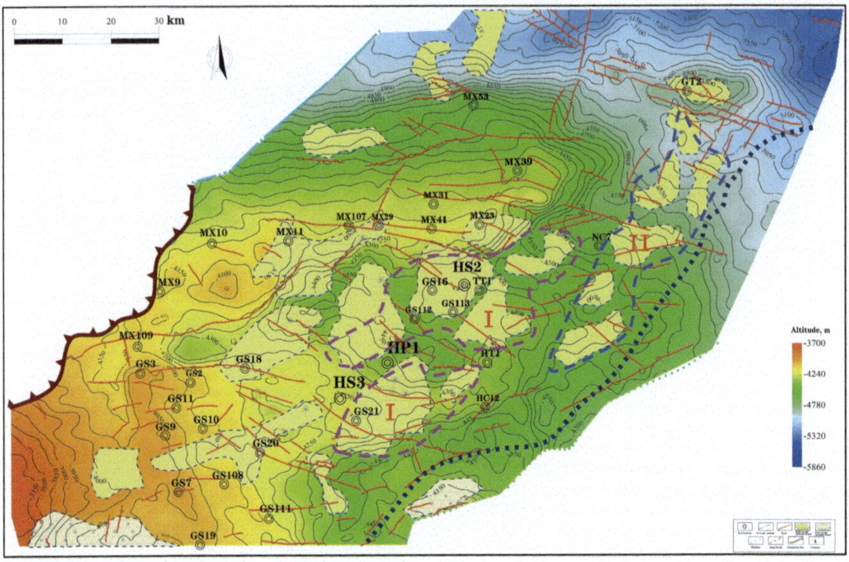

Fig. 9 Comprehensive evaluation map of favorable zones of Xixiangchi Formation in Central Sichuan

according to ① the development of mound-beach bodies in Longwangmiao Formation; ② the tectonic background located in the high part and periphery of continuous structure; ③ the mound-beach bodies adjacent to the thick-layer area of source rocks in Qiongzhusi Formation, and the potential source rocks developed in Doushantuo Formation and pre-Sinian. Based on the tectonic background, Well Tongtan 2 is located in a tectonic trap. According to the seismic profile, the Longwangmiao Formation showed highlight characteristic (Fig. 10), with the development of granular beaches. According to the seismic inversion profile (Fig. 11), the reservoirs are developed.

3 Key Exploration Fields and Zones of Permian Strata

3.1 Key Exploration Areas and Zones

The reservoir accumulations in reservoirs of Permian Qixia-Maokou Formation are developed with lower-generation and upper storage. The major source rocks are developed in Silurian, Lower Permian and Upper Permian. The reservoirs are granular-beach reservoirs in Qixia Formation and granular-beach and karst reservoirs in Maokou Formation. Based on the current exploration results and the main understanding of the controlling factors of the gas reservoir, focusing on the key factors controlling the gas reservoir. Three favorable exploration zones of Qixia

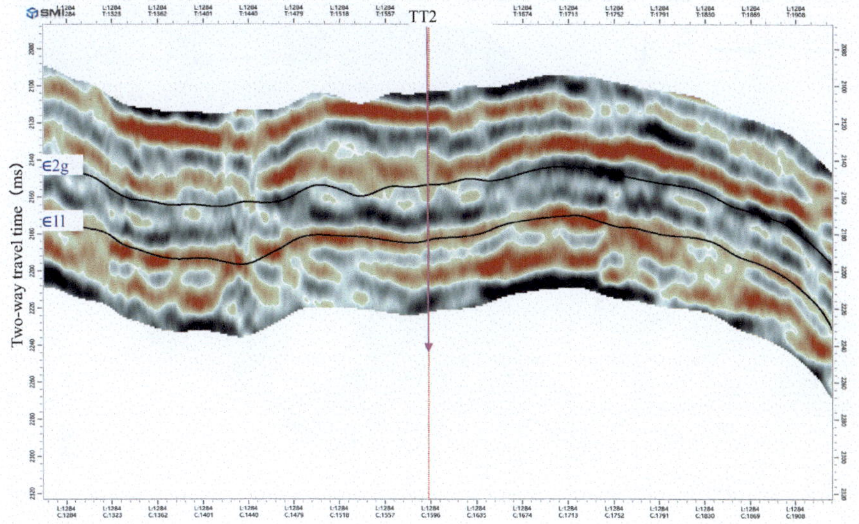

Fig. 10 Profile cross Well Tongtan 2 (L1284)

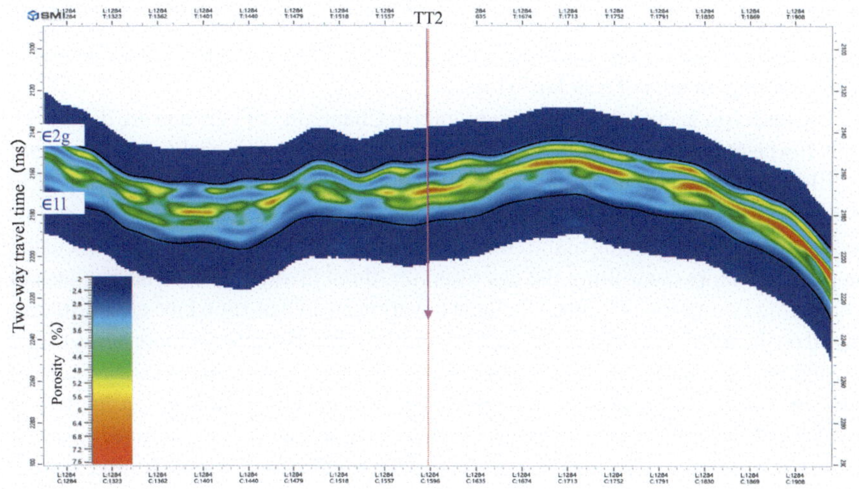

Fig. 11 Inversion profile cross Well Tongtan 2 (L1284)

Formation in Central Sichuan covering an area of 1630 km^2 are evaluated, based on the development of mound-beach bodies, the connection between faults and granular beaches and fault-karst bodies, the tectonic background and the source rock conditions, including two Class I favorable zone covering an area of 1150 km^2, and one Class II favorable area covering an area of 480 km^2 (Fig. 12). There are two favorable exploration zones in Maokou Formation covering an area of 2014 km^2, including

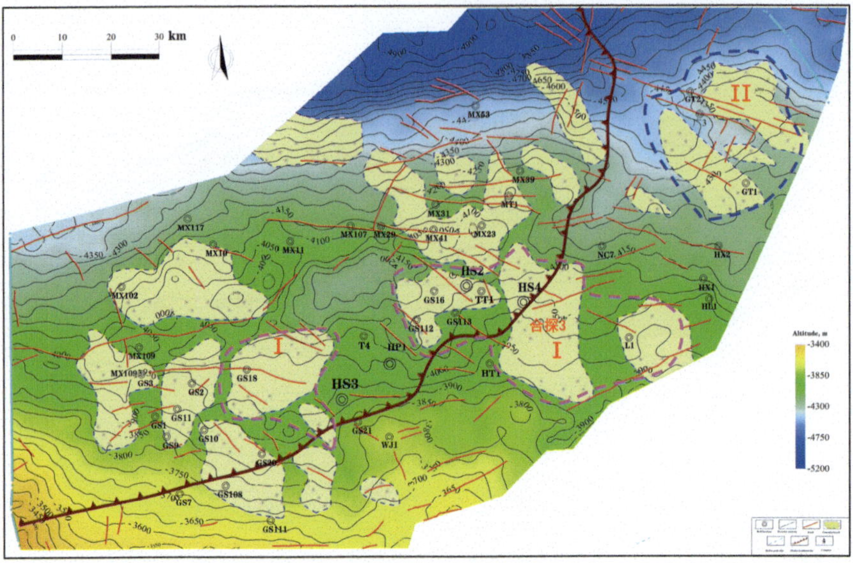

Fig. 12 Comprehensive evaluation map of favorable zones in Qixia Formation in Central Sichuan

one Class I favorable zone covering an area of 1495 km^2, and one Class II favorable zone covering an area of 519 km^2 (Fig. 13).

The reservoir accumulations of reservoirs in Changxing of Permian are developed with lower-generation and upper storage. The major source rocks are developed in Silurian, Lower Permian and Upper Permian. The reservoirs are developed in Changxing Formation mound-beach body. Based on the current exploration results and the main understanding of the controlling factors of the gas reservoir and focusing on the key factors controlling the gas reservoir, three favorable exploration zones of Changxing Formation in Central Sichuan covering an area of 1866 km^2 are evaluated. Based on the relatively development of mound-beach bodies, the tectonic background located on the continuous high points of the structure, the relative development of Permian source rocks, and the oil source faults connecting the Silurian, including two Class I favorable zones (Suining I high-zone and Suning II high-zone) covering an area of 1666 km^2, and one Class II favorable area (Guang'an high-zone) covering an area of 120 km^2 (Fig. 14).

3.2 *Favorable Exploration Targets*

In the fault-karst bodies found in Maokou Formation, four favorable exploration targets are selected (such as Well Gaoshen 1, Hetan 3, Hetan 5 and Hetan 6) according to ① the development of mound-beach bodies and fault-karst bodies; ② the tectonic

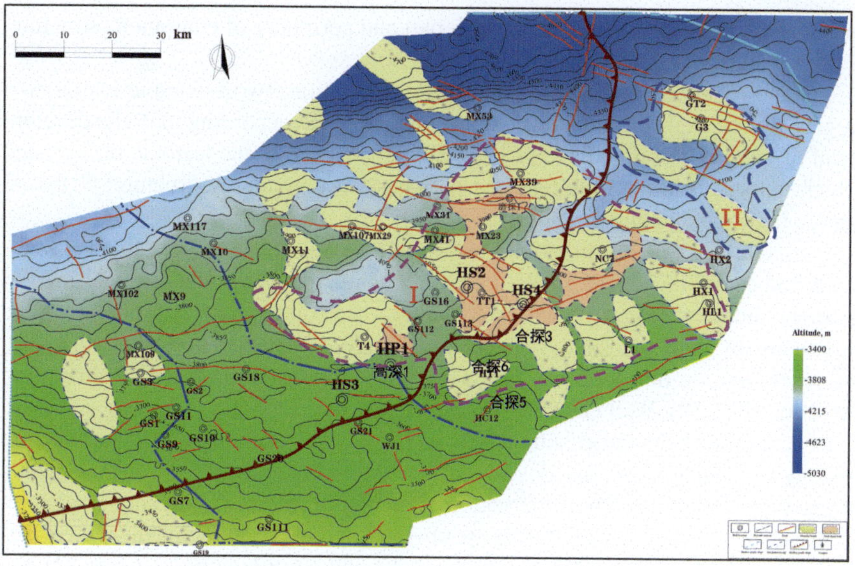

Fig. 13 Comprehensive evaluation map of favorable zones in Maokou Formation in Central Sichuan

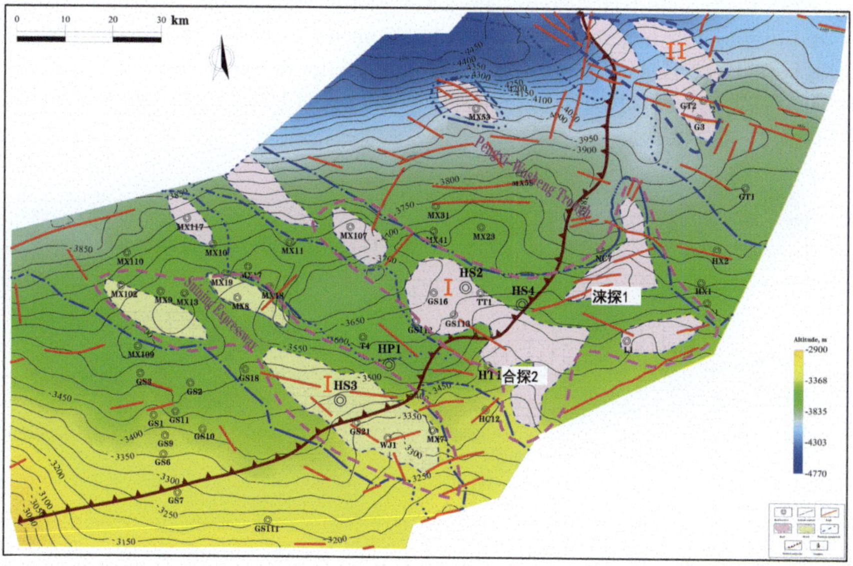

Fig. 14 Comprehensive evaluation map of favorable zones in Changxing Formation in Central Sichuan

background which is located in the high part and periphery of continuous structure; ③ the oil source faults connecting the Silurian.

For the reef-beach bodies in Changxing Formation, two favorable exploration targets are selected, such as Well Laitan 1 and Hetan 2, according to ① the development of mound-beach bodies in Changxing Formation; ② the tectonic background which is located in the high part and periphery of continuous structure; ③ the oil source faults connecting the Silurian.

For example, in the profile of strong event removal, it can be seen that the fault-karst bodies are developed in Well Gaoshen (Fig. 15). From the attribute map of gradient modulus (Figs. 16 and 17), fault-karst bodies' characteristics are more prominent in Well Gaoshen 1. From the profile of seismic inversion (Fig. 18), the reservoirs are developed in Well Gaoshen 1. From the profile of hydrocarbon detection (Fig. 19), Well Gaoshen 1 is developed with high gas potential.

For example, from the seismic profile of Well Laitan 1 (Fig. 20), faults connecting underlying source rocks are developed near Well Laitan 1, which are conducive to reservoir formation. From the bottom flattened profile of Changxing Formation (Fig. 21), the characteristics of reefs are obvious, and the stratum thickness increases obviously where reefs are developed. From the phase attribute profile (Fig. 22), the characteristics of reef are also obvious, and the location where the reef is developed is characterized by mound-shaped chaotic reflection. From the amplitude attribute (Fig. 23), the mound-beach bodies are developed in Well Laitan 1, and are characterized by mound-shaped weak amplitude. From the plan and inversion profile of wave impedance (Figs. 24 and 25), the reservoirs are developed in Well Laitan 1. The hydrocarbon detection profile shows that the Lotus 1 well has a high probability of containing gas (Fig. 26).

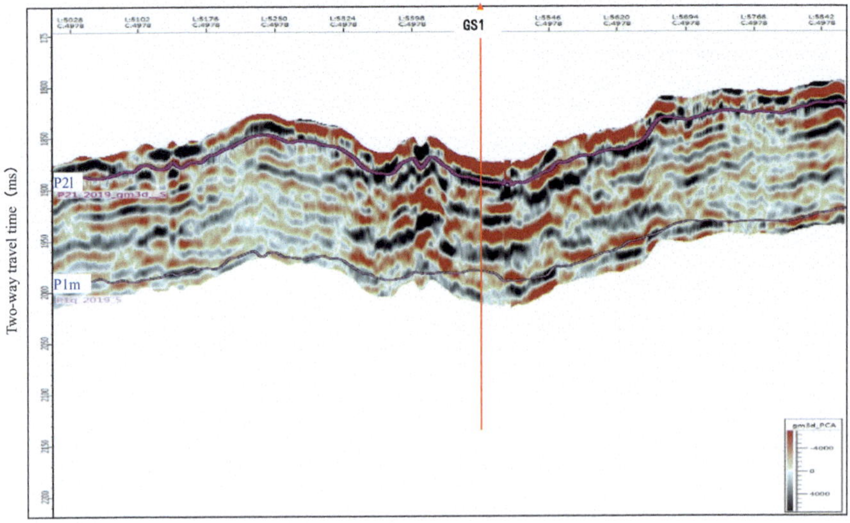

Fig. 15 Seismic profile of strong event removal cross Well Gaoshen 1

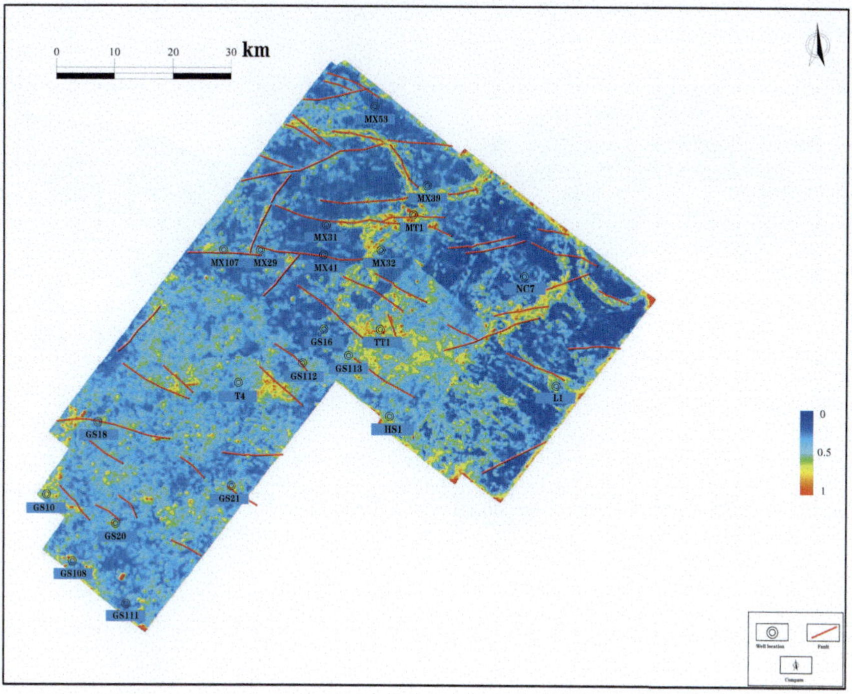

Fig. 16 Plan map of gradient modulus in Maokou Formation of 3D area

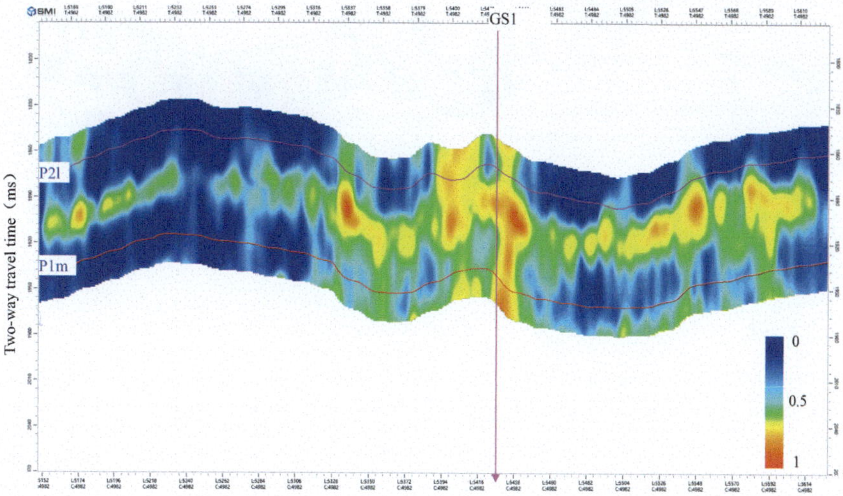

Fig. 17 Attribute profile of gradient modulus cross Well Gaoshen 1

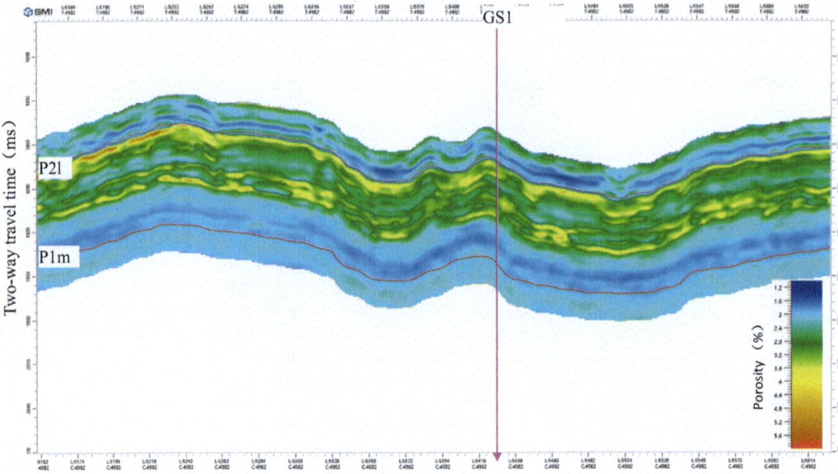

Fig. 18 Inversion profile of wave impedance cross Well Gaoshen 1

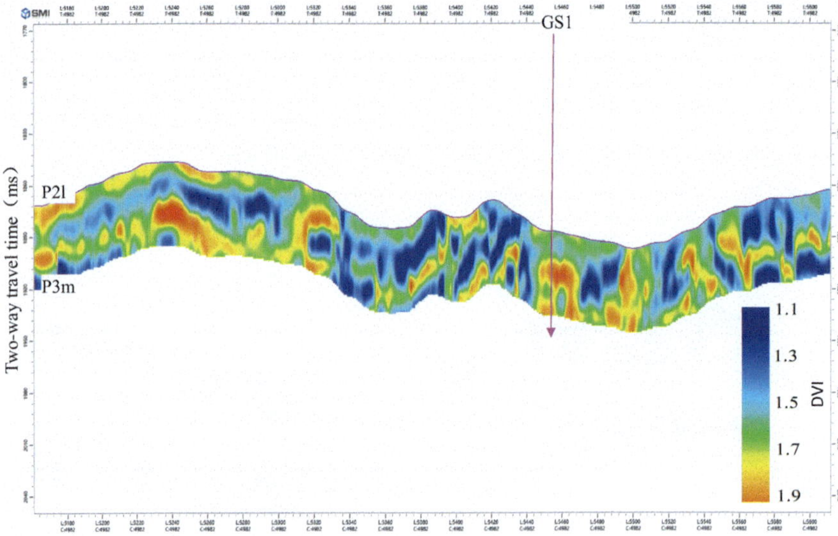

Fig. 19 Profile of hydrocarbon detection cross Well Gaoshen 1

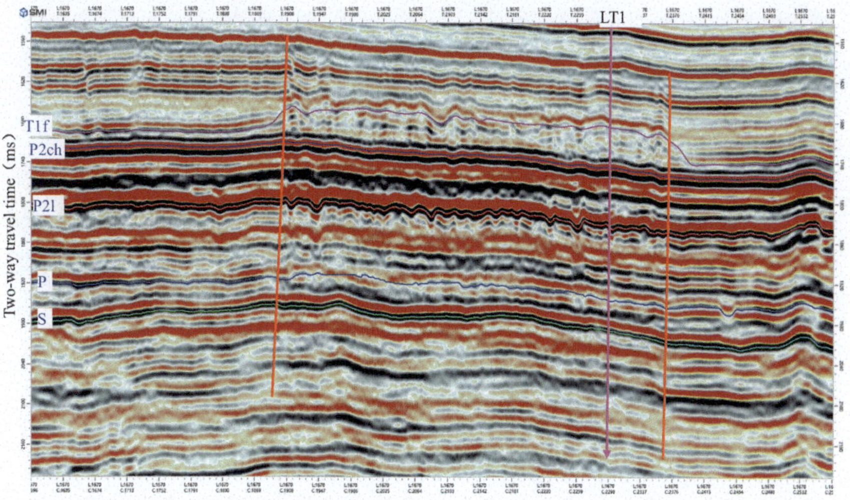

Fig. 20 Seismic profile cross Well Laitan 1

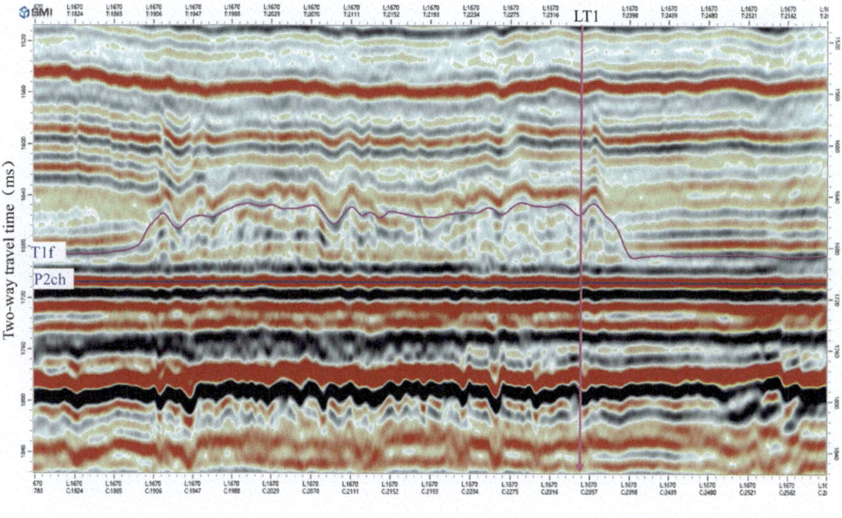

Fig. 21 Seismic profile cross Well Laitan 1 (P2ch flattened)

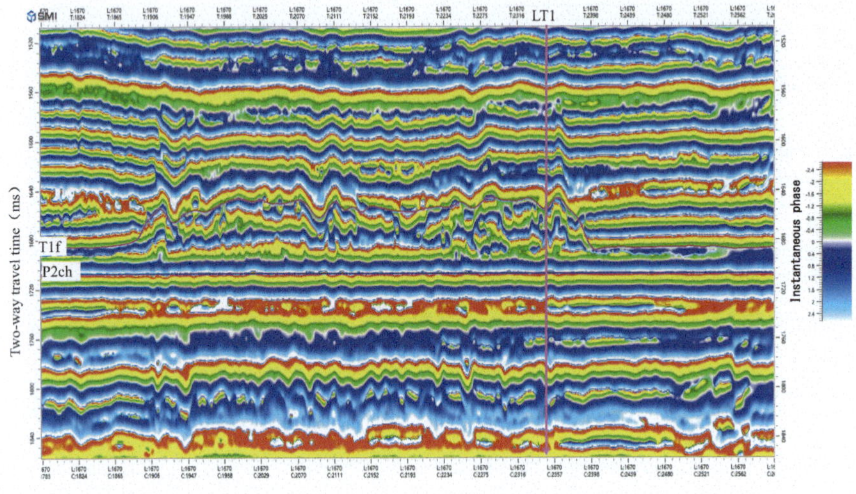

Fig. 22 Phase attribute profile cross Well Laitan 1 (P2ch flattened)

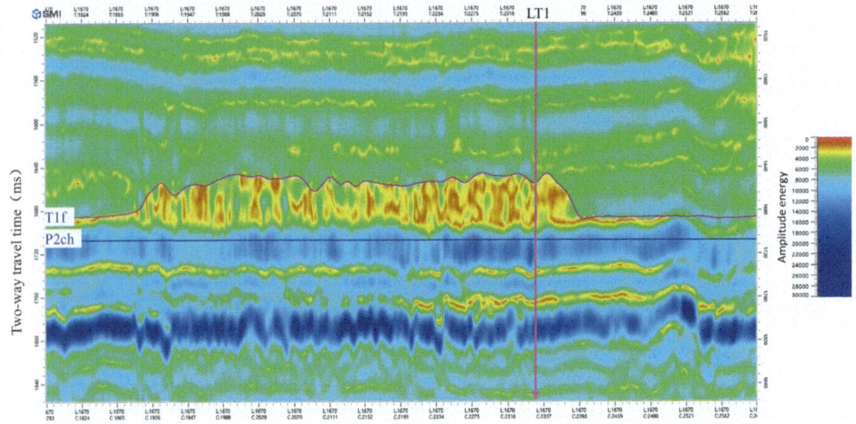

Fig. 23 Amplitude attribute profile of Well Laitan 1

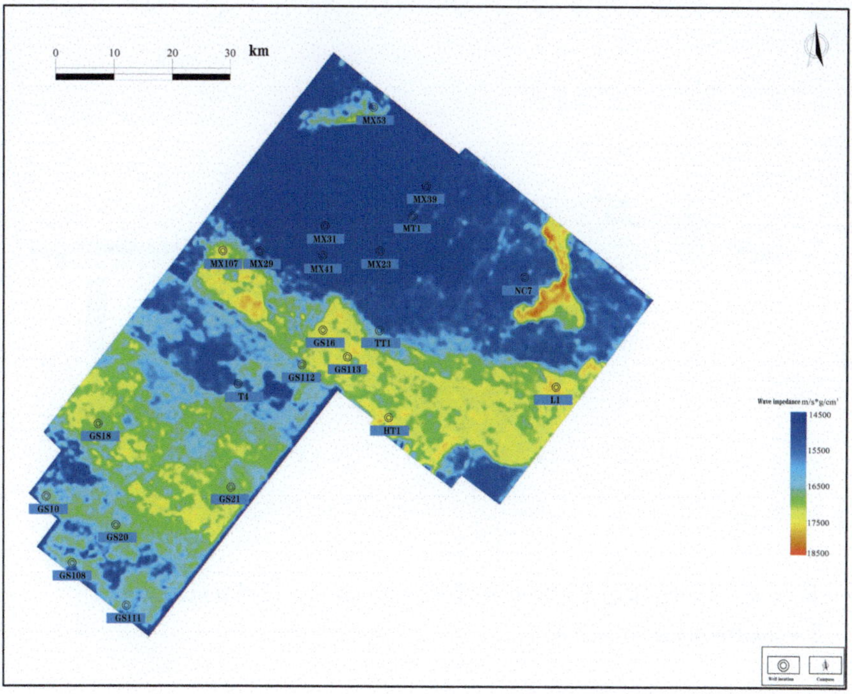

Fig. 24 Plan map of wave impedance in Changxing Formation of 3D area

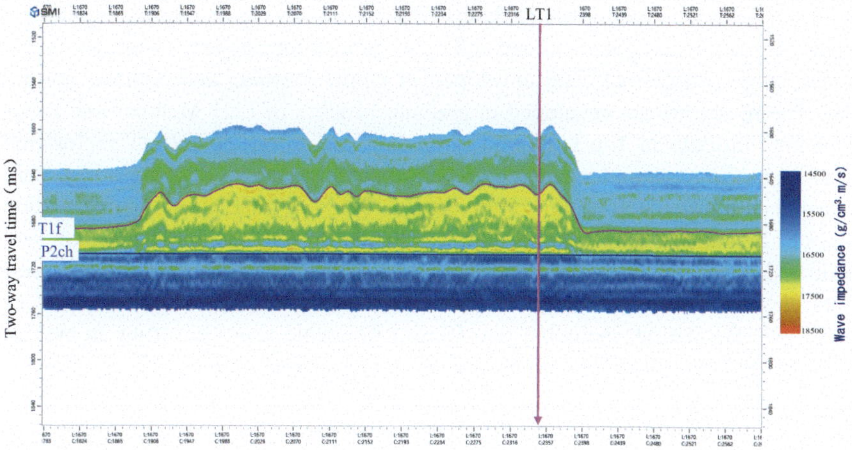

Fig. 25 Inversion profile of wave impedance cross Well Laitan 1

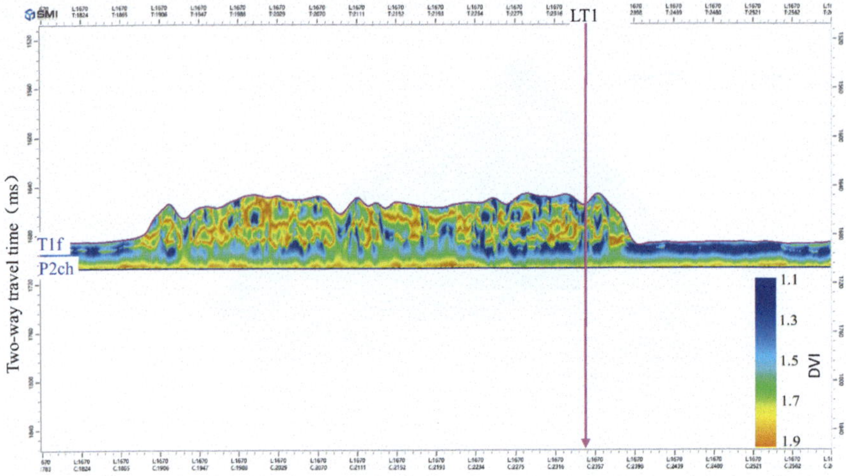

Fig. 26 Profile of hydrocarbon detection cross Well Laitan 1

4 Key Exploration Fields and Zones of the Triassic Leikoupo Formation

4.1 Key Exploration Areas and Zones

The reservoir accumulations in Leikoupo Formation of Triassic are developed with lower-generation and upper storage and upper-generation and lower-storage.

The main source rocks are developed in Upper Permian and Xujiahe Formation. The reservoirs are developed in granular beaches of Lei1 Member and karst-remaining hillocks in Top Leikoupo Formation. Based on the current exploration results and the main understanding of the controlling factors of the gas reservoir, focusing on the key factors controlling the gas reservoir, three favorable exploration zones of Leikoupo Formation in Central Sichuan covering an area of 1260 km^2 are evaluated, based on the geological conditions such as the development of karst reservoirs, the connection of oil source faults, and the thickness area of source rocks in Xujiahe Formation, including two Class I favorable zones covering an area of 947 km^2, and one Class II favorable area covering an area of 313 km^2 (Fig. 27).

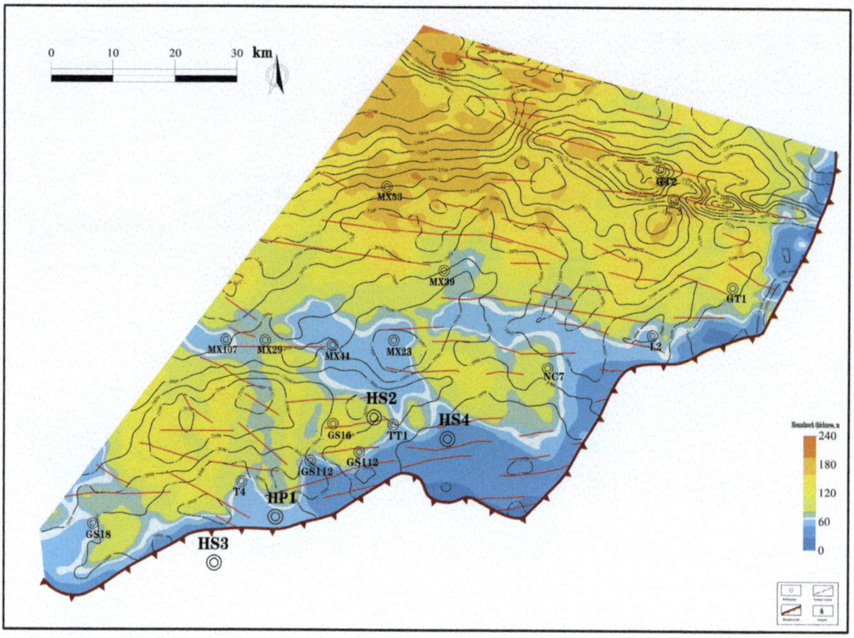

Fig. 27 Comprehensive evaluation map of favorable zones in Leikoupo Formation in Central Sichuan

4.2 Favorable Exploration Targets

Three favorable exploration targets are selected in residual hills at the top of Leikou Formation, such as Well Longqian 1, Tongqian 1 and Anqian 1, according to ① the development of remained hillocks in Lei3^3 Submember; ② the development of source rocks in overlying Xujiahe Formation; ③ the development of faults connecting Xu1 Member and Lei3 Member.

For example, it can be seen from the flattened seismic profile of Lei3 Member (Fig. 28) that karst-remaining hills are developed in the location of Well Tongqian 1; from the seismic inversion section (Fig. 29), the reservoirs are developed in the location Well Tongqian 1; from the hydrocarbon detection profile (Fig. 30), Well Tongqian 1 is developed with high gas potential.

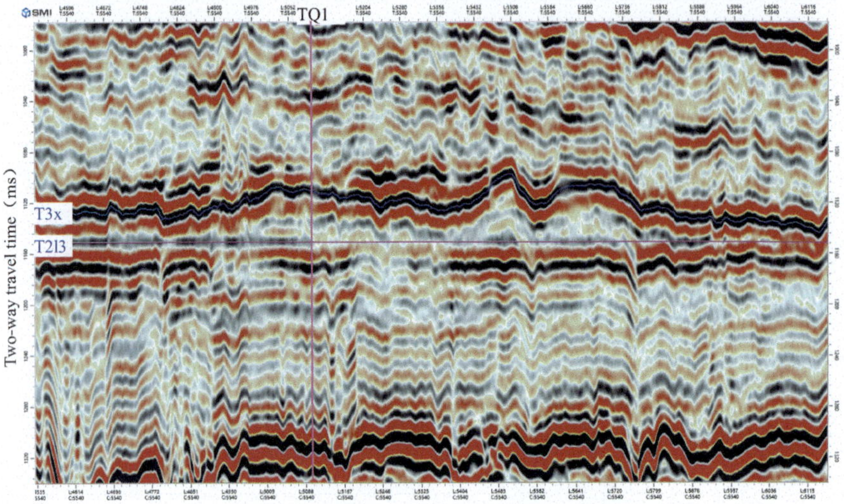

Fig. 28 Seismic profile cross Well Tongqian 1 (Lei3 Member flattened)

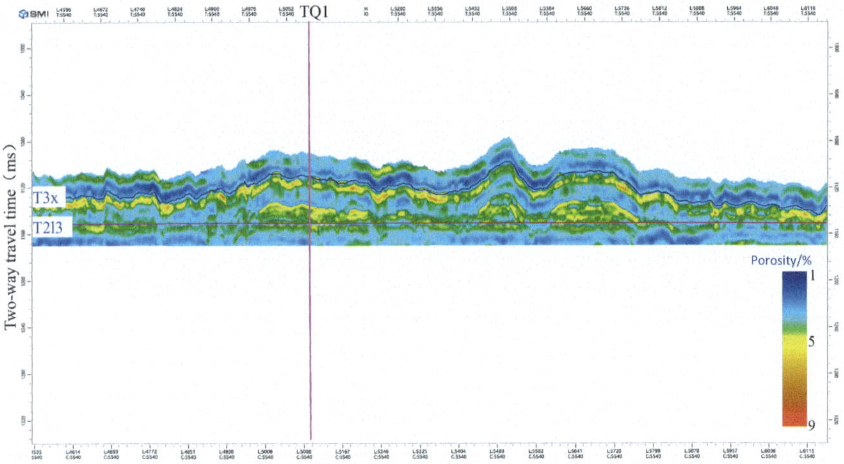

Fig. 29 Inversion profile cross Well Tongqian 1

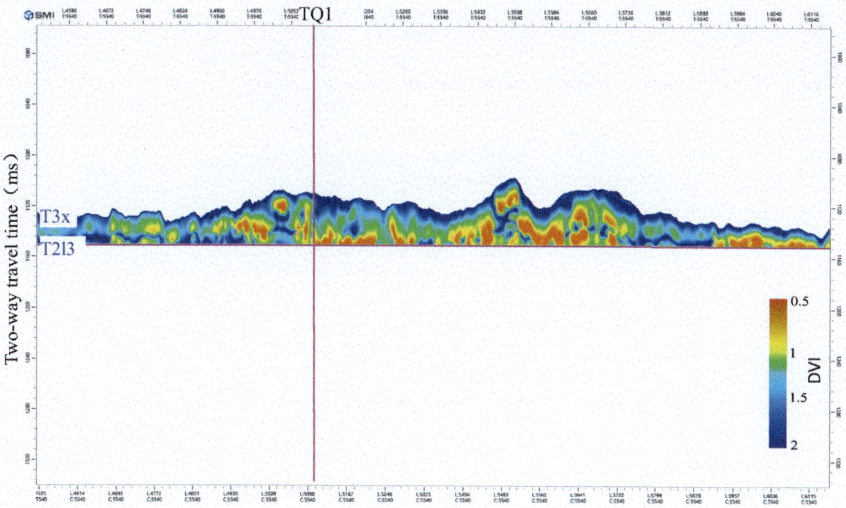

Fig. 30 Hydrocarbon detection profile cross Well Tongqian 1

The manufacturer's authorised representative in the EU is Springer Nature Customer Service Centre GmbH, Europaplatz 3, 69115 Heidelberg, Germany. If you have any concerns regarding our products, please contact ProductSafety@springernature.com

Printed and bound by CPI Group (UK) Ltd, Croydon, CR0 4YY
01/10/2026
02225935-0009